# A Computational Logic

This is a volume in the
ACM MONOGRAPH SERIES

Editor: THOMAS A. STANDISH, *University of California at Irvine*

A complete list of titles in this series appears at the end of this volume.

# A Computational Logic

**ROBERT S. BOYER and J STROTHER MOORE**
SRI International
Menlo Park, California

ACADEMIC PRESS
A Subsidiary of Harcourt Brace Jovanovich, Publishers
New York   London   Toronto   Sydney   San Francisco

COPYRIGHT © 1979, BY ACADEMIC PRESS, INC.
ALL RIGHTS RESERVED.
NO PART OF THIS PUBLICATION MAY BE REPRODUCED OR
TRANSMITTED IN ANY FORM OR BY ANY MEANS, ELECTRONIC
OR MECHANICAL, INCLUDING PHOTOCOPY, RECORDING, OR ANY
INFORMATION STORAGE AND RETRIEVAL SYSTEM, WITHOUT
PERMISSION IN WRITING FROM THE PUBLISHER.


ACADEMIC PRESS, INC.
111 Fifth Avenue, New York, New York 10003


*United Kingdom Edition published by*
ACADEMIC PRESS, INC. (LONDON) LTD.
24/28 Oval Road, London NW1 7DX


**Library of Congress Cataloging in Publication Data**

Boyer, Robert S
   A computational logic.

   (ACM monograph series)
   Includes bibliographical references and index.
   1.  Automatic theorem proving.  I.  Moore,
J Strother, Date     joint author.  II.  Title.
III.  Series:  Association for Computing Machinery.
ACM monograph series.
QA76.9.A96B68    519.4    79–51693
ISBN 0–12–122950–5


PRINTED IN THE UNITED STATES OF AMERICA

79 80 81 82    9 8 7 6 5 4 3 2 1

To our wives,
Anne and Liz

# Contents

*Preface* xi

## I. Introduction
  A. Motivation   2
  B. Our Formal Theory   2
  C. Proof Techniques   3
  D. Examples   3
  E. Our Mechanical Theorem-Prover   4
  F. Artificial Intelligence or Logic?   6
  G. Organization   6

## II. A Sketch of the Theory and Two Simple Examples
  A. An Informal Sketch of the Theory   8
  B. A Simple Inductive Proof   16
  C. A More Difficult Problem   19
  D. A More Difficult Proof   21
  E. Summary   26
  F. Notes   26

## III. A Precise Definition of the Theory
  A. Syntax   28
  B. The Theory of IF and EQUAL   30
  C. Well-Founded Relations   31
  D. Induction   33
  E. Shells   35
  F. Natural Numbers   40
  G. Literal Atoms   41
  H. Ordered Pairs   42
  I. Definitions   43
  J. Lexicographic Relations   51

K. LESSP and COUNT     52
L. Conclusion     54

## IV. The Correctness of a Tautology-Checker
A. Informal Development     57
B. Formal Specification of the Problem     59
C. The Former Definition of TAUTOLOGY.CHECKER     63
D. The Mechanical Proofs     67
E. Summary     84
F. Notes     85

## V. An Overview of How We Prove Theorems
A. The Role of the User     87
B. Clausal Representation of Conjectures     88
C. The Organization of Our Heuristics     89
D. The Organization of Our Presentation     91

## VI. Using Type Information to Simplify Formulas
A. Type Sets     92
B. Assuming Expressions True or False     95
C. Computing Type Sets     96
D. Type Prescriptions     97
E. Summary     101
F. Notes     102

## VII. Using Axioms and Lemmas as Rewrite Rules
A. Directed Equalities     103
B. Infinite Looping     104
C. More General Rewrite Rules     105
D. An Example of Using Rewrite Rules     107
E. Infinite Backwards Chaining     109
F. Free Variables in Hypotheses     111

## VIII. Using Definitions
A. Nonrecursive Functions     114
B. Computing Values     114
C. Diving in to See     116

## IX. Rewriting Terms and Simplifying Clauses
A. Rewriting Terms     120
B. Simplifying Clauses     124
C. The REVERSE Example     126
D. Simplification in the REVERSE Example     127

## X. Eliminating Destructors
  A. Trading Bad Terms for Good Terms    130
  B. The Form of Elimination Lemmas    133
  C. The Precise Use of Elimination Lemmas    134
  D. A Nontrivial Example    135
  E. Multiple Destructors and Infinite Looping    139
  F. When Elimination Is Risky    139
  G. Destructor Elimination in the REVERSE Example    141

## XI. Using Equalities
  A. Using and Throwing Away Equalities    145
  B. Cross-Fertilization    146
  C. A Simple Example of Cross-Fertilization    147
  D. The Precise Use of Equalities    149
  E. Cross-Fertilization in the REVERSE Example    150

## XII. Generalization
  A. A Simple Generalization Heuristic    151
  B. Restricting Generalizations    153
  C. Examples of Generalizations    154
  D. The Precise Statement of the Generalization Heuristic    156
  E. Generalization in the REVERSE Example    157

## XIII. Eliminating Irrelevance
  A. Two Simple Checks for Irrelevance    159
  B. The Reason for Eliminating Isolated Hypotheses    160
  C. Elimination of Irrelevance in the REVERSE Example    162

## XIV. Induction and the Analysis of Recursive Definitions
  A. Satisfying the Principle of Definition    164
  B. Induction Schemes Suggested by Recursive Functions    171
  C. The Details of the Definition-Time Analysis    180
  D. Recursion in the REVERSE Example    183

## XV. Formulating an Induction Scheme for a Conjecture
  A. Collecting the Induction Candidates    185
  B. The Heuristic Manipulation of Induction Schemes    189
  C. Examples of Induction    197
  D. The Entire REVERSE Example    202

## XVI. Illustrations of Our Techniques via Elementary Number Theory
  A. PLUS.RIGHT.ID    209
  B. COMMUTATIVITY2.OF.PLUS    211
  C. COMMUTATIVITY.OF.PLUS    216

### x / CONTENTS

      D. ASSOCIATIVITY.OF.PLUS   221
      E. TIMES   221
      F. TIMES.ZERO   222
      G. TIMES.ADD1   223
      H. ASSOCIATIVITY.OF.TIMES   227
      I. DIFFERENCE   233
      J. RECURSION.BY.DIFFERENCE   233
      K. REMAINDER   242
      L. QUOTIENT   242
      M. REMAINDER.QUOTIENT.ELIM   243

### XVII. The Correctness of a Simple Optimizing Expression Compiler

    A. Informal Development   253
    B. Formal Specification of the Problem   257
    C. Formal Definition of the Compiler   263
    D. The Mechanical Proof of Correctness   266
    E. Notes   279

### XVIII. The Correctness of a Fast String Searching Algorithm

    A. Informal Development   283
    B. Formal Specification of the Problem   291
    C. Developing the Verification Conditions for the Algorithm   292
    D. The Mechanical Proofs of the Verification Conditions   301
    E. Notes   305

### XIX. The Unique Prime Factorization Theorem

    A. The Context   309
    B. Formal Development of the Unique Prime Factorization Theorem   311
    C. The Mechanical Proofs   315

### Appendix A. Definitions Accepted and Theorems Proved by Our System   329

### Appendix B. The Implementation of the Shell Principle   376

### Appendix C. Clauses for Our Theory   380

### References   385

*Index*   389

# Preface

Mechanical theorem-proving is crucial to the automation of reasoning about computer programs. Today, few computer programs can be mechanically certified to be free of "bugs." The principal reason is the lack of mechanical theorem-proving power.

In current research on automating program analysis, a common approach to overcoming the lack of mechanical theorem-proving power has been to require that the user direct a proof-checking program. That is, the user is required to construct a formal proof employing only the simplest rules of inference, such as *modus ponens*, instantiation of variables, or substitution of equals for equals. The proof-checking program guarantees the correctness of the formal proof. We have found proof-checking programs too frustrating to use because they require too much direction.

Another approach to overcoming the lack of mechanical theorem-proving power is to use a weak theorem-proving program and to introduce axioms freely. Often these axioms are called "lemmas," but they are usually not proved. While using a proof checker is only frustrating, introducing axioms freely is deplorable. This approach has been abused so far as to be ludicrous: we have seen researchers "verify" a program by first obtaining formulas that imply the program's correctness, then running the formulas through a simplifier, and finally assuming the resulting slightly simplified formulas as axioms. Some researchers admit that these "lemmas" ought to be proved, but never get around to proving them because they lack the mechanical theorem-proving power. Others, however, believe that it is reasonable to assume lots of "lemmas" and never try to prove them. We are strongly opposed to this latter attitude because it so completely undermines the spirit of proof, and we therefore reply to the arguments we have heard in its defense.

(1) It is argued that the axioms assumed are obvious facts about the concepts involved. We say that a great number of mistakes in computer programs arise from false "obvious" observations, and we have already seen researchers present proofs based on false lemmas. Furthermore, the concepts involved in the complicated computer systems one hopes eventually to certify are so insufficiently canonized that one man's "obvious" is another man's "difficult" and a third man's "false."

(2) It is argued that one must assume some axioms. We agree, but observe that mathematicians do not contrive their axioms to solve the problem at hand. Yet often the "lemmas" assumed in program verification are remarkably close to the main idea or trick in the program being checked.

(3) It is argued that mathematicians use lemmas. We agree. In fact, our theorem-proving system relies heavily on lemmas. But no proof is complete until the lemmas have been proved too. The assumption of lemmas in program proving often amounts to sweeping under the rug the hard and interesting inferences.

(4) It is argued that the definition of concepts necessarily involves the addition of axioms. But the axioms that arise from proper definitions, unlike most "lemmas," have a very special form that guarantees two important properties. First, adding a definition never renders one's theory inconsistent. Second, the definition of a concept involved in the proof of a subsidiary result (but not in the statement of one's main conjecture) can safely be forgotten. It does not matter if the definition was of the "wrong" concept. But an ordinary axiom (or "lemma"), once used, always remains a hypothesis of any later inference. If the axiom is "wrong," the whole proof may be worthless and the validity of the main conjecture is in doubt.

One reason that researchers have had to assume "lemmas" so freely is that they have not implemented the principle of mathematical induction in their theorem-proving systems. Since mathematical induction is a fundamental rule of inference for the objects about which computer programmers think (e.g., integers, sequences, trees), it is surprising that anyone would implement a theorem-prover for program verification that could not make inductive arguments. Why has the mechanization of mathematical induction received scant attention?

Perhaps it has been neglected because the main research on mechanical theorem-proving, the resolution theorem-proving tradition (see Chang and Lee [15] and Loveland [29]), does not handle axiom schemes, such as mathematical induction.

We suspect, however, that the mechanization of mathematical induction has been neglected because many researchers believe that the only need for induction is in program semantics. Program semantics enables one to obtain from a given program and specification some conjectures ("verification conditions") which imply that the program is correct. The study of program semantics has produced a plethora of ways to use induction. Because some programs do not terminate, the role of induction in program semantics is fascinating and subtle. Great effort has been invested in mechanizing induction in program semantics. For example, the many "verification condition generation" programs implicitly rely on induction to provide the semantics of iteration.

But program semantics is not the only place induction is necessary. The conjectures that verification condition generators produce often require inductive proofs because they concern inductively defined concepts such as the integers, sequences, trees, grammars, formulas, stacks, queues, and lists. If you cannot make an inductive argument about an inductively defined concept, then you are doomed to assume what you want to prove.

This book addresses the use of induction in proving theorems rather than the use of induction in program semantics.

We will present a formal theory providing for inductively constructed objects, recursive definitions, and inductive proofs. Readers familiar with programming languages will see a strong stylistic resemblance between the language of our theory and that fragment of the programming language LISP known as "pure LISP" (see McCarthy et al. [35]). We chose pure LISP as a model for our language because pure LISP was designed as a mathematical language whose formulas could easily be represented within computers. Because of its mathematical nature (e.g., one cannot "destructively transform" the ordered pair $\langle 7, 3 \rangle$ into $\langle 8, 3 \rangle$), pure LISP is considered a "toy" programming language. It is an easy jump to the *non sequitur:* "The language and theory presented in this book are irrelevant to real program analysis problems because they deal with a toy programming language." But that statement misses the point. It is indeed true that our theory may be viewed as a programming language. In fact, many programs are naturally written as functions in our theory. But our theory is a mathematical tool for making precise assertions about the properties of discrete objects. As such, it can be used in conjunction with any of the usual program specification methods to state and prove properties of programs written in any programming language whatsoever.

When we began our research into proving theorems about recursive functions [7, 38], we thought of ourselves as proving theorems only about pure LISP and viewed our work as an implementation of McCarthy's [34] functional style of program analysis. However, we now also regard recursion as a natural alternative to quantification when making assertions about programs. Using recursive functions to make assertions about computer programs no more limits the programming language to one that implements recursion than using the ordinary quantifiers limits the programming language to one that implements quantification! In this book we use both the functional style and Floyd's inductive assertion style [18] of program specification in examples. (For the benefit of readers not familiar with the program verification literature, we briefly explain both ideas when they are first used.) We have relegated the foregoing remarks to the preface because we are not in general interested in program semantics in this book. We are interested in how one proves theorems about inductively constructed objects.

Our work on induction and theorem-proving in general has been deeply influenced by that of Bledsoe [3, 4]. Some early versions of our work have been previously reported in [7, 38, 39, 40, 8]. Work closely related to our work on induction has been done by Brotz [11], Aubin [2], and Cartwright [14].

We thank Anne Boyer, Jacqueline Handley, Paul Gloess, John Laski, Greg Nelson, Richard Pattis, and Jay Spitzen for their careful criticisms of this book. We also thank Jay Spitzen for showing us how to prove the prime factorization theorem. We thank Bernard Meltzer for the creative atmosphere in the Metamathematics Unit of the University of Edinburgh, where we began our collaboration. Finally, we thank our wives and children for their usually cheerful long-suffering through the years of late hours behind this book.

Our work has been supported by the National Science Foundation under Grant MCS-7681425 and by the Office of Naval Research under Contract N00014-75-C-0816. We are very grateful to these agencies for their support.

# I
# Introduction

Unlike most texts on logic and mathematics, this book is about how to prove theorems rather than the proofs of specific results. We give our answers to such questions as

When should induction be used?

How does one invent an appropriate induction argument?

When should a definition be expanded?

We assume the reader is familiar with the mathematical notion of equality and with the logical connectives "and," "or," "not," and "implies" of propositional calculus. We present a logical theory in which one can introduce inductively constructed objects (such as the natural numbers and finite sequences) and prove theorems about them. Then we explain how we prove theorems in our theory.

We illustrate our proof techniques by using them to discover proofs of many theorems. For example, we formalize a version of the propositional calculus in our theory, and, using our techniques, we formally prove the correctness of a decision procedure for that version of propositional calculus. In another example, we develop elementary number theory from axioms introducing the natural numbers and finite sequences through the prime factorization theorem.

Since our theory is undecidable, our proof techniques are not perfect. But we know that they are unambiguous, well integrated, and successful on a large number of theorems because we have programmed a computer to follow our rules and have observed the program prove many interesting theorems. In fact, the proofs we describe are actually those discovered by our program.

## A. MOTIVATION

Suppose it were practical to reason, mechanically and with mathematical certainty, about computer programs. For example, suppose it were practical to prove mechanically that a given program satisfied some specification, or exhibited the same output behavior as another program, or executed in certain time or space bounds.[1] Then there would follow a tremendous improvement in the reliability of computer programs and a subsequent reduction of the overall cost of producing and maintaining programs.

To reason mechanically about programs, one must have a formal program semantics, a formal logical theory, and a mechanical theorem-prover for that theory. The study of formal program semantics has provided a variety of alternative methods for specifying and modeling programs. But all the methods have one thing in common: they reduce the question, Does this program have the desired property? to the question, Are these formulas theorems? Because of the nature of computers, the formulas in question almost exclusively involve inductively constructed mathematical objects: the integers, finite sequences, n-tuples, trees, grammars, expressions, stacks, queues, buffers, etc. Thus, regardless of which program semantics we use to obtain the formulas to be proved, our formal theory and mechanical theorem-prover must permit definition and proof by induction. This book is about such a theory and a mechanical theorem-prover for it.

## B. OUR FORMAL THEORY

We will present a logical theory that we have tailored to the needs of thinking about computer programs. It provides for the introduction of new "types" of objects, a general principle of induction on well-founded relations (Noetherian Induction [6]), and a principle permitting the definition of recursive functions. Recursive functions offer

---

[1] See Manna and Waldinger [31] for a description of the many other ways that formal reasoning can be usefully applied in computer programming.

such a powerful form of expression when dealing with discrete mathematics (such as underlies computer programs) that we do not use any additional form of quantification.[2]

## C. PROOF TECHNIQUES

After defining our formal theory, we describe many techniques we have developed for proving theorems in it. We devote eleven chapters to the description of these techniques and how, when, and where they should be applied to prove theorems. The most important of these techniques is the use of induction. The formulation of an induction argument for a conjecture is based on an analysis of the recursive definitions of the concepts involved in the conjecture. Thus the use of recursively defined functions facilitates proving theorems about inductively defined objects. Many of the other proof techniques are designed to support our induction heuristics.

## D. EXAMPLES

All the techniques are illustrated with examples. Most of our techniques are first illustrated with simple theorems about functions on lists and trees. These elementary functions are simple to define and are worth knowing if one is interested in mechanical theorem-proving (as we assume many readers will be). In addition, it is more fun to work through the proofs of novel theorems than through the proofs of, say, the familiar theorems of elementary number theory.

We have also included four complicated examples, chosen from several different subject domains, to illustrate the general applicability of the theory and our proof methods.

In the first such example, we write a tautology-checker as a recursive function on trees representing formulas in propositional calculus. We exercise the theory and proof techniques in an interesting way by stating and proving that the tautology-checker always returns an answer, recognizes only tautologies, and recognizes all tautologies.

---

[2] The program of using recursive functions and induction to understand computer programs, and the use of computers to aid the generation of the proofs, were begun by McCarthy [33, 34]. See also Burstall [12]. The idea of using recursive functions and induction but no other form of quantification in the foundations of mathematics (or at least of arithmetic) was first presented by Skolem in 1923 [52]. See also Goodstein [22].

This example serves two important purposes: it illustrates the theory and proof techniques in use, and it gives the reader a precise definition of a simple mechanical theorem-prover (i.e., the tautology-checker) without requiring a digression into programming languages or computational issues.

In the second major example, we prove the correctness of a simple algorithm that "optimizes" and "compiles" arithmetic expressions into sequences of instructions for a hand-held calculator. In order to specify the algorithm and the calculator, we use (and briefly explain) McCarthy's "functional semantics" [34] for programs. This example is the first part of the book that deals explicitly with computational (rather than mathematical) ideas. Because the example is simple (compared to real compilers and real hardware) and because it is ideally suited to our mathematical language, the reader unfamiliar with computing should be able to read this chapter comfortably (and even learn the basic ideas behind compiling expressions and one style of program specification).

In the third major example, we prove the correctness of a fast string searching algorithm. The algorithm finds the first occurrence of one character sequence in another (if such an occurrence exists), and, on the average, is the fastest such algorithm currently known. In this example we explain and use a second program specification style, called the "inductive assertion" method (Floyd [18]).

Finally, we demonstrate that the theory and proof techniques can be used to prove theorems that are generally considered difficult (rather than just theorems that have not been generally considered) by proving our statement of the unique prime factorization theorem: (a) any positive integer can be represented as the product of a finite sequence of primes, and (b) any two finite sequences of primes with the same product are in fact permutations of one another. We derive this theorem starting from the axioms of Peano arithmetic and finite sequences.

## E. OUR MECHANICAL THEOREM-PROVER

It is one thing to describe a loosely connected set of heuristics that a human might use to discover proofs and quite a different thing to formulate them so that a machine can use them to discover proofs.[3] All of the heuristics described have been implemented and together com-

---

[3] For a survey of nonresolution theorem-proving, see [5].

prise our automatic theorem-proving program. Our description of the heuristics makes little or no reference to the fact that they can be mechanized. However, we want competent readers to be able to reproduce and build upon our results. Thus, we are more precise than we would be had we desired only to teach a student how we prove theorems.

All of the example proofs discussed in this book are actually produced by our program. We present the theorem-prover's proofs in an informal style. While describing proof techniques we present small inference steps, but as we move on to the interesting examples we ascend to the level upon which humans usually deal with proofs. By the time we reach the prime factorization theorem, the proofs we describe are very much like those in number theory textbooks: we make large inference leaps, use lemmas without describing their proofs, and dismiss whole theorems with phrases like "the proof is by induction on X." However, our high-level descriptions of the machine's proofs should not be confused with what the machine does: before pronouncing a conjecture proved, the machine discovers a complete sequence of applications of our proof techniques establishing the conjecture from axioms and previously proved lemmas. Furthermore, given the detailed presentation of our proof techniques and their orchestration, the reader should also be able to discover the proofs mechanically.

It is perhaps easiest to think of our program much as one would think of a reasonably good mathematics student: given the axioms of Peano, he could hardly be expected to prove (much less discover) the prime factorization theorem. However, he could cope quite well if given the axioms of Peano and a list of theorems to prove (e.g., "Prove that addition is commutative,". . . , "Prove that multiplication distributes over addition,". . . , "Prove that the result returned by the GCD function divides both of its arguments,". . . , "Prove that if the products over two sequences of primes are identical, then the two sequences are permutations of one another").

The examples discussed are just part of a standard sequence of approximately 400 definitions and theorems the program is expected to reestablish whenever we incorporate a new technique. The sequence contains theorems some readers will have trouble proving before they read the heuristics. In Appendix A we list the entire sequence as evidence that the heuristics are generally useful and well integrated. It should be noted that when the latter theorems of the sequence are proved the theorem-prover is aware of the earlier theorems. The fact that previously proved results are remembered permits their use as lemmas in later proofs. The theorem-prover would fail to prove many of its most interesting theorems in the absence of such lemmas. However, the more a theorem-proving program knows, the more difficult it becomes for it to prove theorems because the program is often tempted to consider using theorems that have no relevance to the task at hand. That our theorem-proving program does prove the entire list of theorems sequentially is a measure of its capacity to avoid being confused by what it knows.

**F. ARTIFICIAL INTELLIGENCE OR LOGIC?**

While drawing heavily upon important facts of mathematical logic, our research is really more artificial intelligence than logic. The principal question we ask (and sometimes answer) is, How do we discover proofs? It has been argued that mechanical theorem-proving is an impossible task because certain theories are known to be undecidable or super-super-exponential in complexity. Such metamathematical results are, of course, no more of an impediment to mechanical theorem-proving than to human theorem-proving.[4] They only make the task more interesting.

**G. ORGANIZATION**

This book is structured as follows. We begin informally in Chapter II by sketching our logical theory, formulating a simple conjecture, and proving that conjecture by using many of the techniques we will discuss. We present the theory formally in Chapter III.

In Chapter IV we state and prove in our theory the correctness of a function for recognizing tautologies.

In Chapter V through Chapter XV we develop and explain our heuristics for proving theorems. These heuristics are illustrated with simple proof steps taken from many theorems.

In Chapter XVI, our theorem-prover develops an interesting initial segment of elementary number theory.

Finally, in Chapter XVII through Chapter XIX we discuss three complex examples: the proof of correctness of a simple optimizing compiler for arithmetic expressions, the correctnesss of a fast string

---

[4] Nils Nilsson, private communication.

searching algorithm, and the proof of the unique prime factorization theorem.

Readers interested in ascertaining the power of our automatic theorem-proving program and in seeing how recursive functions can be used to formalize a variety of problems should first read the informal overview (Chapter II) and then look at the chapters on the tautology-checker (Chapter IV), compiler (Chapter XVII), string searching algorithm (Chapter XVIII), and prime factorization theorem (Chapter XIX).

The book has three appendixes. The first lists all the definitions and theorems the theorem-prover routinely proves. The second appendix presents implementation details concerning the introduction of new types. The third appendix exhibits, in clausal form, the axioms of our theory.

# II

# A Sketch of the Theory and Two Simple Examples

To prove theorems formally one must have in mind a formal theory in which the proofs are to be constructed. We will present our formal theory in Chapter III. Following the precise presentation of the theory, we describe, in great detail, how we discover proofs. However, before descending into detail, we here sketch the theory informally, exhibit and explain several simple recursive function definitions, and (without regard for mechanization) work through several simple inductive proofs.

## A. AN INFORMAL SKETCH OF THE THEORY

### 1. IF and EQUAL

We employ the prefix notation of Church's lambda calculus [16] and McCarthy's LISP [35] when writing down terms. Thus, we write

    (PLUS (H X) (B))

when others write

    PLUS(H(X),B())

or even

    H(X) + B()

to denote the application of the two-argument function PLUS to (1) the application of H to X and (2) the constant (i.e., function of no arguments) B.

We wish to make it easy to define new functions and predicates on inductively constructed objects. For example, given axioms for the natural numbers we would like to define functions such as addition and multiplication, and predicates such as whether a given number is prime; given the axioms for sequences we would like to define operations such as sequence concatenation and predicates such as whether one sequence is a permutation of another.

We find it most convenient to define new functions with equality axioms of the form:

$$(f\ x_1\ \ldots\ x_n) = body$$

where certain constraints are placed on $f$, the $x_i$, and the term body.

It is often necessary to make conditional definitions. For example, (PLUS X Y) is defined to be one thing if $X=0$, and another thing if $X \neq 0$. In order for definitions to be equality axioms, the right-hand side of the definition, body, must be a term. But in the usual treatment of logic it is not permitted to embed a proposition (such as $X=0$) in a term. Thus, we find it necessary to reproduce the logic of truth functions and equality at the term level.

We add to the usual propositional calculus with variables, function symbols, and equality an axiom supposing the existence of two distinct constants, (TRUE) and (FALSE) (henceforth written T and F), and four axioms defining the new function symbols EQUAL and IF. The axioms are

**Axiom**

$$T \neq F,$$

**Axiom**

$$X = Y \rightarrow (EQUAL\ X\ Y) = T,$$

**Axiom**

$$X \neq Y \rightarrow (EQUAL\ X\ Y) = F,$$

**Axiom**

$$X = F \rightarrow (IF\ X\ Y\ Z) = Z,$$

**Axiom**

$$X \neq F \rightarrow (IF\ X\ Y\ Z) = Y.$$

We can paraphrase the above axioms as follows. T is not F. For all X and Y, (EQUAL X Y) is T if X is Y, and is F if X is not Y. (IF X Y Z) is Y if X is non-F and is Z otherwise. Thus (IF (EQUAL X 0) Y Z) is equal to Y if X is 0 and is equal to Z otherwise.

Strictly speaking, we never define "predicates," for they can only be used in the construction of formulas and thus cannot be used in terms (such as function bodies). Without loss of generality, we restrict our attention to functions. For example, we will later define the *function* PRIME, so that (PRIME X) is T if X is prime and F otherwise.

To permit terms or functions to test logical combinations of expressions, it is convenient to define the functional versions of "not," "and," "or," and "implies." For example, we want (NOT P) to be T if P is F and to be F if P is not F. Similarly, we want (AND P Q) to be T if both P and Q are non-F, and F otherwise. Thus, we define the functions NOT, AND, OR, and IMPLIES as follows:

**Definition**

(NOT P)
=
(IF P F T),

**Definition**

(AND P Q)
=
(IF P (IF Q T F) F),

**Definition**

(OR P Q)
=
(IF P T (IF Q T F)),

**Definition**

(IMPLIES P Q)
=
(IF P (IF Q T F) T).

(We adopt the notational convention of treating AND and OR as though they took an arbitrary number of arguments. For example, (AND p q r) is an abbreviation for (AND p (AND q r)).)

It is easy to show that these definitions capture the semantics of the ordinary logical connectives. For example, it is a theorem that

$$(P \neq F \rightarrow Q \neq F) \leftrightarrow (\text{IMPLIES P Q}) \neq F.$$

Thus, it is also easy to prove that

(IMPLIES (AND P Q) (OR P Q))

is not equal to F. Because of our emphasis on terms rather than formulas we find it convenient to call a term p a "theorem" if it can be proved that p ≠ F. Of course, calling a term a "theorem" is an abuse of terminology, since theorems are in fact understood to be formulas. However, whenever we use a term p as though it were a formula, it is always acceptable to read the formula "p ≠ F" in its place.

## 2. Inductively Constructed Objects

Any theory concerned with the mathematics behind computing must provide inductively constructed objects. For example, it is clear that we must be able to talk formally about the natural numbers, and so we will add to our theory axioms for the natural numbers. In formalizing properties of programs we have found that it is convenient to allow the introduction of "new" types of inductively constructed objects, of which the integers are just a single example. To eliminate the possibility that the axioms for a new type will render the theory inconsistent (or not fully specify the properties of the type) we have included in our theory a general principle under which one can introduce new types. We call the principle the "shell" principle. The name "shell" derives from imagining the new objects to be colored structures encapsulating a fixed number of components (possibly of certain colors).

It is actually with the shell principle that we add the axioms defining the nonnegative integers. We also use the principle to add the set of literal atoms (i.e., atomic symbols such as "NIL" and "X"), and the set of ordered pairs. For example, to axiomatize the set of ordered pairs we incant:

Add the shell CONS,
with recognizer LISTP, and
accessors CAR and CDR that return "NIL" on non-LISTP objects,

which is a shorthand for adding a set of axioms that specifies (CONS X Y) to be the red, say, ordered pair containing X and Y, LISTP to be the function that returns T if its argument is a red pair and F otherwise, and CAR and CDR to be the functions that return the first and second components of red pairs (or "NIL" if given an object other than a red pair).

For example, here are some of the axioms added by the above incantation:

**Axiom** LISTP.CONS:

(LISTP (CONS X1 X2)),

**Axiom** CAR.CONS:

(EQUAL (CAR (CONS X1 X2)) X1),

**Axiom** CDR.CONS:

(EQUAL (CDR (CONS X1 X2)) X2),

**Axiom** CAR/CDR.ELIM:

(IMPLIES (LISTP X)
         (EQUAL (CONS (CAR X) (CDR X)) X)),

**Axiom** CONS.EQUAL:

(EQUAL (EQUAL (CONS X1 X2)
              (CONS Y1 Y2))
       (AND (EQUAL X1 Y1)
            (EQUAL X2 Y2))),

**Axiom** CAR.LESSP:

(IMPLIES (LISTP X)
         (LESSP (COUNT (CAR X)) (COUNT X))),

**Axiom** CDR.LESSP:

(IMPLIES (LISTP X)
         (LESSP (COUNT (CDR X)) (COUNT X))).

LESSP is the "less-than" function on the nonnegative integers. The complete set of axioms for the CONS shell is given schematically in Chapter III. Among the axioms not shown above are, for example, axioms specifying that (CAR X) is "NIL" if X is not a LISTP, and that the set of LISTP objects does not overlap the set of numbers, literal atoms, or other types.

We use ordered pairs in a variety of ways. For example, to talk about finite sequences (sometimes called "lists") we think of "NIL" as the empty sequence, and we think of (CONS X Y) as the sequence whose first element is X and whose remaining elements are those of Y. Thus, we think of

(CONS 1 (CONS 2 (CONS 3 "NIL")))

as the sequence containing 1, 2, and 3.

## 3. Recursively Defined Functions

Our theory includes a principle of definition allowing the introduction of recursive functions. The principle is based on the notion of well-founded relations. In particular, ( f $x_1$ ... $x_n$ ) may be defined to be some term, body, involving "recursive calls" of the form ( f $y_1$ ... $y_n$ ), provided there is a measure and well-founded relation such that in every recursive call the measure of the $y_i$ is smaller, according to the well-founded relation, than the measure of the $x_i$. Since "well-founded" means that there is no infinite sequence of objects, each of which is smaller than its predecessor in the sequence, the above restriction, together with a few simple syntactic requirements, ensures that there exists one and only one function satisfying the definition. The existence of a function satisfying the definition implies that adding the definition as an axiom does not render the theory inconsistent.

We explicitly assume that LESSP is well founded. Furthermore, when the CONS shell is added, the axioms CAR.LESSP and CDR.LESSP, above, inform us that the CAR and CDR of a pair both have smaller size (as measured by the function COUNT) than the pair itself. Thus, if in every recursive call of a function some particular argument is replaced by its own CAR or CDR, and we can establish that in each such case that argument is a pair, then the principle of definition would admit the function.

To illustrate a definition by recursion, suppose we wished to concatenate two finite sequences X and Y. If X is empty, then the result of concatenating X and Y is just Y. If X is nonempty, then X has a first element, ( CAR X ), and some remaining elements, ( CDR X ). The concatenation of X and Y in this case is the sequence whose first element is ( CAR X ) and whose remaining elements are those in the concatenation of ( CDR X ) and Y. Formally, we define the function APPEND so that ( APPEND X Y ) is the concatenation of X and Y:

**Definition**

(APPEND X Y)
=

```
(IF (LISTP X)
    (CONS (CAR X) (APPEND (CDR X) Y))
    Y).
```

APPEND is a particularly simple recursive function. It is easy to see why it is accepted under our principle of definition: the axiom CDR.LESSP, above, establishes that (COUNT X) gets LESSP-smaller in each (i.e., the) recursive call. Later in the book we will introduce more interesting recursive functions—functions for which a measure as obvious as the size of one argument will not suffice to justify their definition.

By the axioms of equality, we can replace any instance of (APPEND X Y) with the corresponding instance of the right-hand side of the definition above. For example, we can show that (APPEND (CONS A D) Y) is equal to (CONS A (APPEND D Y)) as follows. By definition, (APPEND (CONS A D) Y) is equal to the instantiated body

```
(IF (LISTP (CONS A D))
    (CONS (CAR (CONS A D))
          (APPEND (CDR (CONS A D)) Y))
    Y).
```

By the axiom LISTP.CONS, above, (LISTP (CONS A D)) is non-F. But, by the axioms defining IF, we know that when the first argument in an IF-expression is non-F, the IF-expression is equal to its second argument. Thus, the above IF-expression can be replaced by its second argument:

```
(CONS (CAR (CONS A D))
      (APPEND (CDR (CONS A D)) Y)).
```

Finally, applying the axioms CAR.CONS and CDR.CONS, above, we rewrite (CAR (CONS A D)) to A, and (CDR (CONS A D)) to D, obtaining

```
(CONS A (APPEND D Y))
```

as desired.

Similarly, we can prove that (APPEND "NIL" Y) is Y.

We can use a series of such simplifications to "compute" the result of concatenating the sequence containing 1, 2, and 3, to the sequence containing 4 and 5:

```
(APPEND (CONS 1 (CONS 2 (CONS 3 "NIL")))
        (CONS 4 (CONS 5 "NIL")))
=
```

```
  (CONS 1 (APPEND (CONS 2 (CONS 3 "NIL"))
                  (CONS 4 (CONS 5 "NIL"))))
=
  (CONS 1 (CONS 2 (APPEND (CONS 3 "NIL")
                          (CONS 4 (CONS 5 "NIL")))))
=
  (CONS 1 (CONS 2 (CONS 3 (APPEND "NIL"
                                  (CONS 4 (CONS 5 "NIL"))))))
=
  (CONS 1 (CONS 2 (CONS 3 (CONS 4 (CONS 5 "NIL"))))).
```

Using recursion, we can introduce, under our principle of definition, almost all the concepts in which we are interested. Indeed, recursion is a very important tool when dealing with inductively constructed objects such as the integers or sequences. For instance, recursion can be regarded as a form of quantification: we can use recursion to check that all or some elements of a sequence have some property. We do not use any other form of quantification in our formal theory.

## 4. Induction

Because we are interested in proving theorems about inductively constructed objects such as the natural numbers, sequences, pairs, etc., we need a rule of inference that embodies the construction process itself. For example, we know that if X is a pair, then it can be constructed by applying CONS to two "previously" constructed objects, namely, (CAR X) and (CDR X). Thus, we can prove that some property holds for all X by considering two cases. The first case, called the "base case," is to prove that all nonpairs have the property in question. The second case, called the "induction step," is to assume that X is a pair and that (CAR X) and (CDR X) have the desired property, and to prove that X has the property. Such a proof is called a proof by "induction."

The magic idea behind induction, the idea that made it appear unsound to both authors when they first encountered the idea in high school, is that one gets to assume instances of the conjecture being proved during its own proof. Why is induction sound? For example, why can we conclude, after proving the two cases above, that any X must have the property? Suppose some object does not have the property. Then let X be a minimal object not having the property, where we compare two such objects by comparing their COUNTs using the LESSP function. There *is* a minimal object not having the property in question since, by the well-foundedness of LESSP, there is no infinite sequence of objects, each of which has smaller COUNT than the previous one. X must be a pair, because otherwise the base case establishes that X has the property. But if X is a pair, (CAR X) and (CDR X) are both smaller than X (as measured by COUNT and LESSP). Therefore, since everything smaller than X has the property in question, (CAR X) and (CDR X) have the property. But in that case the induction step establishes that X must have it also. Thus the assumption that some object does not have the property has lead to a contradiction.

In general, the induction principle in our theory permits one to assume arbitrary instances of the conjecture being proved, provided those instantiations decrease some measure in some well-founded ordering. Because our induction principle allows the use of arbitrary measures and well-founded relations, we can make inductive arguments that are much more subtle than the "structural induction" illustrated above.[5] We will illustrate more subtle inductions later in the book. Throughout this chapter we confine ourselves to structural induction.

## B. A SIMPLE INDUCTIVE PROOF

One of the hardest problems in discovering an inductive proof is discovering an appropriate application of the principle of induction itself. But the similarity between recursion and induction offers an insight into the problem. For example, suppose we were trying to prove some conjecture involving the expression (APPEND A term). When APPEND was introduced into the theory we were required to exhibit a measure of its arguments that was decreasing. In particular, every time APPEND recurses, the COUNT of its first argument goes down. Thus (APPEND A term) "suggests" an induction: were we to apply the definition of APPEND to "open up" (APPEND A term) we would find ourselves recursively decomposing A into its constituents and would want information about those constituents. But by the observation above, we know those constituents are smaller than A in some well-founded order. Thus, by the induction principle, we can supply ourselves with inductive instances about those constituents.

[5] The use of structural induction to prove programs correct is beautifully described by Burstall in [12].

We illustrate the above reasoning with a simple example. We will prove

**Theorem** ASSOCIATIVITY.OF.APPEND:

  (EQUAL (APPEND (APPEND A B) C)
     (APPEND A (APPEND B C))).

Name the conjecture *1.
The proof is by induction. Three terms—each APPEND expression with a variable in its first argument—suggest "plausible" inductions. Two of these inductions are on A and the third is on B. All occurrences of A are in argument positions being recursively decomposed. Thus, by appropriately opening up APPEND expressions we can reexpress *1 in terms of (CDR A), about which we can supply an inductive hypothesis. (In this case we say that the induction on A is "unflawed.") The induction on B is flawed: B occurs sometimes as the first argument to APPEND and sometimes as the second (which is never changed in recursion). No matter how we expand APPEND expressions, the conjecture will still involve B and (CDR B), and we are unable to supply an induction hypothesis about B while inducting on B.

Thus, we will induct on A, using the following scheme:

  (AND (IMPLIES (NOT (LISTP A)) (p A B C))
    (IMPLIES (AND (LISTP A)
        (p (CDR A) B C))
     (p A B C))).

That is, letting (p A B C) be a schematic representation of *1, we will prove that (p A B C) holds when A is not a LISTP, and we will prove that if A is a LISTP and (p (CDR A) B C) holds, then (p A B C) holds. The induction is sound because the axiom CDR.LESSP establishes that (COUNT A) decreases according to the well-founded relation LESSP in the induction step of the scheme. Instantiating the scheme above with *1 we obtain two new goals:

*Case 1.*
```
(IMPLIES (NOT (LISTP A))
         (EQUAL (APPEND (APPEND A B) C)
                (APPEND A (APPEND B C)))).
```
This is the base case. Since A is non-LISTP, (APPEND A B) is equal to its second argument, B, by the definition of APPEND. Similarly, (APPEND A (APPEND B C)) is equal to its second argument, (APPEND B C). By rewriting these two terms in the conclusion above we obtain:
```
(IMPLIES (NOT (LISTP A))
         (EQUAL (APPEND B C)
                (APPEND B C))),
```
which simplifies, using the axiom X=Y → (EQUAL X Y)=T, to:

   (TRUE).

*Case 2.*
```
(IMPLIES (AND (LISTP A)
              (EQUAL (APPEND (APPEND (CDR A) B) C)
                     (APPEND (CDR A) (APPEND B C))))
         (EQUAL (APPEND (APPEND A B) C)
                (APPEND A (APPEND B C)))).
```
This is the induction step. Since A is a LISTP here, (APPEND A Y), for any Y, is equal to (CONS (CAR A) (APPEND (CDR A) Y)), by the definition of APPEND. Thus, we can "unfold" the two (APPEND A term) expressions in the conclusion above to get:
```
(IMPLIES (AND (LISTP A)
              (EQUAL (APPEND (APPEND (CDR A) B) C)
                     (APPEND (CDR A) (APPEND B C))))
         (EQUAL (APPEND (CONS (CAR A)
                              (APPEND (CDR A) B))
                        C)
                (CONS (CAR A)
                      (APPEND (CDR A) (APPEND B C))))).
```
But, by opening up the definition of APPEND, we know that (APPEND (CONS A D) Y) is equal to (CONS A (APPEND D Y)). Thus, we can expand the first APPEND

term in the conclusion above, to get:

```
(IMPLIES (AND (LISTP A)
              (EQUAL (APPEND (APPEND (CDR A) B) C)
                     (APPEND (CDR A) (APPEND B C))))
         (EQUAL (CONS (CAR A)
                      (APPEND (APPEND (CDR A) B) C))
                (CONS (CAR A)
                      (APPEND (CDR A) (APPEND B C))))).
```

Note that the conclusion above is of the form:

(EQUAL (CONS x y) (CONS u v)).

But according to CONS.EQUAL, two pairs are equal if and only if the components are pairwise equal. That is, the concluding equality may be rewritten to the conjunction of (EQUAL x u) and (EQUAL y v). But in the above application x and u are identical—they are both (CAR A). Thus, we replace the concluding equality above with (EQUAL y v):

```
(IMPLIES (AND (LISTP A)
              (EQUAL (APPEND (APPEND (CDR A) B) C)
                     (APPEND (CDR A) (APPEND B C))))
         (EQUAL (APPEND (APPEND (CDR A) B) C)
                (APPEND (CDR A) (APPEND B C)))).
```

However, this simplifies to:

(TRUE),

because the conclusion is identical to the second hypothesis. In particular, by opening up the correct APPEND expressions we transformed the induction conclusion into the induction hypothesis.

That finishes the proof of *1. Q.E.D.

## C. A MORE DIFFICULT PROBLEM

The proof of the associativity of APPEND illustrates two of the proof techniques we describe later: induction and simplification. Some of the other heuristics we will describe are:

It is sometimes useful to trade "bad" terms for "good" ones by re-representing terms in the conjecture. For example, one might transform a conjecture about I and I−J to one about K+J and K, trading difference for addition, by replacing all occurrences of I by K+J.

One obvious way to use an equality hypothesis, (EQUAL x y), is to substitute x for y throughout the conjecture. But it is sometimes useful to replace only some of the y's by x's and then to "throw away" the equality hypothesis, so as to produce a more general conjecture to prove by induction. We call such use of an equality hypothesis "cross-fertilization."

In proofs by induction, it is easier to prove strong theorems than weak ones, because strong theorems permit one to obtain strong induction hypotheses with which to work. Thus, one should look for ways to generalize a conjecture to be proved by induction.

We illustrate these proof techniques by working through another simple example.

Consider the idea of the "fringe" of a tree of CONSes. Given the tree

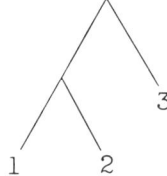

that is, (CONS (CONS 1 2) 3), we wish to return the sequence of tips of the tree, 1, 2, 3, that is, (CONS 1 (CONS 2 (CONS 3 "NIL"))).

An obvious way to "flatten" a tree X is to ask first whether (LISTP X) is true. If so, X is a fork in the tree. The fringe of a fork is the concatenation (as with APPEND) of the fringes of the left and right subtrees (i.e., the recursively obtained fringes of (CAR X) and (CDR X)). If (LISTP X) is false, X is a tip of the tree. The fringe of a tip is the singleton sequence containing the tip, i.e., (CONS X "NIL").[6]

---

[6] Computing the fringe of binary trees brings to mind one of the lesser chestnuts of artificial intelligence: how to compute that two binary trees have the same fringe without using much storage (in particular, without simply computing both fringes and comparing them). We believe the original formulation of the problem was by Carl Hewitt. Burstall and Darlington [13] have investigated the solution of this problem by automatic program transformations. McCarthy (private communication) has written a recursive function, exhibited as SAMEFRINGE in Appendix A, that is more or less the recursive realization of the usual "coroutining" solution to the problem. Our system can prove the correctness of SAMEFRINGE, and we refer the interested reader to Appendix

The above description can be immediately transcribed into the definition of a recursive function that we will call FLATTEN. We exhibit the definition of FLATTEN later, when we formalize the problem. Considered as a recipe for computing the fringe of a tree, FLATTEN is somewhat inefficient. It visits every fork and tip of the tree once, but for every fork, the concatenation process revisits every tip of the left-hand branch. In trees heavily nested to the left, FLATTEN computes in time $n^2$, where n is the number of tips. John McCarthy (private communication) suggested the following more efficient algorithm, which computes in linear time and which we call MC.FLATTEN.

The basic idea is that to collect all the tips in a tree one can initialize a collection site to "NIL", and then sweep the tree adding tips to the site as they are encountered. If the tips are added to the front of the collection site, the answer will be in exactly the reverse of the order in which the tips were visited. If we sweep the tree from right to left (instead of left to right), the result will be in the order desired.

To write the algorithm as a recursive function, we use a second argument, ANS, as the collection site. At a tip, we add the tip to the front of ANS (with CONS) and return the new list as the answer. At a fork, we first collect the tips in the CDR (the right branch), and, using the resulting answer as the collection site, we then collect the tips in the CAR (the left branch). Thus, (MC.FLATTEN X ANS) should append the fringe of X onto ANS.

The formal statement of the relationship between MC.FLATTEN and FLATTEN is

```
(EQUAL (MC.FLATTEN X ANS)
       (APPEND (FLATTEN X) ANS)).
```

In the next section we define FLATTEN and MC.FLATTEN formally, explain why they are admitted under our principle of definition, and then work through the proof of the conjecture above.

## D. A MORE DIFFICULT PROOF

### Definition

```
(FLATTEN X)
   =
(IF (LISTP X)
```

---

A. The problem of computing the fringe of a tree relates to computer science as a whole, since similar tree processing underlies such fundamental algorithms as parsers and compilers.

```
          (APPEND (FLATTEN (CAR X))
                  (FLATTEN (CDR X)))
          (CONS X "NIL")).
```

The lemmas CAR.LESSP and CDR.LESSP establish that (COUNT X) goes down according to the well-founded relation LESSP in each recursive call. Hence, FLATTEN is accepted under the definition principle. Observe that (LISTP (FLATTEN X)) is a theorem.

**Definition**

```
     (MC.FLATTEN X ANS)
       =
     (IF (LISTP X)
         (MC.FLATTEN (CAR X)
                     (MC.FLATTEN (CDR X) ANS))
         (CONS X ANS)).
```

The lemmas CDR.LESSP and CAR.LESSP establish that (COUNT X) decreases according to the well-founded relation LESSP in each recursive call. Hence, MC.FLATTEN is accepted under the definition principle. Note that (LISTP (MC.FLATTEN X ANS)) is a theorem.

**Theorem** FLATTEN.MC.FLATTEN:

```
     (EQUAL (MC.FLATTEN X ANS)
            (APPEND (FLATTEN X) ANS)).
```

Name the conjecture *1.

Let us appeal to the induction principle. There are two plausible inductions. However, they merge into one likely candidate induction. We will induct according to the following scheme:

```
     (AND (IMPLIES (NOT (LISTP X)) (p X ANS))
          (IMPLIES (AND (LISTP X)
                        (p (CAR X)
                           (MC.FLATTEN (CDR X) ANS))
                        (p (CDR X) ANS))
                   (p X ANS))).
```

The inequalities CAR.LESSP and CDR.LESSP establish that the measure (COUNT X) decreases according to the well-founded relation LESSP in the induction step of

## D. A MORE DIFFICULT PROOF / 23

the scheme. Note, however, the inductive instances chosen for ANS. The above induction scheme produces two new formulas:

*Case 1.* (IMPLIES (NOT (LISTP X))
                  (EQUAL (MC.FLATTEN X ANS)
                         (APPEND (FLATTEN X) ANS))),

which simplifies, applying CDR.CONS and CAR.CONS, and expanding the definitions of MC.FLATTEN, FLATTEN and APPEND, to:

    (TRUE).

*Case 2.*
(IMPLIES (AND (LISTP X)
              (EQUAL (MC.FLATTEN (CAR X)
                                 (MC.FLATTEN (CDR X)
                                             ANS))
                     (APPEND (FLATTEN (CAR X))
                             (MC.FLATTEN (CDR X)
                                         ANS)))
              (EQUAL (MC.FLATTEN (CDR X) ANS)
                     (APPEND (FLATTEN (CDR X)) ANS)))
         (EQUAL (MC.FLATTEN X ANS)
                (APPEND (FLATTEN X) ANS))).

This simplifies, unfolding the definitions of MC.FLATTEN and FLATTEN, to:

(IMPLIES (AND (LISTP X)
              (EQUAL (MC.FLATTEN (CAR X)
                                 (MC.FLATTEN (CDR X)
                                             ANS))
                     (APPEND (FLATTEN (CAR X))
                             (MC.FLATTEN (CDR X)
                                         ANS)))
              (EQUAL (MC.FLATTEN (CDR X) ANS)
                     (APPEND (FLATTEN (CDR X)) ANS)))
         (EQUAL (MC.FLATTEN (CAR X)
                            (MC.FLATTEN (CDR X) ANS))
                (APPEND (APPEND (FLATTEN (CAR X))
                                (FLATTEN (CDR X)))
                        ANS))).

Appealing to the lemma CAR/CDR.ELIM, we now replace X

## 24 / II. A SKETCH OF THE THEORY AND TWO SIMPLE EXAMPLES

by (CONS Z V) to eliminate (CAR X) and (CDR X). This generates:

```
(IMPLIES (AND (LISTP (CONS Z V))
              (EQUAL (MC.FLATTEN Z (MC.FLATTEN V ANS))
                     (APPEND (FLATTEN Z)
                             (MC.FLATTEN V ANS)))
              (EQUAL (MC.FLATTEN V ANS)
                     (APPEND (FLATTEN V) ANS)))
         (EQUAL (MC.FLATTEN Z (MC.FLATTEN V ANS))
                (APPEND (APPEND (FLATTEN Z)
                                (FLATTEN V))
                        ANS))),
```

which further simplifies, clearly, to:

```
(IMPLIES (AND (EQUAL (MC.FLATTEN Z (MC.FLATTEN V ANS))
                     (APPEND (FLATTEN Z)
                             (MC.FLATTEN V ANS)))
              (EQUAL (MC.FLATTEN V ANS)
                     (APPEND (FLATTEN V) ANS)))
         (EQUAL (MC.FLATTEN Z (MC.FLATTEN V ANS))
                (APPEND (APPEND (FLATTEN Z)
                                (FLATTEN V))
                        ANS))).
```

We use the first equality hypothesis by cross-fertilizing:

```
        (APPEND (FLATTEN Z)
                (MC.FLATTEN V ANS))
```

for (MC.FLATTEN Z (MC.FLATTEN V ANS)) and throwing away the equality. This generates:

```
        (IMPLIES (EQUAL (MC.FLATTEN V ANS)
                        (APPEND (FLATTEN V) ANS))
                 (EQUAL (APPEND (FLATTEN Z)
                                (MC.FLATTEN V ANS))
                        (APPEND (APPEND (FLATTEN Z)
                                        (FLATTEN V))
                                ANS))).
```

We now use the above equality hypothesis by cross-fertilizing (APPEND (FLATTEN V) ANS) for (MC.FLATTEN V ANS) and throwing away the equality. We thus obtain:

## D. A MORE DIFFICULT PROOF / 25

```
(EQUAL (APPEND (FLATTEN Z)
               (APPEND (FLATTEN V) ANS))
       (APPEND (APPEND (FLATTEN Z) (FLATTEN V))
               ANS)),
```

which we generalize by replacing (FLATTEN V) by Y and
(FLATTEN Z) by A. We restrict the new variables by
appealing to the type restriction lemma noted when
FLATTEN was introduced. The result is:

```
(IMPLIES (AND (LISTP Y) (LISTP A))
         (EQUAL (APPEND A (APPEND Y ANS))
                (APPEND (APPEND A Y) ANS))),
```

which we will finally name *1.1.

Let us appeal to the induction principle. The
recursive terms in the conjecture suggest three
inductions. They merge into two likely candidate
inductions. However, only one is unflawed. We will
induct according to the following scheme:

```
(AND (IMPLIES (NOT (LISTP A)) (p A Y ANS))
     (IMPLIES (AND (LISTP A) (p (CDR A) Y ANS))
              (p A Y ANS))).
```

The inequality CDR.LESSP establishes that the measure
(COUNT A) decreases according to the well-founded
relation LESSP in the induction step of the scheme.
The above induction scheme generates two new
conjectures:

*Case 1.*
```
(IMPLIES (AND (NOT (LISTP (CDR A)))
              (LISTP Y)
              (LISTP A))
         (EQUAL (APPEND A (APPEND Y ANS))
                (APPEND (APPEND A Y) ANS))).
```

This simplifies, appealing to the lemmas CDR.CONS,
CAR.CONS and CONS.EQUAL, and expanding APPEND, to:

```
(IMPLIES (AND (NOT (LISTP (CDR A)))
              (LISTP Y)
              (LISTP A))
         (EQUAL (APPEND (CDR A) (APPEND Y ANS))
                (APPEND (APPEND (CDR A) Y) ANS))).
```

However this again simplifies, unfolding APPEND, to:

(TRUE).

*Case 2.*
```
(IMPLIES (AND (EQUAL (APPEND (CDR A) (APPEND Y ANS))
                     (APPEND (APPEND (CDR A) Y) ANS))
              (LISTP Y)
              (LISTP A))
         (EQUAL (APPEND A (APPEND Y ANS))
                (APPEND (APPEND A Y) ANS))).
```

This simplifies, applying the lemmas CDR.CONS, CAR.CONS and CONS.EQUAL, and expanding the definition of APPEND, to:

(TRUE).

That finishes the proof of *1.1, which, consequently, finishes the proof of *1. Q.E.D.

## E. SUMMARY

The purpose of this chapter was to provide an introduction to our function-based theory and to indicate how we prove theorems in the theory. As noted, all our proof techniques have been implemented in an automatic theorem-proving program. In fact, the last section was written, in its entirety, by our automatic theorem-prover in response to three user commands supplying the definitions of FLATTEN and MC.FLATTEN and the statement of the theorem to be proved. This book is about such questions as how function definitions are analyzed by our theorem-proving system to establish their admissibility, how the system discovers that (LISTP (FLATTEN X)) is a theorem when presented with the definition of FLATTEN, why the system chooses the inductions it does, and why some functions are expanded and others are not. We describe our proof techniques in detail after presenting a precise statement of our formal theory.

## F. NOTES

To illustrate several proof techniques, we instructed the theorem-prover to conduct the FLATTEN.MC.FLATTEN proof in an environment in which it was aware only of the axioms of our basic theory and the definitions of APPEND, FLATTEN, and MC.FLATTEN. In the proof, the program derived a version of the associativity of APPEND (formula *1.1) and proved it with a second induction. Had the theorem-prover previously proved the associativity of APPEND (and been instructed to remember it), the proof above would have been much shorter, as the lemma would have been involved in early simplifications. Later in the book, when we deal with more complicated examples such as the system's proof of the fundamental theorem of arithmetic, we will show the system working primarily from previously proved theorems rather than axioms.

The total amount of CPU time required to analyze the two definitions above and produce the proof is five-and-a-half seconds running compiled INTERLISP [53, 41] on a Digital Equipment Corporation KL-10. For an introduction to LISP see Allen's [1].

The English commentary produced during the definition-time analysis of functions and during proofs is typical of the system's output. Examples will be found throughout the book. The steps in a proof are described in real-time, as the proof is developed, so that the user can follow the theorem-prover's progress. To avoid boring the reader with repetitious phrasing, the system varies its sentence structure.

# III

# A Precise Definition of the Theory

In this chapter, we precisely define the mathematical theory underlying our proof techniques, the theory in which our theorem-prover proves theorems. We will present the axioms and principles of definition and induction to which the system appeals. We will not present the usual elementary rules of logic and equality, but instead we assume the reader is familiar with these rules. The parts of this chapter that are important are set off in boxes. The remainder of the text is motivation for the boxed material. In this motivational material we shall speak in a language of "naive" set theory. It is possible to embed our theory within a theory of sets and to derive our "principles" therein. However, we do not regard set theory or even quantification as being a part of our theory. The proofs our system produces depend only upon (a) the propositional calculus with variables and function symbols, (b) equality reasoning, (c) the rule of instantiation which permits us to infer that any instance of a theorem is a theorem, and (d) the boxed material that follows.

---

**A. SYNTAX**

We will use uppercase typewritten words, sometimes with embedded periods, hyphens, slashes, or digits, as variable symbols and function symbols. Examples are X, X1, PRIME.FACTORS, and PLUS. Associated with each function symbol is a nonnegative integer, the number of arguments the function symbol expects. The number of arguments associated with certain function symbols will become apparent as we proceed.

A *term* is a variable or a sequence consisting of a function symbol of n arguments followed by n terms. If the term t is not a variable and begins with the function symbol f, we say that t is a *call of* f.

We depart from the usual notation of F(X,Y) for function application and will instead write (F X Y). Examples of terms are thus: X, (TRUE), and (P (ADD1 X) Y). The first term is just a variable, the second is the application of the 0-ary function symbol TRUE to no arguments (and hence denotes a constant), and the third is the application of the dyadic function symbol P to the term (ADD1 X) and the variable Y.

To talk about terms, it is convenient to use so-called "metavariables" that are understood by the reader to stand for certain variables, function symbols, or terms. We will use only lowercase typewritten words as metavariables, and we will make clear what type of syntactic object the symbol is to denote. For example, if f denotes the function symbol G, and t denotes the term (ADD1 Y), then (f t X) denotes the term (G (ADD1 Y) X).

When we are speaking in naive set theory we use both upper- and lowercase words as variables ranging over numbers, sets, functions, etc. Context will make clear the range of these variables.

We imagine that axioms, such as function definitions, are added as "time goes by." Whenever we add a new shell or function definition, we insist that certain function symbols not have been mentioned in any previous axiom. We call a function symbol *new* until an axiom mentioning the function symbol has been added.

If i is an integer, then by an abuse of notation we let Xi denote the variable whose first character is X and whose other characters are the decimal representation of i. Thus, if i is 4, Xi is the variable X4.

A finite set s of ordered pairs is said to be a *substitution* provided that for each ordered pair ⟨v, t⟩ in s, v is a variable, t is a term, and no other member of s has v as its first component. The *result of substituting* a substitution s *into* a term p (denoted p/s) is the term obtained by simultaneously replacing, for each ⟨v, t⟩ in s, each occurrence of v as a variable in p with t.

## B. THE THEORY OF IF AND EQUAL

We find it necessary to reproduce the logic of truth functional propositions and equality at the term level. We assume the existence of two distinguished constants, (TRUE) and (FALSE). We use T and F as abbreviations for (TRUE) and (FALSE), respectively. We never use T or F as a variable. We axiomatize below the function EQUAL, of two arguments, to return T or F, depending on whether its two arguments are equal. We also axiomatize the function IF, of three arguments, to return its third argument if the first is F and otherwise return its second argument.

**Axiom**

$$T \neq F,$$

**Axiom**

$$X = Y \rightarrow (\text{EQUAL } X\ Y) = T,$$

**Axiom**

$$X \neq Y \rightarrow (\text{EQUAL } X\ Y) = F,$$

**Axiom**

$$X = F \rightarrow (\text{IF } X\ Y\ Z) = Z,$$

**Axiom**

$$X \neq F \rightarrow (\text{IF } X\ Y\ Z) = Y.$$

The logical functions are defined with the following equations:

**Definition**

(NOT P)
=
(IF P F T),

**Definition**

(AND P Q)
=
(IF P (IF Q T F) F),

**Definition**

    (OR P Q)

    =

    (IF P T (IF Q T F)),

**Definition**

    (IMPLIES P Q)

    =

    (IF P (IF Q T F) T).

We adopt the notational convention of writing (AND a b c) for (AND a (AND b c)), (AND a b c d) for (AND a (AND b (AND c d))), and so on. We make the same convention for OR.

We also adopt the notational convention of sometimes writing a term where a formula is expected (e.g., we may refer to the "theorem" p, where p is a term). When we write a term p where a formula is expected, it is an abbreviation for the formula $p \neq F$.

If a term p is a theorem, then by the rule of instantiation, the result of substituting any substitution into p is a theorem.

## C. WELL-FOUNDED RELATIONS

In the following sections we state a principle of induction, introduce inductively constructed objects such as the natural numbers and ordered pairs, and state a principle of definition for recursive functions. All of these extensions hinge on the idea of a "well-founded relation."

A function r of two arguments is said to be a *well-founded relation* if there is no infinite sequence $x_1$, $x_2$, $x_3$, ... with the property that $(r\ x_{i+1}\ x_i) \neq F$ for all integers i greater than 0. For example, suppose that (L x y) is T if x and y are nonnegative integers and x is less than y, and that (L x y) is F otherwise. Then L is a well-founded relation because there is no infinite sequence of nonnegative integers with the property that each successive integer is less than the previous one. That is, there is no infinite sequence $x_1$, $x_2$, $x_3$, $x_4$, ... such that

$$\ldots x_4 < x_3 < x_2 < x_1.$$

On the other hand, suppose that (LE x y) is T if x and y are nonnegative integers and $x \leq y$. LE is not a well-founded relation because

## 32 / III. A PRECISE DEFINITION OF THE THEORY

the infinite sequence 1, 1, 1, . . . has the property that

$$\ldots 1 \leq 1 \leq 1 \leq 1.$$

If r is a well-founded relation and ( r x y ) holds, we say that x is r-*smaller than* y.

For the purposes of our theory, functions are known to be well founded only by assumption in one of the following three ways:

(1) Whenever we axiomatize an inductively generated type of object, e.g., the integers, we explicitly assume a certain new function to be a well-founded relation. Such an assumption is inherent in any axiomatization of an inductively generated type. See Section E.

(2) We assume explicitly that the function LESSP is a well-founded relation in Section K. We present there an informal proof that LESSP is well founded.

(3) Whenever we have two previously assumed well-founded relations, we assume that the lexicographic relation induced by them is well founded. In Section J we define "induced" and present an informal proof of the well-foundedness of induced lexicographic relations.

The fact that a function has been assumed to be a well-founded relation is used only in our principles of induction and definition and in the formation of induced lexicographic relations.

It is possible to define formally in a theory of sets (for example, see Morse [43] or Kelley [25]) the concept of well-founded relation, to prove that certain relations are well founded, and to derive as metatheorems our principles of induction and definition. However, such a development is not within the scope of this work.

We say that x is an r-*minimal element of* S provided x is a member of S and no member of S is r-smaller than x. Later in this chapter we use the fact that if r is well founded, then for each nonempty set S, there exists an r-minimal element of S.

PROOF. Suppose that r is well founded and S is a nonempty set with no r-minimal element. Let f be a choice function on the power set of S. That is, suppose that for each nonempty subset s of S that ( f s ) is a member of s. Define the sequence $x_1, x_2, x_3, \ldots$ by letting $x_1$ be ( f S ) and by letting $x_{i+1}$ be ( f $s_i$ ), where $s_i$ is the set of all z in S r-smaller than $x_i$. For each i, $s_{i+1}$ is nonempty (otherwise, $x_i$ would be r-minimal). And for each i, $x_{i+1}$ is r-smaller than $x_i$. Q.E.D.

## D. INDUCTION

A rough sketch of our principle of induction is:

> Suppose that r denotes a well-founded relation, x is a variable, d is a function symbol, q is a term, and (IMPLIES q (r (d x) x)) is a theorem. Then, to prove p it is sufficient to prove the following two things:
>
> (base case): (IMPLIES (NOT q) p), and
>
> (induction step): (IMPLIES (AND q p') p), where p' is the result of substituting (d x) for x in p.

This is a version of the generalized principle of induction or Noetherian induction (see Bourbaki [6] and Burstall [12]).

The induction principle we actually use generalizes the principle sketched above in three ways:

> Instead of limiting the principle to one variable that is getting r-smaller, we use an n-tuple $x_1, \ldots, x_n$ of variables and some function m such that (m $x_1 \ldots x_n$) is getting r-smaller. The function m is called a "measure function."
>
> Instead of case splitting on q, we consider k + 1 cases, of which one is a base case and the remaining k are induction steps.
>
> We permit each of the k induction steps to have several induction hypotheses.

---

**The Induction Principle**

Suppose:
- (a) p is a term;
- (b) r is a function symbol that denotes a well-founded relation;
- (c) m is a function symbol of n arguments;
- (d) $x_1, \ldots, x_n$ are distinct variables;
- (e) $q_1, \ldots, q_k$ are terms;
- (f) $h_1, \ldots, h_k$ are positive integers; and
- (g) for $1 \leq i \leq k$ and $1 \leq j \leq h_i$, $s_{i,j}$ is a substitution and it is a theorem that
  
  (IMPLIES $q_i$ (r (m $x_1 \ldots x_n$)/$s_{i,j}$ (m $x_1 \ldots x_n$))).

*Continued*

Then p is a theorem if

> (IMPLIES (AND (NOT $q_1$) ... (NOT $q_k$))
>          p)

is a theorem and for each $1 \leq i \leq k$,

> (IMPLIES (AND $q_i$ p/$s_{i,1}$ ... p/$s_{i,h_i}$)
>          p)

is a theorem.

Note in particular that we have to prove $k + 1$ things (the "base case" and $k$ "induction steps"). Each induction step distinguishes a given case with one of the $q_i$'s and provides $h_i$ inductive instances of the conjecture being proved.

We now illustrate an application of the induction principle. Imagine that LESSP is well founded, that the axioms CAR.LESSP and CDR.LESSP have been added, and that FLATTEN and MC.FLATTEN have been introduced as in Chapter II. The first induction performed in the proof of the FLATTEN.MC.FLATTEN theorem of Chapter II is obtained by the following instance of our induction principle. p is the term (EQUAL (MC.FLATTEN X ANS) (APPEND (FLATTEN X) ANS)); r is LESSP; m is COUNT; n is 1; $x_1$ is X; k is 1; $q_1$ is the term (LISTP X); $h_1$ is 2; $s_{1,1}$ is {⟨X, (CAR X)⟩,⟨ANS, (MC.FLATTEN (CDR X) ANS)⟩}; $s_{1,2}$ is {⟨X, (CDR X)⟩, ⟨ANS, ANS⟩}; the axioms CAR.LESSP and CDR.LESSP establish the two theorems required by condition (g). The base case and the induction steps produced by this application of the induction principle are those exhibited in Chapter II.

We now prove that our induction principle is sound. Suppose we have in mind particular p, r, m, $x_i$, $q_i$, $h_i$, and $s_{i,j}$ satisfying conditions (a) through (g) above, and suppose the base case and induction steps are theorems. Below is a set theoretic proof of p.

PROOF. Without loss of generality we assume that the $x_i$ are X1, X2, ..., Xn; that r is R; that m is M; that Xn+1, Xn+2, ..., Xz are all of the variables other than X1, X2, ..., Xn in p, the $q_i$, and either component of any pair in any $s_{i,j}$; that p is (P X1 ... Xz); that $q_i$ is (Qi X1 ... Xz); and that $s_{i,j}$ replaces Xv, $1 \leq v \leq z$, with some term $d_{i,j,v}$.

Let RM be the dyadic function on z-tuples defined by (RM ⟨U1, ..., Uz⟩ ⟨V1, ..., Vz⟩) = (R (M U1...Uz) (M V1...Vz)). Note that RM is well founded. If p is not a theorem there must exist a z-tuple ⟨X1, ..., Xz⟩ such that (P X1 ... Xz) = F. Let ⟨X1, ..., Xz⟩ be an RM-minimal such z-tuple.

We now consider the cases on which, if any, of the $q_i$ are true on the chosen z-tuple.

*Case 1.* Suppose no $q_i$ is true, i.e., suppose (Q1 X1 ... Xz) = F, (Q2 X1 ... Xz) = F, ..., and (Qk X1 ... Xz) = F. Then by the base case (P X1 ... Xz) ≠ F, contradicting the assumption that (P X1 ... Xz) = F.

*Case 2.* Suppose that at least one of the $q_i$ is true. Without loss of generality we can assume that (Q1 X1 ... Xz) ≠ F. By condition (g) above we have

$$(R\ (M\ d_{1,1,1}\ ...\ d_{1,1,n})\ (M\ X1\ ...\ Xn)),$$
$$(R\ (M\ d_{1,2,1}\ ...\ d_{1,2,n})\ (M\ X1\ ...\ Xn)),$$
...,

and

$$(R\ (M\ d_{1,h_1,1}\ ...\ d_{1,h_1,n})\ (M\ X1\ ...\ Xn)).$$

Thus, by the definition of RM, we have

$$(RM\ \langle d_{1,1,1},\ ...,\ d_{1,1,z}\rangle\ \langle X1,\ ...,\ Xz\rangle),$$
$$(RM\ \langle d_{1,2,1},\ ...,\ d_{1,2,z}\rangle\ \langle X1,\ ...,\ Xz\rangle),$$
...,

and

$$(RM\ \langle d_{1,h_1,1},\ ...,\ d_{1,h_1,z}\rangle\ \langle X1,\ ...,\ Xz\rangle).$$

Since $\langle X1, ..., Xz\rangle$ is an RM-minimal z-tuple such that (P X1 ... Xz) = F, we have

$$(P\ d_{1,1,1}\ ...\ d_{1,1,z}) \neq F,$$
$$(P\ d_{1,2,1}\ ...\ d_{1,2,z}) \neq F,$$
...,

and

$$(P\ d_{1,h_1,1}\ ...\ d_{1,h_1,z}) \neq F.$$

Hence, by the first induction step, we derive (P X1 ... Xz) ≠ F, contradicting the assumption that (P X1 ... Xz) = F.  Q.E.D.


## E. SHELLS

Thus far the theory is somewhat impoverished in that it does not have any "interesting" objects. It would be convenient, for example, if we could refer to the natural numbers 0, 1, 2, ... and ordered pairs

## 36 / III. A PRECISE DEFINITION OF THE THEORY

from within our theory (as we have several times in discussions of our theory). We could invent appropriate axioms for each individual "type" of object. However, we want to ensure that no natural number is T, F, or an ordered pair. In addition, we want to specify how the primitive functions behave on "unexpected" arguments (e.g., what does the successor function return when given T as an argument?).[7] Because of considerations such as these, we address the general problem of extending the theory by adding a new "type" of object.

Among the most common objects in the applications of our theory are what we will call "shells."[8] A shell can be thought of as a colored n-tuple with restrictions on the colors of objects that can occupy its components. For example, the natural numbers can be thought of as shells: a number is either a blue 0 or a blue 1-tuple containing another blue object (namely, the predecessor of the tuple). Ordered pairs can be red 2-tuples containing arbitrary objects. The type consisting of lists of numbers can be either the green, empty list of numbers or else green 2-tuples ⟨x, y⟩ such that x is a number (blue) and y is a list of numbers (green). The fact that ordered pairs and lists of numbers have different colors prevents an ordered pair consisting of a number and a list of numbers from being confused with a list of numbers.

Because it is useful to be able to extend the theory by adding the axioms defining a new shell class and because the required set of axioms can be represented schematically, we will adopt a notational shorthand for adding new shells. We now specify informally the properties of a shell. The basic function for a shell class is one that "constructs" n-tuples of the appropriate color (e.g., the successor function or the pairing function). It is convenient if each shell class can (optionally) have one object that is in the class but is not such an n-tuple (e.g., 0 and the empty list of numbers). Because we will have many kinds of shells, we will need a function, called the "recognizer" of the class that returns T on objects of the class and F otherwise. We also require n "accessor" (or "destructor") functions associated with the class, that, when given an n-tuple of the right color, return the corresponding components of the n-tuple (e.g., the predecessor function is the "accessor" corresponding to the shell for numbers). Finally, we posit that any object in the class can be generated with a finite number of "con-

---

[7] One way to make sure that T is not a number or to escape from asking what is the successor of T is to employ a typed syntax. Indeed, Aubin [2] and Cartwright [14] have implemented theorem-provers for recursive functions that use typed syntax. However, we have grown so accustomed to the untyped syntax of predicate calculus, set theory, LISP, MACRO-10, and POP-2 that we simply do not like typed syntax.

[8] Our shells are inspired by the "structures" used by Burstall [12]. Recently, Oppen [45] has established the decidability of a theory similar to our shell theory.

structions" starting with the bottom object and objects not in the class. This is arranged by assuming a certain function to be a well-founded relation (e.g., one under which the predecessor of a nonzero number is smaller than the number itself).

Because we wish to have restrictions on the types of objects that can be components of a shell n-tuple, we must adopt some convention for specifying the restriction. We also adopt conventions for specifying what a constructor returns when one of its arguments fails to meet the required restriction and what an accessor returns when given an object of the wrong type as an argument. We require that there be associated with each argument position of each shell constructor a "type restriction" that limits the colors of objects that may occupy that component. The restriction is expressed in one of two ways: (a) that an object must have a color that is a member of an explicitly given set of previously (or currently being) axiomatized colors, or (b) that an object may *not* have a color in such a set. The type restriction is written as a propositional term (see condition (b) below). We also require that each argument have a "default value" that is permitted by the type restriction to occupy the corresponding component of the shell tuple. When one of the arguments to a constructor does not satisfy the corresponding type restriction, the default value for that argument position is used in its place. Finally, we arrange that the accessor for a position return the corresponding default value when given either the bottom object or an object of the wrong color.

---

**The Shell Principle**

To *add the shell* const *of* n *arguments with* (optionally, *bottom object* (btm)),
*recognizer* r,
*accessors* $ac_1$, ..., $ac_n$,
*type restrictions* $tr_1$, ..., $tr_n$,
*default values* $dv_1$, ..., $dv_n$, *and*
*well-founded relation* wfn,

where:

(a) const is a new function symbol of n arguments, (btm is a new function symbol of no arguments, if a bottom object is supplied), r, $ac_1$, ..., $ac_n$ are new function symbols of one argument, wfn is a new function symbol of two arguments, and all the above function symbols are distinct;

(b) each $tr_i$ is a term that mentions no symbol as a variable besides Xi and mentions no symbol as a function symbol besides IF, TRUE, FALSE, previously introduced shell recognizers, and r; and

(c)   if no bottom object is supplied, the $dv_i$ are bottom objects of previously introduced shells, and for each i, (IMPLIES (EQUAL Xi $dv_i$) $tr_i$) is a theorem; if a bottom object is supplied, each $dv_i$ is either (btm) or a bottom object of some previously introduced shell, and for each i,

```
(IMPLIES (AND (EQUAL Xi dv_i)
              (r (btm)))
         tr_i)
```

is a theorem,

means to extend the theory by doing the following (using T for (r (btm)) and F for all terms of the form (EQUAL x (btm)) if no bottom object is supplied):

(1)   assume the following axioms:

(OR (EQUAL (r X) T) (EQUAL (r X) F)),

(r (const X1 ... Xn)),

(r (btm)),

(NOT (EQUAL (const X1 ... Xn) (btm))),

and

```
(IMPLIES (AND (r X)
              (NOT (EQUAL X (btm))))
         (EQUAL (const (ac_1 X) ... (ac_n X))
                X));
```

(2)   for each i from 1 to n, assume the following axiom:

```
(IMPLIES tr_i
         (EQUAL (ac_i (const X1 ... Xn))
                Xi));
```

(3)   for each i from 1 to n, assume the following axiom:

```
(IMPLIES (OR (NOT (r X))
             (EQUAL X (btm))
             (AND (NOT tr_i)
                  (EQUAL X
                         (const X1 ... Xn))))
         (EQUAL (ac_i X) dv_i));
```

> (4) assume the axioms:
>
> (NOT (r T))
>
> and
>
> (NOT (r F));
>
> (5) for each recognizer r' of a shell class previously added to the theory, assume the following axiom:
>
> (IMPLIES (r X) (NOT (r' X)));
>
> (6) assume the axiom:
>
> (wfn X Y)
> =
> (OR t
>     (AND (r Y)
>         (NOT (EQUAL Y (btm)))
>         (OR (EQUAL X ($ac_1$ Y))
>            ...
>            (EQUAL X ($ac_n$ Y)))))),
>
> where t is the term (FALSE) if no shell has been added previously, and otherwise is (wfn' X Y) where wfn' is the well-founded relation name for the last shell previously added; and
>
> (7) assume wfn denotes a well-founded relation.
>
> If the $tr_i$ are not specified, they should each be assumed to be T.

The n axioms in (3) specify what the values of the accessors are when given "unexpected" arguments.

It is possible to prove the consistency of the theory resulting from the addition of a finite number of shells by exhibiting a model. A suitable model may be constructed by representing a (nonbottom) member of a shell class having n accessors as an n+1-tuple whose first component is an integer encoding the "color."

Note that merely because we add a finite number of shells we are not assured that every object in the world is in one of our shell classes. That is, we do not have an axiom that says: for any x, x is either T, or x is F, or x satisfies one of the shell recognizers. Indeed, this is an intended feature of the shell principle; we desire that any extension produced by adding shells can be further extended by additional shells without giving rise to inconsistency.

From the point of view of modeling programming language constructs, shells are valuable. They can play a role in the semantics of the usual kinds of "records" since records are often n-tuples with type restrictions on their components. Shells can also be used to model other kinds of common programming objects, such as the integers, atomic objects (words, capabilities, characters, file names), push down stacks, and character strings. In the next three sections, we will use shells to add the natural numbers, literal atoms, and ordered pairs to our theory.

## F. NATURAL NUMBERS

We now axiomatize the natural numbers, 0, 1, 2, etc., using the shell principle:

> **Shell Definition**
>
> Add the shell ADD1 of one argument
> with bottom object (ZERO),
> recognizer NUMBERP,
> accessor SUB1,
> type restriction (NUMBERP X1),
> default value (ZERO), and
> well-founded relation SUB1P.

This axiomatizes a new type of object we will call the "numbers." The numbers consist of the new object (ZERO) and all of the objects returned by the new function ADD1. We now informally repeat the axioms added by the shell principle. The numbers in parentheses indicate the corresponding clause of the definition of the shell addition principle. (1) The function NUMBERP (which recognizes numbers), always returns either T or F, and returns T on (ADD1 X) and (ZERO). (ZERO) is never returned by ADD1. If X is a non-(ZERO) number, then (ADD1 (SUB1 X)) is X. (2) If X is a number, then (SUB1 (ADD1 X)) is X. (3) SUB1 returns (ZERO) if applied to a nonnumber, (ZERO), or (ADD1 X1) when X1 is a nonnumber. (4) T and F are nonnumeric. (5) Because no other shells have been added, clause (5) does not contribute any axioms. (6) We define the function SUB1P so that (SUB1P X Y) is T if Y is a number other than (ZERO) and X is (SUB1 Y), and (SUB1P X Y) is F otherwise. (7) Finally, we assume that SUB1P is a well-founded relation.

Note the fundamental nature of the assumption that SUB1P is a well-founded relation: by virtue of this assumption (and our principle of induction), (P X) can be proved, for all X, by proving (P X) when X is not a number, proving (P (ZERO)), and proving that if X is a number other than (ZERO), then (P (SUB1 X)) implies (P X).

Among the theorems that can be derived from the above axioms is the theorem that if X and Y are numeric, then (ADD1 X) is equal to (ADD1 Y) if and only if X is equal to Y. See Appendix B.

We will abbreviate (ZERO) as 0, and any well-formed nest of ADD1's around a 0 as the decimal numeral expressing the number of ADD1 terms in the nest. Thus, 1 is an abbreviation for (ADD1 0), 2 is an abbreviation for (ADD1 (ADD1 0)), etc.

## G. LITERAL ATOMS

We want to be able to prove theorems about functions that manipulate symbols. For example, in Chapter XVII we prove the correctness of a function that translates from symbolic arithmetic expressions to sequences of instructions for a hand-held calculator. We write symbols as sequences of characters delimited by quotation marks (e.g., "X" and "ABC"). We could adopt the convention that "X", for example, was an abbreviation for 24. Such a convention is part of the logician's standard method for representing terms, known as Gödelization. However, we want to arrange for the symbols to be different from the integers. To this end we add a new shell class called the "literal atoms," and we adopt a syntactic convention that translates from quoted character sequences to literal atoms. The shell class is recognized by the new Boolean function LITATOM. The new function PACK, of one argument, takes an arbitrary object and returns a literal atom. (The name PACK derives from the INTERLISP operation for constructing literal atoms by concatenating sequences of characters.) (PACK x) is the same literal atom as (PACK y) if and only if x is the same object as y. The new function UNPACK, given a literal atom, returns the object used to construct it.

> **Shell Definition**
>
> Add the shell PACK of one argument
> with bottom object (NIL),
> recognizer LITATOM,
> accessor UNPACK,
> default value 0, and
> well-founded relation UNPACKP.

## 42 / III. A PRECISE DEFINITION OF THE THEORY

Note that since ADD1 was the last shell added and the well-founded relation for it was SUB1P, the new well-founded relation UNPACKP holds between X and Y either if SUB1P holds between X and Y or if X is the result of UNPACKing (the non-(NIL) literal atom) Y.

We now adopt a convention for abbreviating literal atoms as symbols. We suppose an enumeration $s_1$, $s_2$, $s_3$, ... of all symbols (character sequences) except "NIL". When we write "NIL" in a term position, it is an abbreviation for (NIL). When we write $s_i$ delimited by quotation marks in a term position, it is an abbreviation for (PACK i).

### H. ORDERED PAIRS

We axiomatize ordered pairs as follows:

---
**Shell Definition**

Add the shell CONS of two arguments
with recognizer LISTP,
accessors CAR and CDR,
default values "NIL" and "NIL", and
well-founded relation CAR.CDRP.

---

We sometimes think of ordered pairs as sequences, binary trees, or terms. For example,

(CONS 1 (CONS 2 (CONS 3 "NIL")))

may be thought of as the sequence 1, 2, 3.

(CONS (CONS 1 2) 3) may be thought of as the binary tree:

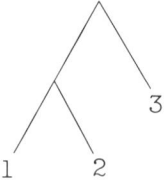

Finally,

(CONS "PLUS"
    (CONS "X" (CONS 3 "NIL"))).

may be thought of as the term (PLUS X 3).

---
Because nests of CARs and CDRs are frequently used, we provide a definition for each function symbol beginning with the letter C,

ending with the letter R, and containing only A's and D's in between. The body of the definition is just the appropriate nest of CARs and CDRs. For example,

**Definition**

>  ( CADDR X )
>  =
>  ( CAR ( CDR ( CDR X ) ) ) .

## I. DEFINITIONS

We have already defined certain simple functions, such as AND, OR, NOT, and IMPLIES. For example,

**Definition**

>  ( AND P Q )
>  =
>  ( IF P ( IF Q T F ) F ) .

Another simple function we define is ZEROP; it returns T if its argument is virtually 0 (in the sense that ADD1 and SUB1 treat it as 0) and F otherwise:

**Definition**

>  ( ZEROP X )
>  =
>  ( OR ( EQUAL X 0 ) ( NOT ( NUMBERP X ) ) ) .

In general, if the function symbol $f$ is new, if $x_1, \ldots, x_n$ are distinct variables, if the term body mentions no symbol as a variable other than these $x_i$, and if body does not mention $f$ as a function symbol, then adding the axiom

>  ( f $x_1$ ... $x_n$ ) = body

is a proper way to define a new function. Indeed, any use of the symbol $f$ as a function symbol in a term, such as ( f $t_1$ ... $t_n$ ), can be completely eliminated by replacing the term with the result of substituting $\{\langle x_1, t_1 \rangle, \ldots, \langle x_n, t_n \rangle\}$ into body.

However, one apparently mild generalization of the above scheme results in our being able to define functions that are considerably more interesting. This generalization allows the use of $f$ as a function symbol in the body of its own definition. For example, to define a function that returns the integer sum of its two arguments we could write

**Definition**

(SUM X Y)
= 
(IF (ZEROP X)
    Y
    (ADD1 (SUM (SUB1 X) Y))).

Unlike our previous definitions, the body of SUM mentions the function symbol being defined. That is, SUM is defined recursively. Nevertheless, SUM is well defined. For example, consider (SUM 3 4).

```
(SUM 3 4) = (ADD1 (SUM 2 4))
          = (ADD1 (ADD1 (SUM 1 4)))
          = (ADD1 (ADD1 (ADD1 (SUM 0 4))))
          = (ADD1 (ADD1 (ADD1 4)))
          = 7.
```

If we were to allow the new function symbol to occur arbitrarily in the right-hand side, we could define all "general recursive" functions. However, we could also fall into inconsistency. For example, were we to add the axiom

(RUSSELL X) = (IF (RUSSELL X) F T),

our theory would be inconsistent. To sketch a proof: (RUSSELL X) must be equal to F or not equal to F. If the former, then by the definition of RUSSELL it follows that (RUSSELL X) = T, a contradiction. If the latter, it follows that (RUSSELL X) = F, another contradiction. Q.E.D.

We now present a principle of definition that allows the introduction of recursive functions. The principle will not allow us to introduce all "general recursive" functions or even all "recursive" functions.[9] However, it will permit the definition of almost all the functions in which we are currently interested. And we shall prove that every application of the principle is sound (unlike the axiomatization of RUSSELL above).

---

**The Definition Principle**

In stating our principle of definition below, we say that a term is *f-free* if the symbol f does not occur in the term as a function symbol.

---

[9] See Peter [46] for a thorough treatment of general recursive functions.

# I. DEFINITIONS / 45

We say that a term t *governs* an occurrence of a term s in a term b either if b contains a subterm of the form (IF t p q) and the occurrence of s is in p, or if b contains a subterm of the form (IF t' p q), where t is (NOT t') and the occurrence of s is in q. Thus, P and (NOT Q) govern the first occurrence of S in

```
(IF P
    (IF (IF Q F S)
        S
        R)
    T).
```

Note that P and (IF Q F S) govern the second occurrence of S.

Our principle of definition is: *To define* f *of* $x_1, \ldots, x_n$ *to be* body (usually written "**Definition** (f $x_1 \ldots x_n$) = body"), where:

(a) f is a new function symbol of n arguments;
(b) $x_1, \ldots, x_n$ are distinct variables;
(c) body is a term and mentions no symbol as a variable other than $x_1, \ldots, x_n$; and
(d) there is a well-founded relation denoted by a function symbol r and a function symbol m of n arguments, such that for each occurrence of a subterm of the form (f $y_1 \ldots y_n$) in body and the f-free terms $t_1, \ldots, t_k$ governing it, it is a theorem that:

```
(IMPLIES (AND t₁ ... tₖ)
         (r (m y₁ ... yₙ) (m x₁ ... xₙ))),
```

means to add as an axiom the defining equation:

(f $x_1 \ldots x_n$) = body.

We now illustrate an application of the principle of definition. Imagine that LESSP is well founded and that the axioms CAR.LESSP and CDR.LESSP have been added as in Chapter II. The defining equation for MC.FLATTEN is added to our theory by the following instantiation of our principle of definition. f is the function symbol MC.FLATTEN; n is 2; $x_1$ is X and $x_2$ is ANS; body is the term

```
(IF (LISTP X)
    (MC.FLATTEN (CAR X)
                (MC.FLATTEN (CDR X) ANS))
    (CONS X ANS));
```

r is LESSP; m is the function symbol COUNT1, where (COUNT1 X Y) is defined to be (COUNT X). The two theorems required by (d) are

```
(IMPLIES (LISTP X)
         (LESSP (COUNT1 (CDR X) ANS)
                (COUNT1 X ANS))),
```

and

```
(IMPLIES (LISTP X)
         (LESSP (COUNT1 (CAR X)
                        (MC.FLATTEN (CDR X)
                                    ANS))
                (COUNT1 X ANS))).
```

Both theorems are easily proved from CAR.LESSP, CDR.LESSP, and the definition of COUNT1. Note that the second theorem is proved before any axiom about MC.FLATTEN has been posited, even though MC.FLATTEN is used as a function symbol in the theorem.

If we have defined ( f $x_1$ ... $x_n$) to be body, then we say that body is the *body* of f and that $x_i$ is the ith *formal parameter* of f. If body mentions f as a function symbol, we say that f is *recursive* and otherwise we say that f is *nonrecursive*.[10] If f has not been defined but has been mentioned as a function symbol of n arguments in an axiom, we say that Xi is the ith *formal parameter* of f, for i from 1 to n.

We now offer a justification for admitting recursive definitions. This justification will relieve the fears raised by RUSSELL. Roughly speaking, we shall prove that for each correct application of the definition principle, we can prove that there exists a unique function f that satisfies the defining equation ( f $x_1$ ... $x_n$) = body.

We shall construct the desired function using a standard set-theoretic method of partial functions that "approximate" the desired function. Unfortunately, different set theories supply different answers to the question: What is the value of applying a function to an object not in the function's domain? Instead of adopting a particular theory of sets, we shall instead make sure in the following proof not to apply any function to an object not in the function's domain. Furthermore, we shall assume the existence of some universal set D to be the domain of discourse for all the functions that we axiomatize in our theory. To be

---

[10] This is potentially confusing, since in both cases the function is (general) recursive in the usual mathematical sense. No confusion should arise from our convention—which is derived from everyday usage in computer programming—since we will nowhere discuss in this book functions that are not general recursive in the mathematical sense. Our definition principle does not permit mutually recursive definitions. If f were defined in terms of a new function g, then after the definition, g would no longer be new, and hence g could not be defined.

precise, we assume that:

> There exists a set D such that each function symbol f mentioned as a function symbol in any axiom denotes a function whose domain is $D^n$ and whose range is a subset of D, where n is the number of arguments of f.

If G is a function whose domain is a subset of $D^n$, for some n, and whose range is a subset of D, then the *extension* of G is the function on $D^n$ to D that is defined to be (G X1 ... Xn) if $\langle X1, \ldots, Xn \rangle$ is a member of the domain of g and is defined to be (TRUE) otherwise.

Suppose that we have in mind some specific f, $x_1, \ldots, x_n$, body, r, and m and suppose they satisfy the conditions (a) through (d) of the definition principle. Before we add the defining axiom ($fx_1 \ldots x_n$) = body, we wish to prove that there exists a unique function f defined on $D^n$ to D satisfying the equation. Without loss of generality, suppose that f, $x_1, \ldots, x_n$, r, and m are the symbols FN, X1, ..., Xn, R, and M, respectively.

Let us adopt the notational convention that b[s] is an abbreviation for the term obtained by replacing every occurrence of FN as a function symbol in the term b with the symbol s. (For example, if term is (ADD1 (FN X1)), then term[G] is an abbreviation for (ADD1 (G X1)).)

Let RM be the well-founded relation defined on n-tuples by (RM $\langle$U1, ..., Un$\rangle$ $\langle$V1, ... Vn$\rangle$) = (R (M U1 ... Un) (M V1 ... Vn)).

Let us say that a subset S of $D^n$ is RM-*closed* if and only if every member of $D^n$ RM-smaller than a member of S is itself a member of S.

Let us say that a function H is *partially correct* if (a) its domain is an RM-closed subset of $D^n$, (b) its range is a subset of D, and (c) if H' is the extension of H, then for each $\langle X1, \ldots, Xn \rangle$ in the domain of H, (H X1 ... Xn) = body[H'].

We now prove a lemma that is used frequently below. Its proof is quite tedious.

**Lemma** Suppose

> F1 is a function whose domain is a subset of $D^n$ and whose range is a subset of D,
>
> F1' is the extension of F1,
>
> G1 is partially correct,
>
> G1' is the extension of G1,
>
> $\langle X1, \ldots, Xn \rangle$ is in $D^n$, and

## 48 / III. A PRECISE DEFINITION OF THE THEORY

F1 and G1 are defined and agree upon every member of $D^n$ that is RM-smaller than $\langle X1, \ldots, Xn \rangle$.

Then body[F1'] = body[G1'].

PROOF. Let $\text{subterm}_1, \ldots, \text{subterm}_k$ be an enumeration of the occurrences of the subterms of body, and suppose that if the term $\text{subterm}_i$ is a proper subterm of the term $\text{subterm}_j$, then $i < j$. Let $\text{tests}_i$, for $1 \leq i \leq k$, be the conjunction of the FN-free terms governing $\text{subterm}_i$.

We prove, for i from 1 to k,

*1      (IMPLIES tests$_i$
              (EQUAL subterm$_i$[F1'] subterm$_i$[G1'])).

Suppose that we have proved *1 for all $j < i$. To prove *1 for i, we consider the form of subterm$_i$.

*Case 1.* The term subterm$_i$ is a variable. The proof in this case is immediate.

*Case 2.* The term subterm$_i$ has function symbol IF. Then subterm$_i$ is (IF subterm$_a$ subterm$_b$ subterm$_c$) for some a, b, and c all less than i. Hence we have previously proved

*2      (IMPLIES tests$_a$
              (EQUAL subterm$_a$[F1'] subterm$_a$[G1'])),

*3      (IMPLIES tests$_b$
              (EQUAL subterm$_b$[F1'] subterm$_b$[G1'])),

and

*4      (IMPLIES tests$_c$
              (EQUAL subterm$_c$[F1'] subterm$_c$[G1'])).

If subterm$_a$ uses FN as a function symbol, then tests$_i$ = tests$_a$ = tests$_b$ = tests$_c$ because the FN-free terms governing the occurrence of subterm$_i$ are the same as those governing subterm$_a$, subterm$_b$, and subterm$_c$. So *1 follows from *2, *3, and *4. If subterm$_a$ does not use FN as a function symbol, tests$_i$ = tests$_a$, tests$_b$ = (AND tests$_i$ subterm$_a$), and tests$_c$ = (AND tests$_i$ (NOT subterm$_a$)). If subterm$_a$ ≠ F, then subterm$_i$[F1'] = subterm$_b$[F1'] and subterm$_i$[G1'] = subterm$_b$[G1'], so *1 follows from *3. If subterm$_a$ = F, then subterm$_i$[F1'] = subterm$_c$[F1'] and subterm$_i$[G1'] = subterm$_c$[G1'], so *1 follows from *4.

*Case 3.* The term subterm$_i$ has a function symbol other than IF. Suppose subterm$_i$ has the form (g subterm$_{j_1}$ ... subterm$_{j_b}$) for some function symbol g, where $j_a < i$, $1 \leq a \leq b$. Note that $tests_i$ = $tests_{j_a}$, for $i \leq a \leq b$. We have previously proved, for $i \leq a \leq b$,

*a        (IMPLIES tests$_i$
                    (EQUAL subterm$_{j_a}$[F1'] subterm$_{j_a}$[G1'])).

If g is not the function symbol FN, then *1 follows immediately from the *a.

So suppose g is FN and b is thus n. If tests$_i$ = F, then *1 is true. If tests$_i \neq$ F, then by condition (d) of the application of the definition principle, it is a theorem that

*d        (IMPLIES tests$_i$
                    (R (M subterm$_{j_1}$ ... subterm$_{j_b}$)
                      (M X1 ... Xn))).

When *d was proved no axioms about FN had been posited. Hence, a proof of

             (IMPLIES tests$_i$
                    (R (M subterm$_{j_1}$[F1'] ... subterm$_{j_b}$[F1'])
                      (M X1 ... Xn)))

and a proof of

             (IMPLIES tests$_i$
                    (R (M subterm$_{j_1}$[G1'] ... subterm$_{j_b}$[G1'])
                      (M X1 ... Xn)))

may be similarly produced. Hence, $\langle$subterm$_{j_1}$[F1'], ..., subterm$_{j_b}$[F1']$\rangle$ is RM-smaller than $\langle$X1, ..., Xn$\rangle$ and so is $\langle$subterm$_{j_1}$[G1'], ..., subterm$_{j_b}$[G1']$\rangle$. But F1 and G1 are defined and agree on members of $D^n$ RM-smaller than $\langle$X1, ..., Xn$\rangle$. Thus, *1 follows from the hypothesis that tests$_i \neq$ F and the *a.    Q.E.D.

We now turn to the construction of the unique function satisfying the defining equation (FN X1 ... Xn) = body.

Let F0 be the union of all partially correct functions. We will prove that F0 is the desired function by demonstrating that (a) F0 is a function, (b) F0 is partially correct, and (c) the domain of F0 is $D^n$. The uniqueness of F0 follows from (a), (b), (c), and the fact that any other function defined on $D^n$ to D satisfying the defining equation is partially correct and hence a subset of F0.

PROOF THAT F0 IS A FUNCTION. If F0 is not a function it is multiply defined somewhere. Let $\langle$X1, ..., Xn$\rangle$ be an RM-minimal member of $D^n$ such that for some two distinct values, Z1 and Z2 say, both $\langle\langle$X1, ..., Xn$\rangle$, Z1$\rangle$ and $\langle\langle$X1, ..., Xn$\rangle$, Z2$\rangle$ are members of F0. Let F1

## 50 / III. A PRECISE DEFINITION OF THE THEORY

and G1 be partially correct functions that contributed $\langle\langle X1, \ldots, Xn\rangle, Z1\rangle$ and $\langle\langle X1, \ldots, Xn\rangle, Z2\rangle$ to F0. Let F1' and G1' be the extensions of F1 and G1. Both F1 and G1 are defined upon all members of $D^n$ RM-smaller than $\langle X1, \ldots, Xn\rangle$ because both are partially correct. F1 and G1 have the same values on all members of $D^n$ RM-smaller than $\langle X1, \ldots, Xn\rangle$ because $\langle X1, \ldots, Xn\rangle$ is RM-minimal. Therefore, by the lemma, body[F1'] = body[G1']. But Z1 = (F1 X1 ... Xn) = body[F1'] = body[G1'] = (G1 X1 ... Xn) = Z2 because both F1 and G1 are partially correct. Q.E.D.

PROOF THAT F0 IS PARTIALLY CORRECT. The domain of F0 is an RM-closed subset of $D^n$ because it is the union of RM-closed subsets of $D^n$. The range of F0 is a subset of D. Let F0' be the extension of F0. Let $\langle X1, \ldots, Xn\rangle$ be any member of the domain of F0 such that (F0 X1 ... Xn) $\neq$ body[F0']. Let G be a partially correct function with $\langle X1, \ldots, Xn\rangle$ in its domain such that (G X1 ... Xn) = (F0 X1 ... Xn). Let G' be the extension of G. By applying the lemma to F0, F0', G, G', and $\langle X1, \ldots, Xn\rangle$ we infer that body[F0'] = body[G']. But (F0 X1 ... Xn) = (G X1 ... Xn) = body[G'] = body[F0']. Q.E.D.

Before proving that the domain of F0 is $D^n$ we adopt the notational convention that body' is an abbreviation for the result of substituting Yi for Xi in body[F0']. For example, if body is the term (IF (LISTP X1) (FN X1 X2) F), then body' is (IF (LISTP Y1) (F0' Y1 Y2) F).

PROOF THAT THE DOMAIN OF F0 IS $D^n$. Suppose that F0 is not defined on every element of $D^n$. Let $\langle Y1, \ldots, Yn\rangle$ be an RM-minimal element of $D^n$ not in the domain of F0. Let F0' be the extension of F0. Let G be the function that results from adding to F0 the ordered pair $\langle\langle Y1, \ldots, Yn\rangle, \text{body}'\rangle$. If we can show that G is partially correct a contradiction will arise because then G would be a subset of F0 by the definition of F0. The domain of G is an RM-closed subset of $D^n$ because it was formed by adding to an RM-closed subset of $D^n$ an RM-minimal element of $D^n$ not in that subset. Let G' be the extension of G. We need to show that for every n-tuple $\langle X1, \ldots, Xn\rangle$ in the domain of G that (G X1 ... Xn) = body[G']. For every $\langle X1, \ldots, Xn\rangle$ in the domain of G, we may apply the lemma for G, G', F0, F0', and $\langle X1, \ldots, Xn\rangle$ to infer that body[G'] = body[F0']. If $\langle Y1, \ldots, Yn\rangle = \langle X1, \ldots, Xn\rangle$, then (G X1 ... Xn) = body[F0'] = body[G']. If $\langle Y1, \ldots, Yn\rangle \neq \langle X1, \ldots, Xn\rangle$, then (G X1 ... Xn) = (F0 X1 ... Xn) = body[F0'] = body[G']. Q.E.D.

That concludes the proof that the definition principle is sound. No constructivist would be pleased by the foregoing justification of recursive definition because of its freewheeling, set-theoretic character. The truth is that induction and inductive definition are more basic than the truths of high-powered set theory, and it is slightly odd to justify a fundamental concept such as inductive definition with set theory.

We have presented this proof only to provide the careful reader with some clear talk about our definition principle. The only other kind of discussion we might have presented would have consisted of examples and the truly hand-waving phrase "and so on." One of our teachers, Paul Lorenzen, once proclaimed that the correct way to introduce induction to a student in an ideal society was simply to draw strokes: |, ||, |||, ||||, and so on until the student "caught on."

## J. LEXICOGRAPHIC RELATIONS

Our theory requires one more concept: the idea of lexicographic relations.

> To define f to be the *lexicographic relation induced by* r and s, where
> (a)  f is a new function symbol of 2 arguments,
> (b)  r and s are function symbols of 2 arguments, and
> (c)  neither r nor s is f,
> means to add as an axiom the following defining equation:
> ```
> (f P1 P2) = (OR (r (CAR P1) (CAR P2))
>                 (AND (EQUAL (CAR P1) (CAR P2))
>                      (s (CDR P1) (CDR P2)))).
> ```

That is, f orders pairs of objects by first comparing their CAR components using r, but using s on their CDR components if the test with r fails and the CARs are equal. (The name "lexicographic" is inspired by the alphabetic order used by lexicographers.)

> If r and s denote well-founded relations and f is defined to be the lexicographic relation induced by r and s, then f denotes a well-founded relation.

PROOF. Suppose that $x_1$, $x_2$, ... were an infinite sequence and that for all positive i, $(f\ x_{i+1}\ x_i) \neq F$. By the definition of f, $(r\ (CAR\ x_{i+1})\ (CAR\ x_i)) \neq F$ or $(CAR\ x_{i+1}) = (CAR\ x_i)$. But since r is well founded, the sequence (CAR $x_1$), (CAR $x_2$), ... cannot be infinitely descending in r. Hence, for some j, for all positive p, (CAR $x_j$) = (CAR $x_{j+p}$). But the sequence (CDR $x_j$), (CDR $x_{j+1}$), ... must then be infinitely descending in s, a contradiction.  Q.E.D.

## K. LESSP AND COUNT

We have now finished defining the formal theory that we use as the logical basis of our theorem-proving system. We now use the theory to define two functions, LESSP and COUNT, that play central roles in our proofs (but have no role in the formal definition of the theory).

LESSP and COUNT are the well-founded relation and measure function we use most often in applying our principles of induction and definition.

(LESSP X Y) returns T if X is less than Y and F otherwise. LESSP treats nonnumeric arguments as 0. LESSP determines whether X is less than Y by counting them both down to 0, seeing which gets there first.

**Definition**

        (LESSP X Y)
          =
        (IF (ZEROP Y)
            F
            (IF (ZEROP X)
                T
                (LESSP (SUB1 X) (SUB1 Y)))).

Since (IMPLIES (NOT (ZEROP X)) (SUB1P (SUB1 X) X)) is a thorem, the first argument of LESSP gets SUB1P smaller in the recursive call. (In fact, so does the second argument.) Thus, LESSP is admitted under the principle of definition.

We claim that LESSP is a well-founded relation. That is, we claim there is no infinite sequence $x_1$, $x_2$, ... such that $x_{i+1}$ is LESSP-smaller than $x_i$. It is easy to see how to prove that LESSP is well founded in a suitable theory of sets, since SUB1P is well founded, and x is LESSP-smaller than y if and only if a finite number of SUB1s will reduce y to x (when both are numbers). We cannot state or prove the well-foundedness of LESSP within our theory.

## K. LESSP AND COUNT

> We assume LESSP to be a well-founded relation.

By virtue of this assumption, it is permitted to make induction arguments in which some measure gets LESSP-smaller in the induction hypotheses. Similarly, it is permitted to define recursive functions in which some measure gets LESSP-smaller in the recursive calls.

A particularly useful measure is the "size" of a shell object obtained by adding one to the sum of the sizes of its components. We first define the addition function for the nonnegative integers. We could use the function SUM defined above. However, SUM suffers from the disadvantage of sometimes returning a nonnumber (it returns Y, whatever that is, when X is 0). SUM is thus not commutative (e.g., (SUM 0 T) = T, while (SUM T 0) = 0). We thus make the following definitions:

**Definition**
```
(FIX X)
   =
(IF (NUMBERP X) X 0).
```

**Definition**
```
(PLUS X Y)
   =
(IF (ZEROP X)
    (FIX Y)
    (ADD1 (PLUS (SUB1 X) Y))).
```

We adopt the notational convention of writing (PLUS a b c) for (PLUS a (PLUS b c)), etc.

Now assume we have added all the shells we will use. We define the function COUNT to return 0 on bottom objects, to return 1 plus the sum of the COUNTs of the components of a nonbottom shell object, and to return 0 on any nonshell object. For example, if the only shells we were ever to add were ADD1, PACK, and CONS, we would define COUNT as

**Definition**
```
(COUNT X)
   =
(IF (NUMBERP X)
    (IF (EQUAL X 0) 0
        (ADD1 (COUNT (SUB1 X))))
```

*Continued*

## 54 / III. A PRECISE DEFINITION OF THE THEORY

```
(IF (LITATOM X)
    (IF (EQUAL X "NIL")
        0
        (ADD1 (COUNT (UNPACK X))))
    (IF (LISTP X)
        (ADD1 (PLUS (COUNT (CAR X))
                    (COUNT (CDR X))))
        0))).
```

The immediately preceding definition of COUNT would be admitted under the principle of definition since at each stage the argument is CAR.CDRP-smaller. In general, the definition of COUNT is admitted under the principle of definition, because at each stage the argument is smaller according to the well-founded relation of the last shell.

To permit the illusion that shells may be added at any time, our theorem-proving program does not actually employ the full definition of COUNT, but instead records (a) for each shell const the theorems

```
(EQUAL (COUNT (const X1 ... Xn))
       (ADD1 (PLUS (IF tr₁ (COUNT X1) 0)
                    ...
                    (IF trₙ (COUNT Xn) 0))))
```

and

```
(EQUAL (COUNT (btm)) 0)
```

(omitting the latter if const has no bottom object), and (b) for each $i$ from 1 to n, the theorem

```
(IMPLIES (AND (r X) (NOT (EQUAL X (btm))))
         (LESSP (COUNT (acᵢ X)) (COUNT X)))
```

(using T for (NOT (EQUAL X (btm))) if const has no bottom object). These theorems may be proved from the shell axioms and the definition of COUNT.

For example, the formula CAR.LESSP of Chapter II is recorded when the shell CONS is added. Strictly speaking, CAR.LESSP is not an axiom but a theorem since it can be proved, using the definition of COUNT. However, we have occasionally referred to CAR.LESSP as an axiom in Chapter II and the examples of this chapter.

## L. CONCLUSION

This concludes the discussion of our formal theory. We recap the topics presented:

We defined with axioms certain functions including IF and EQUAL.

We introduced the idea of well-founded relations.

We introduced a principle of induction.

We introduced a general mechanism for adding "new" types of "colored" n-tuples called "shells."

We used the shell principle to add axioms for the natural numbers, literal atoms, and ordered pairs.

We introduced a principle of definition which allows the introduction of recursive functions.

We introduced the concept of a lexicographic relation.

We used the principle of definition to introduce the usual "less than" function, assumed it was well-founded, and defined the measure function COUNT that computes the size of an object.

# IV

# The Correctness
# of a Tautology-Checker

Before we describe how we prove theorems in the theory just presented, it is important that the reader be familiar with the theory itself. In addition, it is useful to go through the proofs of some difficult theorems, so that the reader gets a feel for what is coming in subsequent chapters. Finally, readers unfamiliar with mechanical theorem-proving may be curious about how one proves theorems mechanically. Since all three of these objectives should be addressed before we begin to present our proof techniques, we have chosen to illustrate them all in a rather novel example: the mechanical proof of the correctness of a simple mechanical theorem-prover. In particular, we prove the correctness of a decision procedure for the propositional calculus.

In the standard development of logic, the propositional calculus is presented first. As in our theory, it often forms part of the logical underpinnings of richer theories. In addition, it offers a simple way of introducing certain important ideas such as soundness, completeness, and decision procedures. Because of its foundational role, discussions of the propositional calculus are usually carried on in the informal "metalanguage." For example, a common definition of the value of the formula "$p \wedge q$" is that it is "true if both p and q have the value true, and false otherwise." In this chapter we exercise the expressive power of our theory, and clarify certain aspects of it, by formalizing the semantics of a version of propositional calculus in our theory. We then introduce certain very simple theorem-proving ideas (such as how to apply a theorem as a rewrite rule, and how to keep track of what assumptions govern a subformula of a formula) by writing, as a recursive function in our theory, a decision procedure for the propositional calculus. Finally, we illustrate some of our proof techniques by proving that the decision procedure is well defined, recognizes only tautologies, and recognizes all tautologies. The proofs described are actually carried out by our own mechanical theorem-prover, and the discussion of the proofs illustrates the role of the human user in our automatic theorem-proving system.

## A. INFORMAL DEVELOPMENT

Throughout this chapter we will be concerned with the set of terms constructed entirely from variables, T, F, and the function symbol IF. We call such terms "propositional IF–expressions." Examples of propositional IF–expressions are

(IF A B C),

(IF T F (IF A B C)),

and

(IF (IF P Q F) (IF P T Q) T).

Note that the first of these expressions sometimes has the value F (when A is T and B is F, for example) but sometimes does not have the value F (when A is T and B is T, for example). On the other hand, the second expression always has the value F, and the third expression never has the value F. We call a propositional IF-expression that never has the value F a "tautology." Note that any formula of the ordinary propositional calculus can be converted to an equivalent propositional IF–expression by using the definitions of AND, OR, NOT, and IMPLIES presented in Chapter III. (Throughout the remainder of this chapter we shall use "expression" as a shorthand for "propositional IF–expression.")

It is our aim in this chapter to construct a procedure for deciding whether an expression is a tautology. Our first step is to indicate more precisely what we mean by "the value of an expression." Informally, let us say that v is an *assignment* provided that v is a function, its domain includes T, F, and the variables, and v maps T to T and F to F. If v is an assignment, then by "the assignment of x under v" we mean v(x). Then the *value of the expression* x *under the assignment* v is defined recursively as

if x has the form (IF p q r),
  then if the value of p under v is F
    then return the value of r under v,
    else return the value of q under v;
  else return the assignment of x under v.

We want to define a mechanical procedure that when given an expression x returns non-F if for every assignment v, the value of x under v is non-F, and returns F if for some assignment v, the value of x under v is F. We will call our procedure TAUTOLOGY.CHECKER.

There are many ways to write TAUTOLOGY.CHECKER. The question. What is the most efficient way to write TAUTOLOGY.-CHECKER? is actually one of the most important unsolved problems in computer science. One method, called the "truth table" method, considers all possible assignments of T and F to the variables in the given expression. The truth table method requires execution time exponential in the number of variables in the expression. No one knows a method that does not require exponential time in the worst case. Furthermore, no one has yet proved that all algorithms require exponential time in the worst case.

The version of TAUTOLOGY.CHECKER that we present is more efficient than the truth table method on one important class of expressions, namely, those in "IF-normal form."[11] An expression x is said to be in *IF-normal form* provided that no subterm of x beginning with an IF has as its first argument another term beginning with an IF. Of the three example formulas above, the first two are in IF-normal form and the last is not.

When a formula is in IF-normal form, we can decide whether it is a tautology very easily: we consider each "branch" through the expression and require either that the tests through which we pass are contradictory or that the tests through which we pass force the output to be something other than F. Consider, for example, the expression

(IF P (IF T T
           F)
      (IF Q (IF P F
                  Q)
            T)).

There are five branches through this expression (one output per line).

---

[11] "This nomenclature is an excellent example of the time-honored custom of referring to a problem we cannot handle as abnormal, irregular, improper, degenerate, inadmissible, and otherwise undesirable." From Kelley [25], on "normal" spaces.

The first branch returns T. The second branch returns F but can never be taken because the second test, on T, never fails. The third branch can never be taken because the first test on P must have returned F so the second must also. The fourth branch returns Q, which is not F because the earlier test on Q determined that Q was not F. And the last branch returns T. So the expression is a tautology.

Informally, then, we have a method for deciding which expressions in IF-normal form are tautologies. To use the method on expressions in general (rather that just those in IF-normal form), we convert the given expression into an equivalent one (that is, one that always has the same value) in IF-normal form. We achieve this normalization by applying the theorem

```
(EQUAL (IF (IF P Q R) LEFT RIGHT)
       (IF P (IF Q LEFT RIGHT)
             (IF R LEFT RIGHT)))
```

repeatedly, as a rewrite rule from left to right. That is, whenever we find an expression of the form (IF (IF p q r) left right), we replace it with the equivalent (IF p (IF q left right) (IF r left right)). Normalizing an expression may produce a formula that is exponentially larger than the one with which we started. So in the worst case, our procedure is at least exponential.

The foregoing sketch of a decision procedure for the tautology problem is very informal. Below we reconsider both the problem and its solution very formally—in fact we formalize both the problem and its solution using the theory presented in Chapter III.

One way to view the formal presentation is as an interaction between four participants: a "buyer" who wants to purchase a recursive function satisfying his specification of a tautology-checker; an "implementor" who encodes his knowledge of theorem-proving in a recursive function claimed to meet the specifications; a "theorem-prover" that proves that the implementor did his job; and a "mathematician user" who aids the theorem-prover by suggesting that it prove certain lemmas.

## B. FORMAL SPECIFICATION OF THE PROBLEM

In this section we play the role of the buyer: we specify our requirements by formally defining what a propositional IF-expression is, what the value of such an expression is, and what it means for a function to be a decision procedure.

## 1. Representing Expressions

Since we want to define functions on IF—expressions, we must represent IF-expressions as objects in our theory. From the point of view of general-purpose theorem-proving programs, the most natural and convenient way to represent terms is to represent variables as literal atoms and to represent the term ( f $x_1$ ... $x_n$ ) as the sequence whose CAR is the literal atom f and whose CDR is the sequence of objects representing the terms $x_1$ through $x_n$. This is the representation we use in our theorem-prover. However, this representation makes it awkward to refer to subterms. For example, if x represented (IF test left right), then in order to refer to the third argument one would write (CAR (CDR (CDR (CDR x)))).

With ease of reading in mind, we represent IF—expressions in this chapter by employing a new shell class (the green triples, say), called the IF.EXPRPs, which we introduce with

## Shell Definition

Add the shell CONS.IF of three arguments
with recognizer IF.EXPRP,
accessors TEST, LEFT.BRANCH, and RIGHT.BRANCH,
default values "NIL", "NIL", and "NIL", and
well-founded relation TEST.LEFT.BRANCH.RIGHT.BRANCHP.

Thus, we represent the term

(IF x y z)

as

(CONS.IF x' y' z')

where x', y', and z' are the representations of x, y, and z, respectively.

We use T and F to represent T and F, respectively. For the purposes of this example we agree that any object other than T, F, or an IF.EXPRP represents a variable.

## 2. Formal Definitions of ASSIGNMENT and VALUE

To represent assignments we use the standard "association list" technique from LISP programming. An association list (or simply "alist") is a sequence of pairs interpreted as associating with the CAR of each pair the item of information in the CDR. The recursive function ASSIGNMENT interprets alists as assignments of values to terms. ASSIGNMENT returns the value assigned to a given term in a given alist (or F if it is not explicitly assigned). ASSIGNMENT assigns T and F to themselves. Here is the definition of ASSIGNMENT:

**Definition**

```
(ASSIGNMENT VAR ALIST)
    =
(IF (EQUAL VAR T)
    T
    (IF (EQUAL VAR F)
        F
        (IF (NLISTP ALIST)
            F
            (IF (EQUAL VAR (CAAR ALIST))
                (CDAR ALIST)
                (ASSIGNMENT VAR
                            (CDR ALIST)))))).
```

(NLISTP X) is defined to be (NOT (LISTP X)).

The formal definition of the value of the expression X under the assignment ALIST is

**Definition**

```
(VALUE X ALIST)
    =
(IF (IF.EXPRP X)
    (IF (VALUE (TEST X) ALIST)
        (VALUE (LEFT.BRANCH X) ALIST)
        (VALUE (RIGHT.BRANCH X) ALIST))
    (ASSIGNMENT X ALIST)).
```

### 3. The Formal Correctness Specifications

As the buyer we want a decision procedure for the propositional calculus. We now specify what we require of a decision procedure.

First of all, we require that the decision procedure be introduced as a function. Let us call it TAUTOLOGY.CHECKER.

Second,* we require that if TAUTOLOGY.CHECKER recognizes an expression (in the sense that TAUTOLOGY.CHECKER returns something other than F when given the expression), then the expression must

## IV. THE CORRECTNESS OF A TAUTOLOGY-CHECKER

have a value other than F under all assignments.[12] Stated formally, this requirement is

**Theorem** TAUTOLOGY.CHECKER.IS.SOUND:
  (IMPLIES (TAUTOLOGY.CHECKER X)
     (VALUE X A)).

When specifying requirements, one must be very careful to say enough. The careless buyer might think that we have fully specified TAUTOLOGY.CHECKER. However, a function that always returned F (i.e., recognized nothing) would satisfy the above specification. We require more than that of TAUTOLOGY.CHECKER. In fact, we require that it recognize all tautologies; when TAUTOLOGY.CHECKER fails to recognize an expression there must exist an assignment for which the VALUE of the expression is F. Since we do not use existential quantification, how can we express the proposition that when TAUTOLOGY.CHECKER fails to recognize an expression there exists a falsifying assignment?

The answer is that we require that somebody define a recursive function that explicitly constructs a falsifying assignment for a nontautological expression. We call the function FALSIFY. Then the statement that the tautology-checker recognizes all tautologies is

**Theorem** TAUTOLOGY.CHECKER.IS.COMPLETE:
  (IMPLIES (NOT (TAUTOLOGY.CHECKER X))
     (EQUAL (VALUE X (FALSIFY X)) F)).

That is, if the tautology-checker fails to recognize an expression, then there is an assignment (namely (FALSIFY X)) for which the value of the expression is F. From our perspective as the buyer, the definition of FALSIFY is irrelevant. That is, if somebody were to supply us with a legal definition of TAUTOLOGY.CHECKER (and also one of FALSIFY) such that the above two conjectures were theorems, then we would believe that TAUTOLOGY.CHECKER was a decision procedure.

---

[12] The purist may note that in a freewheeling set-theoretic approach to validity, one would consider all assignments rather than merely the finite assignments to which we limit ourselves when we represent assignments as finite alists. Of course, no real damage is done, because (in a suitable theory of sets) one can prove that it is sufficient to restrict one's attention to assignments that assign a meaning to the finite number of variables in the term in which one is interested.

## C. THE FORMAL DEFINITION OF TAUTOLOGY.CHECKER

We now take on the role of the implementor. The buyer's definitions and conjectures in the last section specify what is required of TAUTOLOGY.CHECKER. As the implementor, we now define a function we claim has the desired properties. Since the specified task is to write a simple mechanical theorem-prover, it happens (in this example) that the implementor must appeal to some of the basic ideas in mechanical theorem-proving.

### 1. TAUTOLOGY.CHECKER

The definition of TAUTOLOGY.CHECKER requires two subsidiary concepts:

NORMALIZE, which given an expression returns an equivalent expression in IF-normal form, and

TAUTOLOGYP, which given an expression in IF-normal form and a list of assumptions determines if the expression is never F under those assumptions.

Given NORMALIZE and TAUTOLOGYP we define TAUTOLOGY.CHECKER as

**Definition**

```
(TAUTOLOGY.CHECKER X)
    =
(TAUTOLOGYP (NORMALIZE X) "NIL").
```

*a. NORMALIZE*

Recall that the basic idea behind NORMALIZE is to apply the theorem

```
(EQUAL (IF (IF P Q R) LEFT RIGHT)
       (IF P (IF Q LEFT RIGHT)
             (IF R LEFT RIGHT)))
```

as a rewrite rule until we have removed all IFs from the tests of other IFs. Thus, to normalize an expression x we proceed as follows:

If x is not an IF.EXPRP, we return x.

If x is of the form (IF test left right), then we ask whether test is of the form (IF p q r).

If so, we return the result of normalizing the expression (IF p (IF q left right)
                       (IF r left right)).

If not, we return the expression
(IF test left' right'), where left' and right' are the results of normalizing left and right.

The formal definition of this process is

**Definition**

```
(NORMALIZE X)
   =
(IF
 (IF.EXPRP X)
 (IF
  (IF.EXPRP (TEST X))
  (NORMALIZE (CONS.IF (TEST (TEST X))
                      (CONS.IF (LEFT.BRANCH (TEST X))
                               (LEFT.BRANCH X)
                               (RIGHT.BRANCH X))
                      (CONS.IF (RIGHT.BRANCH (TEST X))
                               (LEFT.BRANCH X)
                               (RIGHT.BRANCH X))))
  (CONS.IF (TEST X)
           (NORMALIZE (LEFT.BRANCH X))
           (NORMALIZE (RIGHT.BRANCH X))))
 X).
```

*b. TAUTOLOGYP*

Now that we can put an expression into normal form, we consider TAUTOLOGYP, which determines whether an expression x in IF–normal form is never F. Recall that the basic idea is to explore every branch through the IF-expression and check that either the tests along the branch are contradictory or else that the output of the branch is forced by the tests to be non-F. Our definition requires that TAUTOLOGYP have a second argument, alist, used to remember the tests that have been seen and whether they are being assumed true or false on the current branch. Since all the tests in a normalized IF-expression are variables (or else the constants T or F) the alist of

## C. THE FORMAL DEFINITION OF TAUTOLOGY.CHECKER

assumptions can also be thought of as an assignment to some of the variables in x. Here is how TAUTOLOGYP determines that x is never F under the assumptions in alist:

If x is not an IF.EXPRP, then it is either T, F, or a variable.

If x is T, then it is never F.

If x is F, then clearly it is "sometimes" F.

If x is neither T nor F, then it is a variable. If x is currently assumed non-F, then x is never F under alist; otherwise x is sometimes F.

If x is an IF.EXPRP, say representing (IF test left right), then there are three possibilities:

If test (which must be a variable, T, or F) is T or assumed non-F in alist, then x is never F under alist if and only if left never is.

If test is F or assumed F in alist, then x is never F under alist if and only if right never is.

Otherwise, x is never F under alist if and only if both:

left is never F under the assumptions in alist plus the additional assumption that test is non-F, and

right is never F under the assumptions in alist plus the additional assumption that test is F.

To define TAUTOLOGYP formally we use four auxiliary functions:

ASSIGNEDP, which determines whether a given variable is explicitly assumed F or non-F in a given alist,

ASSIGNMENT (defined above), which returns the assumed value of a variable in an alist,

ASSUME.TRUE, which adds to an alist the pair that indicates that a given variable is being assumed non-F, and

ASSUME.FALSE, which adds to an alist the pair that indicates that a given variable is being assumed F.

The definitions of these functions are in Appendix A. ASSIGNEDP is very similar to ASSIGNMENT. ASSUME.TRUE and ASSUME.FALSE simply CONS the appropriate pair onto the alist.

The formal definition of TAUTOLOGYP is

## Definition

```
(TAUTOLOGYP X ALIST)
    =
(IF
 (IF.EXPRP X)
 (IF (ASSIGNEDP (TEST X) ALIST)
     (IF (ASSIGNMENT (TEST X) ALIST)
         (TAUTOLOGYP (LEFT.BRANCH X) ALIST)
         (TAUTOLOGYP (RIGHT.BRANCH X) ALIST))
     (AND (TAUTOLOGYP (LEFT.BRANCH X)
                     (ASSUME.TRUE (TEST X) ALIST))
          (TAUTOLOGYP (RIGHT.BRANCH X)
                     (ASSUME.FALSE (TEST X) ALIST))))
 (ASSIGNMENT X ALIST)).
```

### 2. Summary of the Simple Theorem-Proving Ideas

As implementor of the buyer's specifications, we claim that our job is done: TAUTOLOGY.CHECKER is a decision procedure for the propositional calculus. Before discussing the proof of this assertion we review the simple theorem-proving techniques illustrated:

Terms can (and must) be represented as objects. In our case, we used the shell facility to define a new class of objects, called the IF-expressions.

Both NORMALIZE and TAUTOLOGYP illustrate how terms can be explored mechanically.

The function NORMALIZE shows how a theorem can be exhaustively applied as a rewrite rule in a mechanical way.

TAUTOLOGYP illustrates the use of alists of assumptions (manipulated by the functions ASSUME.TRUE, ASSUME.FALSE, ASSIGN-EDP, and ASSIGNMENT) to remember one's context while exploring a formula.

The use of NORMALIZE to produce an equivalent expression that is amenable to exploration by TAUTOLOGYP illustrates the value of "simplification" even when it produces a larger expression.

Our own proof techniques (in contrast to those of an implementor concerned only with the propositional calculus), involve ideas such as those above, but usually in much more elaborate form. In fact, the next five chapters of this book are concerned entirely with how we represent formulas, how we remember the assumptions governing the subexpressions in an expression, and how we use rewrite rules to simplify expressions.

## D. THE MECHANICAL PROOFS

We now describe the proofs of TAUTOLOGY.CHECKER.IS.SOUND and TAUTOLOGY.CHECKER.IS.COMPLETE. The proofs are constructed entirely by our mechanical theorem-prover. However, the mathematician user of the system plays an important role by suggesting that the theorem-prover prove certain lemmas first, thus making the system cognizant of truths that were not evident to it previously. It is important to understand from the outset that an incompetent human user may not be able to get the theorem-prover to admit that a valid conjecture is a theorem. On the other hand, the user does not have to be trusted: if a monkey were to cause the theorem-prover to announce that a conjecture were a theorem, the conjecture would indeed be a theorem.

In this section we primarily play the role of the mathematician user. However, occasionally we take the role of the theorem-prover to illustrate how the proofs go. The precise user commands to our theorem-prover can be found in Appendix A. Events CONS.IF through TAUTOLOGY.CHECKER.IS.SOUND are what the user had to type to define the necessary concepts and cause the theorem-prover to prove the desired results.

### 1. Complying with the Principle of Definition

Before anything can be proved, the necessary concepts must be introduced. In particular, we must introduce the CONS.IF shell class and the functions derived above. Furthermore, the system must confirm that the shell principle and the definition principle admit the definitions.

The introduction of CONS.IF is trivial. The only restrictions are syntatic in nature and the system immediately adds the axioms indicated in Chapter III.

As for the function definitions, the system must confirm that in each definition some measure of the arguments is decreasing according to a well-founded relation in every recursive call. All but one of the functions easily pass the test because either they have no recursive calls (e.g., ASSUME.TRUE) or they do simple recursion on components of LISTPs or IF.EXPRPs (e.g., ASSIGNMENT and TAUTOLOGYP) so that the COUNT of some argument is always decreasing. However, one function, namely NORMALIZE, provides a nontrivial challenge because in one of its recursive calls the COUNT of its only argument increases. Before the theorem-prover will accept the definition of NORMALIZE, we must lay a certain amount of groundwork. If the reader has not yet discovered a measure and well-founded relation that decrease when NORMALIZE recurses, he is encouraged to do so before reading further.

We know of several measures and well-founded relations justifying the definition of NORMALIZE. We discuss only one of them, namely, the first one we discovered.

In general, one has to justify all the recursive calls simultaneously. That is, it will not do to find one measure that goes down in one call and a different one that goes down in another unless they can be lexicographically combined to account for all the recursions. We will eventually exhibit such a lexicographic combination. To derive it, we will start with the first recursive call in NORMALIZE, the one in which the COUNT of its argument increases as a result of transforming (IF (IF p q r) left right) to (IF p (IF q left right) (IF r left right)).

Consider the function IF.DEPTH:

**Definition**

>     (IF.DEPTH X)
>     =
>     (IF (IF.EXPRP X)
>         (ADD1 (IF.DEPTH (TEST X)))
>         0).

IF.DEPTH is admitted under the principle of definition since it recurses on components of IF.EXPRPs. In particular, the lemma

>     (IMPLIES (IF.EXPRP X)
>              (LESSP (COUNT (TEST X)) (COUNT X)))

(added by the shell mechanism when the CONS.IF shell was axiomatized) is precisely the theorem required by the definition principle.

Now that IF.DEPTH has been admitted, consider what it does: it counts the depth of IF-nesting in the TEST component of a propositional IF-expression. Thus, when a function (such as NORMALIZE) recurses by changing (IF (IF p q r) left right) to (IF p (IF q left right) (IF r left right)), it drives down the IF.DEPTH of its argument, according to the well-founded relation LESSP. Thus, the lemma

**Theorem** IF.DEPTH.GOES.DOWN:

```
(IMPLIES
    (AND (IF.EXPRP X) (IF.EXPRP (TEST X)))
    (LESSP (IF.DEPTH (CONS.IF (TEST (TEST X)) Y Z))
           (IF.DEPTH X)))
```

suggests a justification of NORMALIZE. The system proves IF.-DEPTH.GOES.DOWN using the definitions of LESSP and IF.DEPTH.

But now let us consider the other two recursive calls in NORMALIZE, those that recurse on the LEFT.BRANCH and RIGHT.BRANCH of IF.EXPRPs. If IF.DEPTH decreased according to LESSP in those recursions, we would be finished: IF.DEPTH would be a measure that got LESSP-smaller in all recursions, and LESSP is well founded. But IF.DEPTH does not necessarily decrease in the last two recursions. For example, the IF.DEPTH of the expression

```
(IF X (IF (IF A B C) D E) Y)
```

is 1, while the IF.DEPTH of its LEFT.BRANCH is 2. In general, the IF.DEPTH of a branch of an IF.EXPRP is unrelated to that of the expression itself and might be arbitrarily bigger.

We remedy the situation by defining a second measure, called IF.COMPLEXITY, and by proving that it gets LESSP-smaller on the latter two recursive calls while not changing on the first recursive call. Given such results it is clear that the measure

```
(CONS (IF.COMPLEXITY X) (IF.DEPTH X))
```

gets lexicographically smaller on each call (using the well-founded relation induced by LESSP and LESSP).

Defining IF.COMPLEXITY so that it decreases on the branches of IF.EXPRPs is easy. (For example, COUNT goes down.) But defining IF.COMPLEXITY so that it also stays unchanged when (IF (IF p q r) left right) is transformed to (IF p (IF q left right) (IF r left right)) is more difficult. The definition of IF.COMPLEXITY that we use is

**Definition**

```
(IF.COMPLEXITY X)
  =
(IF (IF.EXPRP X)
    (TIMES (IF.COMPLEXITY (TEST X))
```

```
                (PLUS (IF.COMPLEXITY (LEFT.BRANCH X))
                      (IF.COMPLEXITY (RIGHT.BRANCH X))))
     1).
```

The three theorems

**Theorem** IF.COMPLEXITY.GOES.DOWN1:

```
(IMPLIES (IF.EXPRP X)
         (LESSP (IF.COMPLEXITY (LEFT.BRANCH X))
                (IF.COMPLEXITY X))),
```

**Theorem** IF.COMPLEXITY.GOES.DOWN2:

```
(IMPLIES (IF.EXPRP X)
         (LESSP (IF.COMPLEXITY (RIGHT.BRANCH X))
                (IF.COMPLEXITY X))),
```

**Theorem** IF.COMPLEXITY.STAYS.EVEN:

```
(IMPLIES
 (AND (IF.EXPRP X) (IF.EXPRP (TEST X)))
 (EQUAL
   (IF.COMPLEXITY
          (CONS.IF (TEST (TEST X))
                   (CONS.IF (LEFT.BRANCH (TEST X))
                            (LEFT.BRANCH X)
                            (RIGHT.BRANCH X))
                   (CONS.IF (RIGHT.BRANCH (TEST X))
                            (LEFT.BRANCH X)
                            (RIGHT.BRANCH X))))
   (IF.COMPLEXITY X)))
```

establish that IF.COMPLEXITY has the desired properties. The system proves these theorems using the definitions of LESSP and IF.COMPLEXITY, together with the inductively proved lemma that IF.COMPLEXITY is never 0 and about twenty well-known lemmas about PLUS, TIMES, and LESSP and the relations between them. The user must have the theorem-prover prove these twenty arithmetic lemmas before it proves the IF.COMPLEXITY lemmas. We do not descend into the proofs here. The statements of the lemmas are among those preceding the tautology theorems in Appendix A. (In fact, we had the system prove these elementary arithmetic lemmas long before we even considered the tautology problem; the system recalled them automatically during the IF.COMPLEXITY proofs.)

Once the IF.COMPLEXITY theorems have been proved by the theorem-prover (and thus brought to its attention), the theorem-prover accepts the definition of NORMALIZE under our principle of definition and responds with

    The lemmas IF.COMPLEXITY.GOES.DOWN1,
    IF.COMPLEXITY.GOES.DOWN2, IF.COMPLEXITY.STAYS.EVEN
    and IF.DEPTH.GOES.DOWN can be used to prove that:

        (CONS (IF.COMPLEXITY X) (IF.DEPTH X))

    decreases according to the well-founded lexicographic
    relation induced by LESSP and LESSP in each recursive
    call. Hence, NORMALIZE is accepted under the
    principle of definition. Observe that:

        (OR (IF.EXPRP (NORMALIZE X))
            (EQUAL (NORMALIZE X) X))

    is a theorem.
    CPU time (devoted to theorem-proving): 1.388 seconds

It is important to note that the admissibility of NORMALIZE has been *proved* and in no way assumed. In particular, we defined certain functions (IF.DEPTH and IF.COMPLEXITY) that were admitted to the theory because of shell axioms. Then we proved certain theorems about those functions, namely, that the IF.DEPTHs (or IF.COMPLEXITYs) of certain expressions were LESSP-smaller than those of others. Some of these theorems required induction to prove—inductions justified by shell axioms. Finally, invoking the theorems just proved, the well-foundedness of LESSP, and the principle of lexicographic relations, we exhibited a measure and well-founded relation justifying the definition of NORMALIZE. Note further that the newly invented measure and well-founded relation permit an induction not permitted by the shell axioms alone—an induction in which we may assume an instance about (IF p (IF q left right) (IF r left right)) while trying to prove a conjecture about (IF (IF p q r) left right).

## 2. Mechanical Proof of TAUTOLOGY.CHECKER.IS.SOUND

Now we turn our attention to the proofs of our main theorems. To enable the theorem-prover to prove TAUTOLOGY.CHECKER.IS.-SOUND we decompose the problem into three main lemmas establishing properties of NORMALIZE and TAUTOLOGYP, the two "subroutines" of TAUTOLOGY.CHECKER. To prove one of these main lemmas the theorem-prover needs several subsidiary lemmas about assignments. Below we present the decomposition, the subsidiary lemmas, and the proofs of the main lemmas. Finally, we combine the main lemmas to prove TAUTOLOGY.CHECKER.IS.SOUND.

*a. Decomposition*

We can establish that TAUTOLOGY.CHECKER recognizes only tautologies by proving that

when TAUTOLOGYP returns non-F on an expression in IF-normal form, the value of the expression is non-F under all assignments,

NORMALIZE produces expressions in IF-normal form, and

NORMALIZE produces an expression with the same VALUE as its input (so that if one is a tautology, the other is also).

Note that this was exactly the decomposition of the problem employed when we, in the role of the implementor, defined TAUTOLOGY.CHECKER.

We define NORMALIZED.IF.EXPRP to be the recursive function that determines whether or not an expression is in IF-normal form (see Appendix A). The formal statements of the three main lemmas are

**Theorem** TAUTOLOGYP.IS.SOUND:

(IMPLIES (AND (NORMALIZED.IF.EXPRP X)
              (TAUTOLOGYP X A1))
         (VALUE X (APPEND A1 A2))),

**Theorem** NORMALIZE.NORMALIZES:

(NORMALIZED.IF.EXPRP (NORMALIZE X)),

**Theorem** NORMALIZE.IS.SOUND:

(EQUAL (VALUE (NORMALIZE X) A)
       (VALUE X A)).

Note that the first lemma is more general than required by the decomposition. The decomposition calls for

(IMPLIES (AND (NORMALIZED.IF.EXPRP X)
              (TAUTOLOGYP X "NIL"))
         (VALUE X A)).

That is, if X is in IF-normal form and (TAUTOLOGYP X "NIL") is non-F, then the VALUE of X is non-F under all assignments. The more general TAUTOLOGYP.IS.SOUND, which can be proved by induction, says that if X is in IF-normal form and (TAUTOLOGYP X A1) is non-F, then the VALUE of X is non-F under any assignment having A1 as an initial segment. TAUTOLOGYP.IS.SOUND reduces to the desired lemma when A1 is "NIL", since (APPEND "NIL" A2) is A2.

*b. Subsidiary Lemmas*

Before proving the three main lemmas, we make explicit four facts about assignments and VALUE.

First, since ASSIGNMENT returns F if the variable in question is not explicitly assigned in A, we conclude that when (ASSIGNMENT X A) is non-F, X must be explicitly assigned:

**Theorem** ASSIGNMENT.IMPLIES.ASSIGNEDP:

    (IMPLIES (ASSIGNMENT X A)
         (ASSIGNEDP X A)).

Second, since ASSIGNMENT returns the first assignment it finds for X in an alist, the assignment of X in the alist (APPEND A B) is either the assignment of X in A (if X is explicitly assigned in A) or else is the assignment of X in B:

**Theorem** ASSIGNMENT.APPEND:

    (EQUAL (ASSIGNMENT X (APPEND A B))
       (IF (ASSIGNEDP X A)
          (ASSIGNMENT X A)
          (ASSIGNMENT X B))).

Third, if an alist contains a pair that can be deleted without changing the assignment of the variable involved, then the value of any expression under the original alist is equal to that under the shorter alist. In particular, if VAR already has value VAL in A, then the value of any given expression under (CONS (CONS VAR VAL) A) is the same as its value under A. Since we are interested only in whether VALUEs of expressions are F or non-F, we can generalize the theorem to

    (IMPLIES (IFF VAL (ASSIGNMENT VAR A))
        (IFF (VALUE X (CONS (CONS VAR VAL) A))
          (VALUE X A))),

where (IFF X Y) is T if X and Y are both F or both non-F and F otherwise. In order to make the lemma useful as a rewrite rule, we actually express it as two implications:

**Theorem** VALUE.CAN.IGNORE.REDUNDANT.ASSIGNMENTS:
```
(AND
 (IMPLIES (AND (IFF VAL (ASSIGNMENT VAR A))
               (VALUE X A))
          (VALUE X (CONS (CONS VAR VAL) A)))
 (IMPLIES (AND (IFF VAL (ASSIGNMENT VAR A))
               (NOT (VALUE X A)))
          (NOT (VALUE X (CONS (CONS VAR VAL) A))))).
```

Fourth, and finally, if X is an IF.EXPRP and is in IF–normal form, then the VALUE of (TEST X) under A is just the assignment of (TEST X) in A:

**Theorem** VALUE.SHORT.CUT:
```
        (IMPLIES (AND (IF.EXPRP X)
                      (NORMALIZED.IF.EXPRP X))
                 (EQUAL (VALUE (TEST X) A)
                        (ASSIGNMENT (TEST X) A))).
```

The proofs of these four lemmas are all straightforward. The first three are proved by induction and the fourth is immediate from the definitions of NORMALIZED.IF.EXPRP and VALUE. We do not discuss the proofs.

*c. Main Lemmas*

We now return to the main lemmas in the decomposition of TAUTOLOGY.CHECKER.IS.SOUND. The hardest of these lemmas is the first:

**Theorem** TAUTOLOGYP.IS.SOUND:
```
        (IMPLIES (AND (NORMALIZED.IF.EXPRP X)
                      (TAUTOLOGYP X A1))
                 (VALUE X (APPEND A1 A2))).
```

We now sketch the machine's proof. Let (p X A1 A2) be a schematic representation of the above conjecture. The machine describes its induction analysis as follows:

```
Let us appeal to the induction principle. The
recursive terms in the conjecture suggest four
inductions. They merge into three likely candidate
```

inductions, none of which is unflawed. However, one
is more likely than the others. We will induct
according to the following scheme:

```
(AND (IMPLIES (NOT (IF.EXPRP X))
              (p X A1 A2))
     (IMPLIES (AND (IF.EXPRP X)
                   (p (RIGHT.BRANCH X)
                      (CONS (CONS (TEST X) F) A1)
                      A2)
                   (p (LEFT.BRANCH X)
                      (CONS (CONS (TEST X) T) A1)
                      A2)
                   (p (RIGHT.BRANCH X) A1 A2)
                   (p (LEFT.BRANCH X) A1 A2))
              (p X A1 A2))).
```

The inequalities LEFT.BRANCH.LESSP and RIGHT.BRANCH.LESSP establish that the measure (COUNT X) decreases according to the well-founded relation LESSP in the induction step of the scheme. Note, however, the inductive instances chosen for A1.

(The inequalities LEFT.BRANCH.LESSP and RIGHT.BRANCH.LESSP are added by the addition of the CONS.IF shell.)

The base case (in which X is not an IF.EXPRP) can be simplifed to

```
            (IMPLIES (AND (NOT (IF.EXPRP X))
                          (ASSIGNMENT X A1))
                     (ASSIGNMENT X (APPEND A1 A2)))
```

by applying the definitions of NORMALIZED.IF.EXPRP, TAUTOLOGYP, and VALUE, and further reduces to T using ASSIGNMENT.APPEND and ASSIGNMENT.IMPLIES.ASSIGNEDP.

The induction step is considerably more complicated. We break our analysis of it into two cases according to the VALUE of (TEST X) under (APPEND A1 A2), which, by VALUE.SHORT.CUT and ASSIGNMENT.APPEND, gives rise to many cases depending on whether (TEST X) is assigned in A1 or A2 and what the assignment is. We sketch just one of these cases to indicate how the proof goes.

Suppose that (TEST X) is unassigned in A1 and has a non-F assignment in A2. The conclusion, (p X A1 A2), of the induction step simplifies to

```
*concl
(IMPLIES (AND (NOT (IF.EXPRP (TEST X)))
              (NORMALIZED.IF.EXPRP (LEFT.BRANCH X))
              (NORMALIZED.IF.EXPRP (RIGHT.BRANCH X))
              (TAUTOLOGYP (LEFT.BRANCH X)
                          (CONS (CONS (TEST X) T)
                                A1))
              (TAUTOLOGYP (RIGHT.BRANCH X)
                          (CONS (CONS (TEST X) F)
                                A1)))
         (VALUE (LEFT.BRANCH X) (APPEND A1 A2)))
```

using the definitions of NORMALIZED.IF.EXPRP, TAUTOLOGYP, and VALUE, the lemmas ASSIGNMENT.APPEND and ASSIGNMENT.IMPLIES.ASSIGNEDP, and the case assumptions about the assignment of (TEST X).

Consider the second induction hypothesis:

(p (LEFT.BRANCH X) (CONS (CONS (TEST X) T) A1) A2),

which, after applying the definition of APPEND, is

```
*hyp
(IMPLIES (AND (NORMALIZED.IF.EXPRP (LEFT.BRANCH X))
              (TAUTOLOGYP (LEFT.BRANCH X)
                          (CONS (CONS (TEST X) T)
                                A1)))
         (VALUE (LEFT.BRANCH X)
                (CONS (CONS (TEST X) T)
                      (APPEND A1 A2)))).
```

By virtue of ASSIGNMENT.APPEND and our case analysis, we know that (TEST X) is already assigned non-F in (APPEND A1 A2). Thus, by VALUE.CAN.IGNORE.REDUNDANT.ASSIGNMENTS, the pair (CONS (TEST X) T) can be deleted from the alist (CONS (CONS (TEST X) T) (APPEND A1 A2)) (occurring in the conclusion of *hyp above) without changing the truth value of any expression. That is, we know that *hyp is equivalent to

```
*hyp'
(IMPLIES (AND (NORMALIZED.IF.EXPRP (LEFT.BRANCH X))
              (TAUTOLOGYP (LEFT.BRANCH X)
                          (CONS (CONS (TEST X) T)
                                A1)))
         (VALUE (LEFT.BRANCH X)
                (APPEND A1 A2))).
```

Since *hyp' propositionally implies the induction conclusion,
*concl, we have completed the proof of the case in which (TEST X)
is unassigned in A1 and assigned non-F in A2.

The remaining cases of the induction step are similar.

The proofs of the two NORMALIZE lemmas, namely, that NORMALIZE
produces expressions in IF-normal form while preserving their
values, are straightforward by induction. However, the inductions are
interesting because of the unusual recursion in NORMALIZE. To illustrate the induction, we exhibit the machine's proof of NORMALIZE.IS.SOUND. The proof of NORMALIZE.NORMALIZES is similar.

**Theorem** NORMALIZE.IS.SOUND:

```
(EQUAL (VALUE (NORMALIZE X) A)
       (VALUE X A))
```

Give the conjecture the name *1.
Perhaps we can prove it by induction. The recursive
terms in the conjecture suggest two inductions,
neither of which is unflawed, and both of which
appear equally likely. So we will choose the one that
will probably lead to eliminating the nastiest
expression. We will induct according to the following
scheme:

```
(AND
 (IMPLIES (NOT (IF.EXPRP X)) (p X A))
 (IMPLIES
     (AND (IF.EXPRP X)
          (IF.EXPRP (TEST X))
          (p (CONS.IF (TEST (TEST X))
                      (CONS.IF (LEFT.BRANCH (TEST X))
                               (LEFT.BRANCH X)
                               (RIGHT.BRANCH X))
                      (CONS.IF (RIGHT.BRANCH (TEST X))
                               (LEFT.BRANCH X)
                               (RIGHT.BRANCH X)))
             A))
     (p X A))
 (IMPLIES (AND (IF.EXPRP X)
               (NOT (IF.EXPRP (TEST X)))
               (p (RIGHT.BRANCH X) A)
               (p (LEFT.BRANCH X) A))
          (p X A))).
```

## IV. THE CORRECTNESS OF A TAUTOLOGY-CHECKER

The inequalities IF.COMPLEXITY.GOES.DOWN1,
IF.COMPLEXITY.GOES.DOWN2, IF.COMPLEXITY.STAYS.EVEN
and IF.DEPTH.GOES.DOWN establish that the measure:

    (CONS (IF.COMPLEXITY X) (IF.DEPTH X))

decreases according to the well-founded lexicographic relation induced by LESSP and LESSP in each induction step of the scheme. The above induction scheme generates three new conjectures:

*Case 1.*
(IMPLIES (NOT (IF.EXPRP X))
       (EQUAL (VALUE (NORMALIZE X) A)
            (VALUE X A))).

This simplifies, unfolding NORMALIZE and VALUE, to:

    (TRUE).

*Case 2.*
(IMPLIES
 (AND
  (IF.EXPRP X)
  (IF.EXPRP (TEST X))
  (EQUAL
   (VALUE
    (NORMALIZE
       (CONS.IF (TEST (TEST X))
            (CONS.IF (LEFT.BRANCH (TEST X))
                  (LEFT.BRANCH X)
                  (RIGHT.BRANCH X))
            (CONS.IF (RIGHT.BRANCH (TEST X))
                  (LEFT.BRANCH X)
                  (RIGHT.BRANCH X))))
    A)
   (VALUE (CONS.IF (TEST (TEST X))
              (CONS.IF (LEFT.BRANCH (TEST X))
                  (LEFT.BRANCH X)
                  (RIGHT.BRANCH X))
              (CONS.IF (RIGHT.BRANCH (TEST X))
                  (LEFT.BRANCH X)
                  (RIGHT.BRANCH X)))
         A)))
 (EQUAL (VALUE (NORMALIZE X) A)
      (VALUE X A))),

which we simplify, applying RIGHT.BRANCH.CONS.IF,
LEFT.BRANCH.CONS.IF and TEST.CONS.IF, and expanding
the definitions of VALUE and NORMALIZE, to:

    (TRUE).

*Case 3.*
(IMPLIES
  (AND (IF.EXPRP X)
      (NOT (IF.EXPRP (TEST X)))
      (EQUAL (VALUE (NORMALIZE (RIGHT.BRANCH X)) A)
           (VALUE (RIGHT.BRANCH X) A))
      (EQUAL (VALUE (NORMALIZE (LEFT.BRANCH X)) A)
           (VALUE (LEFT.BRANCH X) A)))
  (EQUAL (VALUE (NORMALIZE X) A)
     (VALUE X A))).

This simplifies, appealing to the lemmas RIGHT.-
BRANCH.CONS.IF, LEFT.BRANCH.CONS.IF and TEST.CONS.-
IF, and unfolding the functions NORMALIZE and
VALUE, to:

    (TRUE).

That finishes the proof of *1. Q.E.D.
CPU time (devoted to theorem-proving): 8.852 seconds

    The induction performed above is analogous to the recursion in
NORMALIZE. There is a base case for the possibility that X is not a
CONS.IF and there are two induction steps. In the first, X is assumed
to be of the form (IF (IF p q r) left right) and a single inductive
instance is provided, in which X is replaced by (IF p (IF q left
right) (IF r left right)). In the second, X is assumed to be of
the form (IF p left right), where p is not an IF-expression, and
two inductive instances are provided, one in which X is replaced by
left and the other in which X is replaced by right. The induction is
justified by the measure and well-founded relation justifying NOR-
MALIZE. The simplifications of the three cases of the induction rely
upon definitions and the axioms of the CONS.IF shell.

## d. *TAUTOLOGY.CHECKER.IS.SOUND*

    The proof of TAUTOLOGY.CHECKER.IS.SOUND follows immedi-
ately from our three main lemmas, given the following "bridge"
lemma:

**Theorem** TAUTOLOGY.CHECKER.SOUNDNESS.BRIDGE:

(IMPLIES (AND (TAUTOLOGYP Y A1)
              (NORMALIZED.IF.EXPRP Y)
              (EQUAL (VALUE X A2)
                     (VALUE Y (APPEND A1 A2))))
         (VALUE X A2)).

The bridge lemma tells the mechanical theorem-prover that to prove that (VALUE X A2) is true it is sufficient to show that it is equal to (VALUE Y (APPEND A1 A2)), where Y is in IF–normal form and a TAUTOLOGYP under A1. The bridge lemma is trivial to prove: the result of substituting (VALUE Y (APPEND A1 A2)) for (VALUE X A2) in the bridge lemma is an instance of TAUTOLOGYP.IS.SOUND. Without the bridge lemma, the mechanical theorem-prover would not consider "reducing" a problem about (VALUE X A) to one about (VALUE (NORMALIZE X) (APPEND "NIL" A)), which is a necessary step in our decomposition of the problem.

Given the bridge, the proof of

**Theorem** TAUTOLOGY.CHECKER.IS.SOUND:

>   (IMPLIES (TAUTOLOGY.CHECKER X)
>            (VALUE X A))

is trivial. After replacing TAUTOLOGY.CHECKER by its definition, the theorem-prover obtains

>   (IMPLIES (TAUTOLOGYP (NORMALIZE X) "NIL')
>            (VALUE X A)).

But the theorem-prover can derive the above conjecture from the bridge lemma as follows. It instantiates the bridge lemma, replacing Y by (NORMALIZE X), A1 by "NIL", and A2 by A:

(IMPLIES (AND (TAUTOLOGYP (NORMALIZE X) "NIL")
              (NORMALIZED.IF.EXPRP (NORMALIZE X))
              (EQUAL (VALUE X A)
                     (VALUE (NORMALIZE X)
                            (APPEND "NIL" A))))
         (VALUE X A)).

The second hypothesis reduces to T by NORMALIZE.NORMALIZES and the third reduces to T by NORMALIZE.IS.SOUND and the definition of APPEND. After thus removing these true hypotheses the theorem-prover obtains

## D. THE MECHANICAL PROOFS / 81

```
(IMPLIES (TAUTOLOGYP (NORMALIZE X) "NIL")
         (VALUE X A)).
```

Q.E.D.

### 3. Mechanical Proof of TAUTOLOGY.CHECKER.IS.COMPLETE

All of the above proof steps were described from the theorem-prover's point of view. We now turn our attention to the proof of

**Theorem** TAUTOLOGY.CHECKER.IS.COMPLETE:

```
(IMPLIES (NOT (TAUTOLOGY.CHECKER X))
         (EQUAL (VALUE X (FALSIFY X)) F)),
```

and describe it from the user's point of view.

### a. FALSIFY

In our scenario, neither the buyer nor the implementor were required to define FALSIFY. However, before the theorem-prover can prove TAUTOLOGY.CHECKER.IS.COMPLETE the mathematician user must define FALSIFY.

Recall that FALSIFY must return an assignment that falsifies any expression not recognized by our tautology-checker. The definition of FALSIFY is extremely similar to that of TAUTOLOGY.CHECKER. FALSIFY puts the expression into IF–normal form with NORMALIZE and tries to construct a falsifying assignment for that equivalent expression, using a function called FALSIFY1 that is very similar to TAUTOLOGYP.

FALSIFY1 walks through a normalized IF–expression with an alist that assigns values to some variables. The function tries to extend that alist to one that falsifies the current expression. FALSIFY1 returns F if it fails to find such an alist, and otherwise returns the extended alist. The following two observations are basic to how FALSIFY1 works:

An unassigned variable can be falsified by assuming it false.

(IF test left right), where test is an unassigned variable, can be falsified by assuming test true and falsifying left (if possible) or by assuming test false and falsifying right (if possible).

Of course, FALSIFY1 must respect its current assignments (e.g., if the test of an IF is already assigned, FALSIFY1 must try to falsify the appropriate branch).

The definitions of FALSIFY1 and FALSIFY are

## Definition

```
(FALSIFY1 X ALIST)
    =
(IF (IF.EXPRP X)
    (IF (ASSIGNEDP (TEST X) ALIST)
        (IF (ASSIGNMENT (TEST X) ALIST)
            (FALSIFY1 (LEFT.BRANCH X) ALIST)
            (FALSIFY1 (RIGHT.BRANCH X) ALIST))
        (IF (FALSIFY1 (LEFT.BRANCH X)
                      (ASSUME.TRUE (TEST X) ALIST))
            (FALSIFY1 (LEFT.BRANCH X)
                      (ASSUME.TRUE (TEST X) ALIST))
            (FALSIFY1 (RIGHT.BRANCH X)
                      (ASSUME.FALSE (TEST X)
                                    ALIST))))
    (IF (ASSIGNEDP X ALIST)
        (IF (ASSIGNMENT X ALIST) F ALIST)
        (CONS (CONS X F) ALIST))),
```

## Definition

```
        (FALSIFY X)
            =
        (FALSIFY1 (NORMALIZE X) "NIL").
```

Both definitions are accepted immediately by the theorem-prover.

*b. Sketch of the Proof of*
*TAUTOLOGY.CHECKER.IS.COMPLETE*

We prove that if TAUTOLOGY.CHECKER returns F, then there exists a falsifying assignment by proving that

when TAUTOLOGYP fails to recognize a normalized IF-expression, FALSIFY1 returns an assignment (rather than F) on the expression, and

if FALSIFY1 returns an assignment (rather than F) for a normalized IF-expression, then the value of the expression under the assignment is F.

The formal statements of these lemmas are

**Theorem** TAUTOLOGYP.FAILS.MEANS.FALSIFY1.WINS:

```
        (IMPLIES (AND (NORMALIZED.IF.EXPRP X)
```

```
              (NOT (TAUTOLOGYP X A))
              A)
          (FALSIFY1 X A)),
```

**Theorem** FALSIFY1.FALSIFIES:

```
     (IMPLIES (AND (NORMALIZED.IF.EXPRP X)
                   (FALSIFY1 X A))
              (EQUAL (VALUE X (FALSIFY1 X A)) F)).
```

TAUTOLOGY.CHECKER.IS.COMPLETE follows from the above two lemmas and NORMALIZE.NORMALIZES and NORMALIZE.IS.SOUND. To get the theorem-prover to put them together in the desired way, a "bridge" version of FALSIFY1.FALSIFIES is required:

**Theorem** TAUTOLOGY.CHECKER.COMPLETENESS.BRIDGE:

```
     (IMPLIES (AND (EQUAL (VALUE Y (FALSIFY1 X A))
                          (VALUE X (FALSIFY1 X A)))
                   (FALSIFY1 X A)
                   (NORMALIZED.IF.EXPRP X))
              (EQUAL (VALUE Y (FALSIFY1 X A)) F)).
```

The proof of the bridge is trivial using FALSIFY1.FALSIFIES.

All that remains is to prove FALSIFY1.FALSIFIES and TAUTOLOGYP.FAILS.MEANS.FALSIFY1.WINS. These two theorems yield immediately to the correct induction provided the previously proved ASSIGNMENT.IMPLIES.ASSIGNEDP and VALUE.SHORT.CUT are used and the following property of FALSIFY1 is known by the theorem-prover: if VAR is explicitly assigned in A, then the assignment of VAR in (FALSIFY1 X A) is the assignment of VAR in A if (FALSIFY1 X A) is non-F. The formal statement of the relationship between FALSIFY1 and ASSIGNMENT is

**Theorem** FALSIFY1.EXTENDS.MODELS:

```
     (IMPLIES (ASSIGNEDP X A)
              (EQUAL (ASSIGNMENT X (FALSIFY1 Y A))
                     (IF (FALSIFY1 Y A)
                         (ASSIGNMENT X A)
                         (EQUAL X T)))).
```

This property of FALSIFY1 is a crucial aspect of the process of falsification: it is illegal to change the assignment of a variable in midstream. FALSIFY1.EXTENDS.MODELS is proved by induction. The proof appeals to the previously proved ASSIGNMENT.IMPLIES.ASSIGNEDP.

## E. SUMMARY

The theorem-prover (with some guidance from the user) has thus established that TAUTOLOGY.CHECKER is well defined, only recognizes tautologies, and recognizes all tautologies.

It is worthwhile to summarize our objectives in this chapter.

First, we hope the reader better understands our theory and its expressive power. The utility and power of IF and the convenience of T and F as objects was demonstrated. The shell principle was used to introduce IF–expressions as objects. The use of measures and well-founded relations to justify recursive definitions was illustrated. In addition, we demonstrated how we could justify definitions by introducing new measures and proving theorems establishing that the measures decrease. We also illustrated the induction principle in use. Finally, FALSIFY illustrated how we use recursive functions to express existential quantification. Advocates of quantification may feel that our lack of quantification makes it difficult for us to state certain conjectures. We agree; but we observe that the use of explicit existential quantification makes it more difficult to find constructive proofs. Any constructive proof of a conjecture involving existential quantification (such as the proposition that when TAUTOL-OGY.CHECKER returns F on an expression there exists a falsifying assignment) must exhibit a method (such as FALSIFY) for obtaining the objects alleged to exist.

The second objective of this chapter was to expose the reader to our proof techniques and to indicate the role of the user in the theorem-proving process. Perhaps the easiest way to summarize the description of the proofs is to note that the proof of TAUTOL-OGY.CHECKER.IS.COMPLETE is in many ways harder than the proof of TAUTOLOGY.CHECKER.IS.SOUND but that our discussion of it was much more brief. The reason is that we presented a good deal of the proof of TAUTOLOGY.CHECKER.IS.SOUND from the theorem-prover's point of view: we talked about such things as induction schemes, case splitting, and use of function definitions and previously proved theorems. The discussion of TAUTOLOGY.CHECKER.IS.COMPLETE, on the other hand, was almost entirely from the user's point of view: we talked about the concepts and "obvious" relationships that hold between them, and left the proofs to the theorem-prover.

The third objective of this chapter was to illustrate briefly some simple theorem-proving ideas. We illustrated how to represent and manipulate terms mechanically, how to use theorems to rewrite expressions, and how to maintain a representation of one's knowledge about the context of a term in an expression.

Finally, let us note one unstated objective: illustrating the value of definitions and proofs. Consider what the mathematician user of the theorem-prover must tell the buyer when the final proof is completed. He can say, simply, "Here is the function you ordered. Your two conjectures are theorems. Go in peace." The mathematician does not have to mention the decomposition of the problem, because all the lemmas were proved. Furthermore, he does not have to mention NOR-MALIZED.IF.EXPRP or any of the other functions not written by the buyer, because those functions were all introduced under the principle of definition. In contrast, consider the mathematician's duty to the buyer had the proofs required adding nondefinitional axioms about concepts (even concepts, such as NORMALIZED.IF.EXPRP, not involved in the statement of the main results but just in their proofs); he would have had to say, "I'm afraid we did not prove your conjectures. However, I have managed to shift the burden of proof back to you. If you will just accept these few axioms . . ."

## F. NOTES

Church [17, pp. 129–132], discusses a formulation of the propositional calculus very similar to ours. He uses [B, A, C] to denote what we denote by (IF A B C). Interestingly, he suggests the oral reading: "[p, q, r]" is "p or r according as q or not q." Church's version of propositional calculus differs from ours principally because our IF may return arbitrary objects, not merely truth values. It is odd that logicians have not sufficiently seen the utility of IF to include it in their formal theories. To our knowledge, McCarthy [34] first introduced the IF we use.

We are grateful to Greg Nelson, of Stanford University, who first suggested that we try to prove TAUTOLOGYP.IS.SOUND. We tried to do so several months before beginning to write this book and failed, primarily because, as users of the system, we failed to see the importance of VALUE.CAN.IGNORE.REDUNDANT.ASSIGNMENTS. After beginning the book we realized it would be beneficial to take advantage of the theory's expressive power to exhibit a simple theorem-prover in it. Thus, we returned to the TAUTOLOGYP example with the results reported above. We then elaborated it by introducing NORMAL-IZE and TAUTOLOGY.CHECKER, and proving that TAUTOL-

OGY.CHECKER is a decision procedure. The entire effort required about 12 hours of real-time by one of the authors playing the role of the buyer, implementor, and user. Most of that time was spent watching the theorem-prover failing to prove theorems, figuring out the facts that were "obvious" to the user but unknown to the system, and instructing the system to prove those facts as lemmas. The total amount of computer time expended to check all the definitions and prove all the theorems in the final sequence is about 10 minutes.

# V

# An Overview of How We Prove Theorems

In this and the next ten chapters, we describe some techniques for proving theorems in our theory. In fact, we describe how our mechanical theorem-proving program works. The reader familiar with mechanical theorem-proving should be able to reconstruct our theorem-prover from these discussions. However, the reader not familiar with mechanical theorem-proving will be able to follow our presentation without difficulty. No knowledge of computer programming is required, but we describe our proof techniques in complete detail and leave no important decisions to the reader's imagination.

## A. THE ROLE OF THE USER

Suppose that someone, whom we shall call "the user," has conjectures that he wants proved by a very careful but unimaginative device that we shall call "the theorem-prover." The user typically intuits a theorem and perhaps even sketches a proof on paper. We assume the user is willing to help the theorem-prover by formulating the conjectures in the most general way and by laying appropriate groundwork.

We require that when the user brings a conjecture to the theorem-prover's attention (by asking it to prove it or assume it as an axiom), he give the theorem-prover hints regarding how the theorem should be used subsequently. A theorem may be used in any of four ways: as a rewrite, elimination, generalization, or induction lemma. Rewrite lemmas are used to simplify conjectures by replacing terms with other terms. Elimination lemmas are used to remove certain "undesirable" expressions from conjectures. Generalization lemmas point out properties of terms that are good to keep in mind when generalizing formulas. Induction lemmas point out that a given operation drives a given measure down according to some well-founded relation. The precise meaning of these lemma types will be provided when we describe our heuristics for proving theorems. When the user suggests the types for a lemma, he is not merely suggesting that the lemma can be used in the ways indicated (that can be checked mechanically), but that he wants the lemma to be used in the ways indicated.

## B. CLAUSAL REPRESENTATION OF CONJECTURES

We now present the concept of clause as used in our theorem-prover. Through the remainder of this work, a *clause* is a list of terms. The meaning that we attach to a clause is the disjunction of its members. Following Robinson [49], we sometimes delimit clauses with set brackets. For example, if p, q, and r are terms, the clause {p, q, r} is the list of the three terms p, q, and r and means $p \neq F$, $q \neq F$, or $r \neq F$. Sometimes we display clauses as terms to make them easier to read. We may display the clause {p, (NOT q), r} as (IMPLIES (AND (NOT p) q) r), we may display the clause {p, q} as (IMPLIES (NOT p) q), and we may display the clause {p} as p. In the remainder of this book, whenever we refer to a conjecture, theorem, or formula being manipulated by the theorem-proving program, the reader should understand that a clause is being processed.

The members of a clause are called *literals*. If a literal p has the form (NOT atm), then the *atom* of p is atm, and otherwise the *atom* of p is p.

Because our earlier work [7] did not use clauses and because many researchers in mechanical theorem-proving have come to believe that using clauses is counterproductive, we now explain why we choose to use them.

The basic propositional connective in our theory is IF. Consider how we might represent a term such as (IF p T q), where p and q are Boolean (i.e., always return T or F). (IF p T q) is the term representing (OR p q). When we recursively explore (IF p T q) with a process such as TAUTOLOGYP, we naturally have assumed that p is false when we encounter q. But when we encounter p, we have assumed nothing about q, because we have not yet seen q. But (IF p T q) is equal to (IF q T p), and thus, when we encounter p, we could assume q false. The asymmetry is magnified if the q in (IF p T q) is itself an IF-expression, such as (IF r s t), for then when we encounter p we should actually note that either (a) r is true and s is false, or (b) r is false and t is false.

To avoid this asymmetry, we reduce a term containing an IF to an IF-free conjunction of clauses. It is easy to put an expression into this form by using the fact that (IF p q r) is a theorem if and only if the two clauses {(NOT p) q} and {p r} are theorems. The clauses for an IF-expression correspond to the branches through the IF-normal form of the expression.

If we are trying to prove a clause, then when we consider any given literal of it, we assume the remaining literals false. Furthermore, if we ever manage to reduce a literal to T (or any other non-F expression), the clause is true, and if we ever manage to reduce a literal to F, we remove that literal from the clause.

## C. THE ORGANIZATION OF OUR HEURISTICS

When should induction be used? We believe that induction should only be used as a last resort when all other methods of proof have failed. Furthermore, we believe that induction should be applied only to the most simple and generally stated propositions possible.

To this end we have developed a series of heuristics for preparing formulas for induction:

> Simplify the conjecture by applying axioms, rewrite lemmas, and function definitions and by converting the conjecture to a conjunction of IF-free clauses. Sometimes simplification will prove the conjecture. When it does not, at least it will reduce the complexity of the conjecture.
>
> When possible, reformulate the conjecture to eliminate "undesirable" concepts.
>
> Use equalities and then throw them away.
>
> Generalize the conjecture by introducing variables for terms that have "played their role."
>
> Eliminate irrelevant terms from the conjecture.

We apply first the safest operations (those, such as simplification, that usually convert a conjecture into an equivalent one) before trying more daring operations (such as generalization). Induction is applied last for two reasons: it is difficult to invent the right induction argument for anything but the simplest, strongest conjecture available, and induction increases the size and complexity of the conjecture.

Given a conjecture to prove, we apply the above heuristics in the order listed. Each of these heuristics can be regarded as taking a clause as input and returning a set of clauses as output. If each of the output clauses is a theorem, the input clause is a theorem. Thus, each of the output clauses is to be proved instead of the input. To get the process started on a user-supplied term p we start with the unit clause {p}.

Sometimes a clause is proved by one of these heuristics, in which case the empty set of clauses is returned. If a heuristic cannot prove or improve its input clause, it returns the singleton set containing the input, and then the next heuristic is tried. But if a heuristic changes its input, then we recursively start the whole sequence of heuristics over again on each of the output clauses. A clause only emerges from this sequence of heuristics when it can no longer be changed by any of them.

A good metaphor for the organization of these heuristics is an initially dry waterfall. One pours out a clause at the top. It trickles down and is split into pieces. Some pieces evaporate as they are proved. Others are further split up and simplified. Eventually at the bottom a pool of clauses forms whose conjunction suffices to prove the original formula.[13] We clean up this pool by removing any clause that is a substitution instance of another (i.e., we delete subsumed clauses). Having thus obtained the simplest and most general set of clauses we know how, we attempt to prove each by induction.[14]

We choose a clause, remove it from the pool, invent an induction

---

[13] The only thing wrong with this analogy is that our waterfall is recursive: every time a clause splits up, no matter how far down the waterfall, the pieces spill over the top of the fall.

[14] There are theorems that seem to require induction on a conjunction of formulas to permit sufficiently strong induction hypotheses. A classic example is the theorem that ( GCD X Y ) divides both X and Y (rather than just one of them). We do not know mechanical ways of recognizing this situation. Thus, if the user has stated a conjecture that, by the time it first reaches the induction stage, has been split into two or more formulas, we abandon the work we did to split up the conjecture and go into induction on the original formula, assuming the user would not have suggested that we prove a conjunction of two conjectures, each of which required an inductive proof, unless induction on their conjunction is necessary. After the first induction, this restriction is no longer applied.

scheme for it, and generate a conjunction of new clauses to prove (the base case and the induction steps). We then try to prove each of the new clauses by pouring each of them over the waterfall. Descendants of these clauses may also eventually trickle into the same pool from which we drew the one upon which we inducted. If we are successful, then eventually no clause will be in the pool or trickling over the waterfall. We will have proved the original conjecture.

## D. THE ORGANIZATION OF OUR PRESENTATION

In the next ten chapters, we present the details of the various heuristics we have developed. We begin the presentation with a general discussion of how we extract from an expression certain information about the type of its value and how such information is used to help us keep track of what we know in any given context. Then we discuss how rewrite axioms and theorems are used to rewrite terms. Following that, we explain how we use function definitions. After these three topics have been discussed, we describe how they are used to simplify clauses. We thus devote four chapters to simplification.

Recall that simplification is the first of the heuristics we apply when trying to prove a theorem. After simplification, there follow the elimination of "undesirables," utilization of equalities, generalization, and the elimination of irrelevant terms. A chapter is devoted to each of these heuristics.

Finally comes induction. We have devoted two chapters to that topic. The first deals with observations about recursive functions that enable the recognition of appropriate induction schemes. The second describes how we choose an induction scheme for a conjecture, given the previously computed information about how the functions in it recurse.

We illustrate each heuristic with particularly appropriate examples. We touch upon a great many theorems and problem domains during this discussion. To tie all the heuristics together, we have also chosen one simple example about list processing. At the end of the discussion on simplification (Chapter IX) and at the end of each subsequent chapter, we advance the proof of this simple theorem.

When we have completed our presentation of our heuristics, we discuss several much more complicated examples.

# VI

## Using Type Information to Simplify Formulas

The ability to simplify formulas is central to the ability to prove theorems by induction; a principal activity during an inductive proof is transforming the induction conclusion into an expression resembling the induction hypothesis so that the induction hypothesis can be used. To simplify a formula, we simplify its subterms under the assumptions we can glean from the context of the subterms in the formula. In this chapter, we describe how we represent such assumptions so that they can be readily used. We also explain how we derive from an expression or a recursive function definition a superset of the types of objects the expression or function returns. The ability to look at a term and immediately see that it may return certain types of objects is used in almost all our theorem-proving heuristics.

### A. TYPE SETS

In the chapter on the tautology-checker we arranged to remember assumptions that certain terms were F or non-F by maintaining a sequence of pairs. As we explored an expression to determine whether it was a tautology under such a sequence (alist) of assumptions, we both added new assumptions and searched through the current assumptions to determine whether any had been made about a certain term. In the simplification techniques of our theorem-proving system, we use a refinement of the idea of alists of assumptions. Instead of simply

## A. TYPE SETS

pairing terms with T or F, we pair terms with "type sets," which indicate that the terms have certain "types."

In this chapter, we define type set, and we describe an algorithm that computes a type set for an expression. The algorithm can itself be viewed as a refinement of the function TAUTOLOGYP, which determined whether an expression (in IF-normal form) never returned F. The type set algorithm recursively explores an expression, sometimes adding new assumptions about the types of certain terms and often asking whether an assumption has been made about a given term. While TAUTOLOGYP only made use of knowledge about the functions IF, TRUE, and FALSE, the type set algorithm also uses knowledge of EQUAL, knowledge of shell constructor, accessor, and recognizer functions, and knowledge of defined functions.

Assume that 3 + n shells have been added to our theory, with the recognizers NUMBERP, LITATOM, LISTP, and $r_1, \ldots, r_n$. Let $\underline{F}$ be the set {F}, let $\underline{T}$ be the set {T}, let NUMBERP be the set of all objects for which NUMBERP returns T, let LITATOM be the set of all objects for which LITATOM returns T, and so on for the sets LISTP, $\underline{r_1}, \ldots, \underline{r_n}$. Let OTHERS be the set of all objects not in any of $\underline{T}$, $\underline{F}$, NUMBERP, LITATOM, LISTP, $\underline{r_1}, \ldots, \underline{r_n}$.

The *types* are the sets $\underline{T}$, $\underline{F}$, NUMERP, LITATOM, LISTP, $\underline{r_1}, \ldots, \underline{r_n}$, and OTHERS. A *type set* is a set of types. Let UNIVERSE be the set of all types. A term t is said to *have type set* s provided s is a type set and the value of t under any interpretation of the variables of t is a member of some member of s. For example, the following terms have the corresponding type sets.

| term | type set |
| --- | --- |
| X | UNIVERSE |
| (ZERO) | {NUMBERP} |
| (ADD1 X) | {NUMBERP} |
| (EQUAL X Y) | {$\underline{T}$ $\underline{F}$} |
| (IF P (ADD1 X) (CONS X Y)) | {NUMBERP LISTP} |
| (IF P X 0) | UNIVERSE |
| (IF (NUMBERP X) X 0) | {NUMBERP} |
| (IF (NUMBERP X) T X) | UNIVERSE−{NUMBERP} |

To say that t has type set {NUMBERP LISTP} is just to say that t is a number or a list. To say that t has type set UNIVERSE − {NUMBERP} is to say that t is anything but a number.

## VI. USING TYPE INFORMATION TO SIMPLIFY FORMULAS

Our algorithm for computing a type set for an expression actually computes a type set for the expression under some assumptions that certain terms have certain type sets.

Our algorithm does not always return the smallest type set for the given expression; any algorithm that did so would be able to decide for any formula p whether it was a theorem by determining whether F was not in the computed type set of p. Instead of computing the smallest type set, our algorithm always returns a superset of the smallest type set.

Before describing how we determine a type set for an expression, we first illustrate with a few examples the kind of reasoning employed in computing and using type sets.

Suppose we knew that x had type set s. What could we infer if we encountered the term (NUMBERP x), while simplifying a formula, say? If s contained NUMBERP and no other element, then we would know that (NUMBERP x) were true. If s did not contain NUMBERP, then we would know that (NUMBERP x) were false. If s contained both NUMBERP and another element, then if we were asked to suppose that (NUMBERP x) were true, we could assume that x had type set {NUMBERP}, and if we were asked to suppose that (NUMBERP x) were false, we could assume that x had type set s − {NUMBERP}.

As another example, suppose that we knew that x had type set {NUMBERP} and y had type set {LISTP}. Then if we encountered the term (EQUAL x y), we would know that it was false because the two type sets had an empty intersection.

Our type set algorithm utilizes an auxiliary algorithm for adding to a set of assumptions a new assumption that a given expression is true or false. More precisely, the auxiliary algorithm accepts an expression together with a collection of assumptions (that certain terms have certain type sets), and returns one of three possible answers. The first possible answer is that the expression must be true (i.e., non-F) under the assumptions given. The second possible answer is that the expression must be false (i.e., F) under the assumptions given. The third possible answer has two parts. The first part is a set of assumptions (that certain terms have certain type sets) equivalent to the conjunction of the input assumptions and the new assumption that the input expression is true. The second part is a set of assumptions equivalent to the conjunction of the input assumptions and the new assumption that the input expression is false.

The algorithm for making assumptions is mutually recursive with the type set algorithm. In the next two sections, we present first the details of the assumption algorithm and then the details of the type set algorithm.

## B. ASSUMING EXPRESSIONS TRUE OR FALSE

To assume an expression p true or false in the context of certain assumptions about type sets of some terms, we consider the form of p.

### 1. Assuming an EQUAL-Expression True or False

If p has the form (EQUAL $t_1$ $t_2$), then we first compute type sets for $t_1$ and $t_2$, say $s_1$ and $s_2$. If $s_1$ and $s_2$ have an empty intersection, then p must be false. If $s_1$ and $s_2$ are equal, $s_1$ has only one member, and that member has only one member, we conclude that p must be true. Examples of such singleton–singleton type sets are $\{\underline{T}\}$, $\{\underline{F}\}$, and the singleton of the type for a shell of no components and no bottom object. Otherwise, we return two sets of assumptions.

To assume p true, we add to our current assumptions that p has type set $\{\underline{T}\}$. "Add" means that the assumption overrides any previous assumption about the type set of p. We also add the assumption that the commuted version of p, (EQUAL $t_2$ $t_1$), has type set $\{\underline{T}\}$. Finally, we add the two assumptions that $t_1$ and $t_2$ each have as type sets the intersection of $s_1$ and $s_2$. Thus, if $t_1$ were known to be a list or a literal atom, and $t_2$ were known to be a nonlist, then in assuming (EQUAL $t_1$ $t_2$) true we would assume that both $t_1$ and $t_2$ were literal atoms.

To assume p false, we add the assumption that p has type set $\{\underline{F}\}$. We also add the assumption that (EQUAL $t_2$ $t_1$) has type set $\{\underline{F}\}$. If $s_1$ is a singleton–singleton, we add the assumption that $t_2$ has type set $s_2 - s_1$. We perform the symmetric addition if $s_2$ is a singleton–singleton.

### 2. Assuming a Recognizer Expression True or False

If p has the form (r t), where r is a shell recognizer, we first compute the type set s of t. If s is $\{\underline{r}\}$, then p must be true. If s does not contain $\underline{r}$, then p must be false.

Otherwise, to assume p true we add the assumption that t has type set $\{\underline{r}\}$, and to assume p false we add the assumption that t has type set $s - \{\underline{r}\}$.

## 3. Assuming Other Expressions True or False

If p is not an equality or recognizer expression, we first compute the type set, s, of p. If s is {F}, then p must be false. If s does not contain F, then p must be true. Otherwise, to assume p true we add the assumption that the type set of p is s − {F}, and to assume p false we add the assumption that the type set of p is {F}.

## C. COMPUTING TYPE SETS

Given the ability to determine that a term must be true, must be false, or to assume it true or false, we now describe how we determine a type set of an expression under a set of assumptions that certain terms have certain type sets.

If the expression is among those terms with an assumed type set, then that type set is the answer. Otherwise, we consider the form of the expression.

### 1. The Type Set of a Variable

If the expression is a variable, then we return UNIVERSE.

### 2. Fixed Type Sets

If the expression is not a variable, we consider the function symbol of the expression.

If the function symbol is TRUE or FALSE, we return {T} or {F}, respectively.

If the function symbol is EQUAL or a shell recognizer, we return {T F}. (If a type set of a term is a subset of {T F}, we say the term is *Boolean*.)

If the function is a shell constructor or the function symbol of a bottom object, we return {r}, where r is the corresponding recognizer.

For each shell accessor, we return the type set determined by the type restrictions on the corresponding shell component (e.g., all SUB1 expressions have type set {NUMBERP} by virtue of the type restriction (NUMBERP X1) on the ADD1 shell, and similarly, all CAR expressions have type set UNIVERSE by virtue of the default type restriction T on the types of arguments to CONS).

## 3. The Type Set of IF-Expressions

The case for IF is more interesting. If we can determine that the test of the IF must be true or must be false under our current assumptions, then we merely return the type set of the appropriate branch of the IF. Otherwise we obtain the type sets of the two branches—assuming the test true on the one and false on the other—and return the union of those two type sets.

## 4. The Type Set of Other Functions

To explain how we compute a type set for other function symbols, we first define

The pair ⟨ts, args⟩ is a *type prescription for the function symbol* f provided (1) ts is a type set, (2) args is a subset of the set of formal parameters of f, and (3) whenever terms $t_1$, ..., $t_n$ have type sets $s_1$, ..., $s_n$, then ( f $t_1$ ... $t_n$) has as a type set the union of ts with the union of those $s_i$ such that the ith formal parameter of f is in args.

In the next section, we explain how we determine a particular type prescription for a function symbol, given its definitional equation or a theorem about the type of the function.

However, given a previously determined type prescription ⟨ts, args⟩ for f, it is easy to compute a type set for ( f $t_1$ ... $t_n$). In particular, we return the union of ts with the type sets of those $t_i$ indicated by args (under the current list of type assumptions).

## D. TYPE PRESCRIPTIONS

We now turn to the problem of computing a type prescription for a newly introduced function. Aside from the primitives and shell functions (whose type sets are described explicitly above), a function symbol can be introduced into our theory in one of two ways. The function symbol can be "declared" to take a given number of arguments and forever remain undefined, or else it can be defined under the principle of definition. In the former case we associate the type prescription ⟨UNIVERSE, {}⟩ with the function symbol. In the latter case, we compute a type prescription from the definition.

## VI. USING TYPE INFORMATION TO SIMPLIFY FORMULAS

If the user adds a rewrite type axiom or proves a rewrite type theorem that is equivalent to a certain function's having a particular type prescription, then we store that axiom or theorem as the type prescription for the function.

For example, in Chapter XVII we add the rewrite type axiom:

**Axiom** NUMBERP.APPLY:

(NUMBERP (APPLY FN X Y)),

where APPLY is an undefined (and otherwise unaxiomatized) function symbol. The axiom is stored as the type prescription ⟨{NUMBERP}, {}⟩ for APPLY.

### 1. Examples of Type Prescriptions

Before describing how we compute the type prescription for a defined function, we first consider a few examples.

Given the definition of APPEND,

**Definition**

```
(APPEND X Y)
  =
(IF (LISTP X)
    (CONS (CAR X) (APPEND (CDR X) Y))
    Y),
```

we conclude that APPEND returns either a LISTP object (due to the CONS expression) or the second argument Y. That is, it is a theorem that

```
(OR (LISTP (APPEND X Y))
    (EQUAL (APPEND X Y) Y)).
```

The type prescription for APPEND is thus ⟨{LISTP}, {Y}⟩. Some examples of type sets computed for APPEND-expressions are

| expression | type set |
|---|---|
| (APPEND X Y) | UNIVERSE |
| (APPEND X (CONS A B)) | {LISTP} |
| (APPEND X "NIL") | {LISTP LITATOM} |

Given the definition of REVERSE,

**Definition**

```
(REVERSE X)
    =
(IF (LISTP X)
    (APPEND (REVERSE (CDR X))
            (CONS (CAR X) "NIL"))
    "NIL"),
```

we observe that REVERSE either returns a LISTP (because we have already observed that (APPEND r (CONS c n)) is always a LISTP) or else returns a LITATOM (because "NIL" is a LITATOM). Thus any term beginning with the function symbol REVERSE has type set {LISTP LITATOM}.

Given the definition of SUM,

**Definition**

```
(SUM X Y)
    =
(IF (ZEROP X) Y (ADD1 (SUM (SUB1 X) Y))),
```

we observe that SUM returns either a NUMBERP or its second argument. Thus, if y is known to be a number, then (SUM x y) has type set {NUMBERP}.

If we were to define TIMES in terms of SUM,[15]

**Definition**

```
(TIMES I J)
    =
(IF (ZEROP I)
    0
    (SUM J (TIMES (SUB1 I) J))),
```

we could observe that TIMES would always return a number. For, when TIMES returned the 0 it would be returning a number. When it returned the value of the SUM expression it either would be returning a number (because SUM may return a number ) or it would be returning the second argument to the SUM expression (because SUM may return its second argument). But the second argument to the SUM expression would be a recursive call of TIMES. Thus, it could be

---

[15] We actually define TIMES in terms of the numeric function PLUS.

100 / VI. USING TYPE INFORMATION TO SIMPLIFY FORMULAS

nothing but a number. The computation of the type prescription of a recursive function involves such an inductive argument.

## 2. Computing the Type Prescription

To compute the type prescription for a function, we use a generalized version of the idea of type set. A *definition type set* is a pair consisting of a type set and a finite set of variables. A term *has definition type set* ⟨ts, s⟩, provided ts is a type set, s is a set of variables, every variable in s occurs in the term, and the value of the term under any interpretation of the variables in the term either is a member of a member of ts or is the value of some member of s under the same interpretation.

For example, a definition type set for the term

```
(IF (LISTP X)
    (CONS (CAR X) (APPEND (CDR X) Y))
    Y)
```

is ⟨{LISTP}, {Y}⟩. Given accurate type prescriptions for the functions concerned, the computation of a definition type set for an expression (under some assumptions about the definition type sets of certain terms) is closely analogous to the computation of a type set for the expression. Because of the similarity, we do not further discuss the definition type set computation. Instead, we describe how we use it to discover a type prescription for a newly defined function.

Note that one way to confirm that ⟨ts, args⟩ is a type prescription for a newly defined function symbol f is to assume that ⟨ts, args⟩ is such a type prescription, to compute a definition type set ⟨ts', s'⟩ for the body of f (under the assumption that each formal parameter x of f has definition type set ⟨{}, {x}⟩), and to check that ts' is a subset of ts and that s' is a subset of args. If so, the definition type set computation above constitutes an inductive proof that f does have type prescription ⟨ts, args⟩.

The type prescription ⟨UNIVERSE, formals⟩, where formals is the set of formal parameters of f, is a type prescription for any function f. However, we prefer to find a more restrictive type prescription. Instead of searching exhaustively through all possible type prescriptions, we adopt the following search strategy. First we assume that f has the type prescription ⟨{}, {}⟩. Then we compute the definition type set of the body of f. We next assume that f has as its type prescription the pairwise union of the previously assumed type prescription and the newly computed definition type set. We iterate until the computed definition type set is a pairwise subset of the type prescription just previously assumed. We then use the type prescription just previously assumed as the final type prescription for f. The iteration always stops because on each iteration the type prescription "grows," but there are only a finite nmber of type prescriptions.

The type prescription computation is performed only after the definition has been accepted by the definition principle, for otherwise the inductive argument inherent in the computation may be invalid.

## E. SUMMARY

In this chapter we explained

what information we glean from assuming a term true or false,

how we store that information as type sets,

how we compute type sets for terms, and

how we compute type prescriptions for recursive functions from their definitions.

The definition-time discovery of a type set for a function is an important and sometimes surprising aspect of our mechanical theorem-prover. Recall, for example, in Chapter II, that when FLATTEN was defined as

### Definition

```
(FLATTEN X)
    =
(IF (LISTP X)
    (APPEND (FLATTEN (CAR X))
            (FLATTEN (CDR X)))
    (CONS X "NIL")),
```

the theorem-prover announced, "Observe that (LISTP (FLATTEN X)) is a theorem." The fact that FLATTEN always returns a list is nonobvious, but it is important in proofs about FLATTEN. By following the recipe given above, the reader should be able to "discover" the theorem (LISTP (FLATTEN X)) for himself.

## F. NOTES

Since type sets are finite sets, we implement them as bit strings. For example, we obtain the union of two type sets by computing their "logical or."

Because type set information is used so extensively by all our heuristics, our theorem-proving program never reports the use of a type set axiom or lemma. The reader may assume that if a proof involved a function symbol f, then the proof may have tacitly employed type set lemmas about f.

# VII

# Using Axioms and Lemmas as Rewrite Rules

We have been discussing the handling of type information and how it can be used to represent our current set of assumptions as we walk through a formula. Let us now move on to the second major aspect of the simplification of expressions: how one can use axioms and lemmas as rewrite rules.

## A. DIRECTED EQUALITIES

If one has a lemma of the form

    (EQUAL lhs rhs),

it is sound to use it to replace any instance of lhs with the corresponding instance of rhs. (Term t *is an instance of* term lhs provided t is the result of substituting some substitution into lhs.)

For example, in Chapter IV we saw how the theorem

    (EQUAL (IF (IF P Q R) X Y)
           (IF P (IF Q X Y) (IF R X Y)))

could be used as a rewrite rule.

In general, we want to use arbitrary user-supplied theorems as rewrite rules. For example, if we proved (and have been instructed to use as a rewrite rule)

(EQUAL (APPEND (APPEND X Y) Z)
        (APPEND X (APPEND Y Z))),

then whenever we encounter a term of the form (APPEND (APPEND a b) c) we replace it by (APPEND a (APPEND b c)).

Treating equalities in this directed way is arbitrary. Given that equality is symmetric, it is just as reasonable to replace (APPEND a (APPEND b c)) with (APPEND (APPEND a b) c) as vice versa. (Indeed, it is just as reasonable to rewrite neither term and merely to note that they are equal.) We do not know how to decide mechanically in which direction to use a rewrite rule.[16] We leave it to the user to declare what theorems are to be used as rewrite rules and to recognize that we use equalities in this asymmetric, left-to-right fashion when he formulates theorems.

## B. INFINITE LOOPING

Even with the proviso that the supplier of theorems be cognizant of our conventions, many useful equality theorems could lead us around in circles. A simple example is the commutativity of PLUS:

*CP1    (EQUAL (PLUS X Y) (PLUS Y X)).

It is definitely advantageous to know that PLUS is commutative. However, using the above lemma as a simple rewrite rule would cause the following sequence of rewrites to occur on (PLUS A B):

(PLUS A B) rewrites to
(PLUS B A), which rewrites to
(PLUS A B), which rewrites to
. . . .

To prevent such loops, one can observe the following simple (indeed, almost mindless) rule: if a rewrite rule is *permutative* (i.e., the left- and right-hand sides are instances of one another), then do not apply the rewrite rule when the application would move a term to the left into a position previously occupied by an alphabetically smaller term.

Thus, while the rule would permit the use of the commutativity of PLUS to rewrite (PLUS B A) to (PLUS A B), it would not permit the

---

[16] But see Knuth and Bendix [26] and Lankford and Ballantyne [28] for interesting work on such questions.

use of the rule to rewrite (PLUS A B) to (PLUS B A). While this may seem a capricious way to prevent loops, it does have a certain normalizing effect on terms. For example, let us consider another permutative rewrite about PLUS:

*CP2 (EQUAL (PLUS X (PLUS Y Z))
    (PLUS Y (PLUS X Z))).

Given the term (PLUS (PLUS A B) C), the following is the only allowed sequence of rewrites involving *CP1 and *CP2:

(PLUS (PLUS A B) C) rewrites to
(PLUS C (PLUS A B)), using *CP1, which rewrites to
(PLUS A (PLUS C B)), using *CP2, which rewrites to
(PLUS A (PLUS B C)), using *CP1.

Note that at each stage the term is getting alphabetically smaller. Furthermore, after the third rewrite neither *CP1 nor *CP2 can be legally applied.

Finally, note that we just proved that PLUS is associative. In fact, using *CP1, *CP2, and the associativity of PLUS (stated to right-associate PLUS), we can rewrite any nest of PLUS expressions to the nest that is right-associated with the arguments in ascending alphabetic order. Thus any two nests of PLUS expressions with the same "bag" of arguments can be rewritten to identical expressions. Moreover, the analogous lemmas about any other function symbol (such as TIMES or GCD) allow us to normalize nests of that function symbol.

## C. MORE GENERAL REWRITE RULES

Up to now we have considered only rewrite rules of the form (EQUAL lhs rhs). The idea of using a theorem to rewrite expressions can be generalized considerably. In the first place, any literal can be interpreted as an equality in our theory. For example, (NOT p) can be regarded as (EQUAL p F). If p is Boolean, then the literal p can be interpreted as (EQUAL p T). If p is not Boolean, then, although the literal p cannot be interpreted as (EQUAL p T), it is sound, knowing that p is nonfalse, to replace p by T in any position of a formula in which only the "truth value" and not the identity of p is of concern. For example, knowing that p is nonfalse is as good as knowing that it is explicitly T in the test of an IF. Thus, requiring a rewrite rule to be of the form (EQUAL lhs rhs) is actually no restriction.

A more substantial generalization is the idea of using multiliteral formulas as rewrite rules. Consider a formula of the form

(IMPLIES (AND $h_1$ ... $h_n$)
 (EQUAL lhs rhs)).

A "natural" interpretation of this is that any instance of lhs can be replaced by the corresponding instance of rhs, provided the corresponding instances of the $h_i$ are true.

A good example of such a rule is

(IMPLIES (NOT (LISTP X))
 (EQUAL (CAR X) "NIL")).

A natural way to interpret this as a rewrite rule is to observe that we can rewrite any expression of the form (CAR x) to "NIL", provided we can establish that (LISTP x) is false (in the context in which the expression occurs).

Once again we see an element of arbitrariness. The above lemma can just as easily be interpreted as a way to establish that (LISTP x) is true: establish that (CAR x) is not "NIL". Rather than try to develop heuristics for guessing the ways the lemma should be used (or, worse, using it in all possible ways), we obtain this information implicitly from the statement of the lemma. In particular, if the user states a lemma as an implication, we take it to mean that the conclusion is to be used as a rewrite rule when the hypotheses are established.

How can we establish the hypotheses? The answer is simple: rewrite them recursively and if each is reduced to true, then the conclusion may be applied. By recursively rewriting hypotheses we are enabling our knowledge of type sets, lemmas, and recursive functions to be applied to them.

Thus, the following scheme for using rewrite lemmas suggests itself. Suppose we have a term (f $t_1$ ... $t_n$) and we have recursively rewritten the arguments $t_i$ and now wish to apply our known lemmas to the term. Then we look for a lemma with a conclusion of the form (EQUAL lhs rhs), where (f $t_1$ ... $t_n$) is an instance of lhs under some substitution s. If one is found, we instantiate the hypotheses of the lemma with s. Then we recursively rewrite each hypothesis. (Note that, like the tests of IFs and the literals of a clause, only the truth value of a hypothesis matters. Thus, if we know p is not false, we rewrite it to T when it appears as a hypothesis.) If each of the hypotheses rewrites to nonfalse, then we replace (f $t_1$ ... $t_n$) by the instantiated rhs and then recursively rewrite that. Of course, the application of the rewrite rule is subject to our rule about infinite looping due to permutative lemmas.

## D. AN EXAMPLE OF USING REWRITE RULES

Suppose we have proved as rewrite lemmas the following theorems from Chapter IV:

```
*T1     (EQUAL (VALUE (NORMALIZE X) A) (VALUE X A)),

*T2     (NORMALIZED.IF.EXPRP (NORMALIZE X)),

*T3     (IMPLIES (AND (NOT (TAUTOLOGYP X A))
                      (NORMALIZED.IF.EXPRP X)
                      A)
                 (FALSIFY1 X A)),
```

and

```
*T4     (IMPLIES (AND (EQUAL (VALUE Y (FALSIFY1 X A))
                             (VALUE X (FALSIFY1 X A)))
                      (NORMALIZED.IF.EXPRP X)
                      (FALSIFY1 X A))
                 (NOT (VALUE Y (FALSIFY1 X A)))).
```

The first three are lemmas discussed carefully in Chapter IV. We have used shorter names here. The fourth theorem, *T4, is the "bridge" lemma mentioned. It is FALSIFY1.FALSIFIES phrased in a way that makes it a more powerful rewrite rule in our system. FALSIFY1.FALSIFIES says that under some conditions we can conclude a certain thing about (VALUE X (FALSIFY1 X A)). *T4 says that under identical conditions we can conclude the same thing about the more general term (VALUE Y (FALSIFY1 X A)) —provided we can prove those two VALUE-expressions equal.

Let us see how our rewrite scheme would prove TAUTOLOGY.CHECKER.IS.COMPLETE. After expanding the definitions of TAUTOLOGY.CHECKER and FALSIFY, we obtain

```
*T5     (IMPLIES (NOT (TAUTOLOGYP (NORMALIZE P)
                                  "NIL"))
                 (NOT (VALUE P (FALSIFY1 (NORMALIZE P)
                                         "NIL")))).
```

We cannot rewrite the hypothesis of *T5. Thus, we start to rewrite

## 108 / VII. USING AXIOMS AND LEMMAS AS REWRITE RULES

the conclusion of *T5. However, in the process we can assume the hypothesis true. That is, we can assume

*A1    (TAUTOLOGYP (NORMALIZE P) "NIL")

has type set {F}.

Under this assumption we rewrite the atom of the conclusion of *T5:

*T5.2   (VALUE P (FALSIFY1 (NORMALIZE P) "NIL")).

We can use *T4 to rewrite this to F (and thus reduce *T5 to a tautology since *T5.2 is negated in *T5) if we can only establish the three hypotheses of *T4 after instantiating them by replacing X by (NORMALIZE P), Y by P, and A by "NIL".

The first instantiated hypothesis of *T4 is

*T4.1   (EQUAL (VALUE P
                       (FALSIFY1 (NORMALIZE P)
                                 "NIL"))
               (VALUE (NORMALIZE P)
                      (FALSIFY1 (NORMALIZE P)
                                "NIL"))).

But, using *T1, we rewrite the right-hand side of this equality to precisely the left-hand side (since the VALUE of (NORMALIZE P) is the VALUE of P). Thus, we have established the first hypothesis of *T4.

The second hypothesis of *T4 is

*T4.2   (NORMALIZED.IF.EXPRP (NORMALIZE P)).

But we recursively rewrite this to T using *T2.

The third hypothesis of *T4 is

*T4.3   (FALSIFY1 (NORMALIZE P) "NIL").

We can rewrite this to T (since it is in a hypothesis position) using *T3, if we can establish the three hypotheses of *T3. The first hypothesis of *T3 is

*T3.1   (NOT (TAUTOLOGYP (NORMALIZE P) "NIL")).

But according to *A1, the type set of the atom of this literal is {F}. Thus, the negation of it is T and we have established *T3.1.

The second hypothesis of *T3 is

*T3.2   (NORMALIZED.IF.EXPRP (NORMALIZE P)),

which once again rewrites to T using *T2.

The third and final hypothesis of *T3 is just

*T3.3 "NIL",

which is non-F (by type set reasoning, for example).

Therefore, we apply *T3 to rewrite *T4.3 to T, and hence, apply *T4 to rewrite *T5.2 to F, and have thus proved *T5.

Of course, we did not have to cite such a complicated example to illustrate the use of rewrite lemmas. In particular, the reader should keep in mind that virtually everything our implementation knows about arithmetic, lists, etc. is represented with rewrite lemmas.

Below are some simple rewrite lemmas added by our implementation of the shell principle (see Appendix B):

>     (EQUAL (CAR (CONS X1 X2)) X1),
>
>     (IMPLIES (NOT (LISTP X))
>              (EQUAL (CAR X) "NIL")),
>
>     (EQUAL (EQUAL (CONS X1 X2)
>                   (CONS Y1 Y2))
>            (AND (EQUAL X1 Y1)(EQUAL X2 Y2))),

and

>     (EQUAL (SUB1 (ADD1 X1))
>            (IF (NUMBERP X1) X1 0)).

Note that the last two examples illustrate the very useful idea of stating a lemma, when possible, as an unconditional rewrite (by using an IF or other propositional construct in the right-hand side) rather than as a collection of multiliteral rewrite rules.

## E. INFINITE BACKWARDS CHAINING

The idea of recursively appealing to lemmas to establish the hypotheses of other lemmas is called "backwards chaining." Implementing it mechanically requires addressing one major problem: it is possible to backwards chain indefinitely.

For example, consider the rewrite lemma

>     (IMPLIES (LESSP X (SUB1 Y))
>              (LESSP X Y)),

which says that we can rewrite (LESSP X Y) to T if we can establish

## 110 / VII. USING AXIOMS AND LEMMAS AS REWRITE RULES

that X is less than Y − 1. Note how we might be tempted to use this theorem to establish that (LESSP I J) is true:

Using the lemma we can prove (LESSP I J),
if only we could prove (LESSP I (SUB1 J)).

Ah ha! We can prove (LESSP I (SUB1 J)),
if only we could prove (LESSP I (SUB1 (SUB1 J))).

Ah ha! We can prove (LESSP I (SUB1 (SUB1 J))),
if only . . .

To prevent this from occurring, we remember what we are trying to do and we give up when certain contraindications arise.

In particular, we keep a list of the negations of the hypotheses we are currently trying to establish. Here is how we use the list. Suppose we are trying to apply a rewrite rule r, one of whose hypotheses is the term new. If we find new on the list, then it is sound to assume new true and spend no further time trying to establish it. (That is, because new is on the list, we know that we are in the process of trying to establish its negation. But it is permitted to assume p while trying to establish (NOT p).) If we find new's negation on the list, then we are looping, and we abandon the attempt to apply r. We also decide we are looping (but in a much more insidious way) when we find that the atom of new is an "elaboration of" the atom of one of the hypotheses we are already trying to establish. If none of these contraindications are present, we store the negation of new on the list, and then recursively rewrite new.

The basic idea of "elaboration" is suggested by the example above. We say that new is an *elaboration of* old if either (1) new is identical to old or (2) the number of occurrences of function symbols in new is greater than or equal to the number in old and new is "worse than" old.

We say that new is *worse than* old if (1) old is a variable and properly occurs in new, or (2) neither old nor new is a variable and either (2a) new and old have different function symbols and some subterm of new is worse than or identical to old, or (2b) some argument of new is worse than the corresponding argument of old but no argument of new is a variable or an "explicit value" unless the corresponding argument of old is, and furthermore no argument of old is worse than the corresponding argument of new. The definition of "explicit value" is given in Chapter VIII.

Thus, (LESSP I (SUB1 (SUB1 J))) is worse than (LESSP I (SUB1

J)). (LESSP I (SUB1 (SUB1 J))) is not considered worse than (LESSP (F I) (SUB1 J)) because I is a variable and (F I) is not.

## F. FREE VARIABLES IN HYPOTHESES

The discussion of how we establish the hypotheses of rewrite lemmas ignored an important point. We said that after discovering that the term to be rewritten is an instance of the left-hand side of the rewrite lemma's conclusion, we instantiate the hypotheses and rewrite them recursively. However, it is possible that not all the variables in the hypotheses are mentioned in the left-hand side of the conclusion, and hence, some variable may not be instantiated. We refer to such a variable as a "free" variable.

An example of a useful rewrite lemma with free variables in it is

```
(IMPLIES (AND (LESSP X Y)
              (LESSP Y Z))
         (LESSP X Z)).
```

That is, we can rewrite (LESSP X Z) to T if we can establish that X is less than Y and Y is less than Z.

The correct way to use such a lemma is to try to find an instantiation for Y that makes one of the two hypotheses true and then to rewrite the appropriate instance of the other hypothesis to establish it.

We try to find an instantiation y for Y by searching through the current type set assumptions, looking for a term (LESSP x y) assumed to be true, where x is the instantiation of X picked up when (LESSP X Z) was initially instantiated. If such a y is found, we use it for Y while trying to establish the remaining hypothesis.

Our approach to handling free variables is very weak. In particular, our system cannot establish a hypothesis with a free variable in it unless an instance of the hypothesis has previously been assumed true.

Here is an example of the inadequacy of our handling of free variables. Assume we have proved the two lemmas

```
(IMPLIES (NOT (ZEROP X))
         (LESSP (SUB1 X) X)),
(IMPLIES (NOT (ZEROP Y))
         (LESSP X (PLUS X Y))),
```

and suppose we wish to prove

```
(IMPLIES (AND (NOT (ZEROP I))
              (NOT (ZEROP J)))
         (LESSP (SUB1 I)
                (PLUS I J))).
```

All that is required is a single application of the transitivity of LESSP, using I as the choice for Y and the two previously mentioned lemmas to establish the hypotheses. Our system is not able to do this.

The system can prove this lemma (and hundreds like it) by induction. But it is painful to the user to state explicitly a theorem that is an easy consequence of substitutions and *modus ponens*.

# VIII

## Using Definitions

We now move on to the third (and last) fundamental activity involved in simplifying expressions: the use of function definitions.

Suppose we have a function definition of the form

$$(f \ v_1 \ \ldots \ v_n) \ = \ \text{body}.$$

By the axioms of equality we can replace a call of $f$, $(f \ t_1 \ \ldots \ t_n)$, by the body of $f$, after substituting the $t_i$ for the corresponding formals in the body. We call this "opening up" or "expanding" the function call.

Since a function definition is just an equation, there is in principle no difference between using function definitions and using other equality theorems. However, following the rules sketched above, we might expand recursive definitions indefinitely, since the recursive calls in the instantiated body could also be expanded.

Making intelligent decisions about whether to use a definition is crucial to proving theorems in a theory with a definition principle (be it a recursive one or not). Unnecessary expansion of definitions will swamp a theorem-prover (mechanical or otherwise) with irrelevant detail. Failure to expand a definition can prevent a proof from being found at all.

In studying many hand proofs we have learned three heuristics that are useful to know when considering whether to expand a recursive definition. We discuss each of these in turn.

## A. NONRECURSIVE FUNCTIONS

If one encounters a function call of f, and f is nonrecursive, it is clear that expanding its definition is safe in the sense that it will not lead to infinite loops.

In general, it is not necessarily wise to open up a function simply because it is safe. For example, if P is defined nonrecursively but the definition requires several pages to write down, then the theorem (IMPLIES (P X) (P X)) will take longer to prove if P is opened up. Nevertheless, our theorem-prover opens up every nonrecursive call it encounters.

This heuristic is tolerable only because most of the nonrecursive functions with which our theorem-prover deals have small bodies. Recall the definitions of such typical nonrecursive functions as NOT, AND, ZEROP, and FALSIFY.

## B. COMPUTING VALUES

It is sometimes possible to open up recursive functions and not introduce new recursive calls. For example, recall the definition of APPEND:

**Definition**
```
(APPEND X Y)
    =
(IF (LISTP X)
    (CONS (CAR X) (APPEND (CDR X) Y))
    Y),
```
and consider the term
```
(APPEND (CONS 1 (CONS 2 "NIL"))
        (CONS 3 "NIL")).
```
If we open up APPEND repeatedly and simplify the result we obtain
```
(CONS 1 (CONS 2 (CONS 3 "NIL"))),
```
completely eliminating APPEND.

The key to this particular example is the use of "explicit values" as arguments. Examples of explicit values are "NIL", 0, 1 (i.e., (ADD1 0)) and (CONS 1 "NIL"). An *explicit value* is defined to be either

T, F, a bottom object of some shell, or else the constructor of some shell applied to explicit values satisfying the type restrictions for the constructor. Thus, while T and (ADD1 0) are explicit values, (ADD1 T) is not. (We will later need the concept of an *explicit value template*, to wit: a nonvariable term composed entirely of shell constructors, bottom objects, and variables. Examples of explicit value templates are (CONS X "NIL") and (ADD1 I).)

The reason explicit values are mathematically interesting is that they constitute a normal form for a certain class of expressions. It can be proved inductively that two explicit values are equal if and only if they are identical terms. If we did not require that each component of an explicit value be of the right type, then the theorem just mentioned would have counterexamples (ADD1 T) and (ADD1 0). For (ADD1 T) and (ADD1 0) are equal but not identical.

We call a function f *explicit value preserving* if and only if f is TRUE, FALSE, IF, EQUAL, one of the functions introduced by the shell principle, or f was defined under our principle of definition and the body of f calls only f or explicit value preserving functions. Most of the functions with which we deal are explicit value preserving. The function EVAL, used in Chapter XVII, is an example of a function in our theory that is not explicit value preserving because it uses the undefined function APPLY. If f is explicit value preserving and $t_1$, ..., $t_n$ are explicit values, then (f $t_1$ ... $t_n$) can be rewritten to a unique equivalent explicit value by literally computing it from the axioms and definitions.

Thus, we know we can open up (APPEND "NIL" (CONS 3 "NIL")), for example, without even worrying that we will indefinitely expand the recursion in the body.

In fact, we do not require that all the arguments be explicit values. Recall that when we define a recursive function, the definition principle requires that we find a measure and well-founded relation justifying the function's definition. In APPEND, we see that the function is accepted because (COUNT X) gets LESSP-smaller on each (i.e., the) recursive call.

We say that {X} is a "measured subset" of the formals of APPEND since a measure of that subset decreases on every recursive call in the definition of APPEND (we will define "measured subset" in Chapter XIV). If some subset of the arguments to a call (f $t_1$ ... $t_n$) of a function are explicit values, and that subset "covers" a measured subset of the function (in the sense that $t_i$ is an explicit value when the ith formal parameter of f is in the subset), then we open up the call.

For example, we open up

```
(APPEND (CONS 1 (CONS 2 "NIL")) y),
```
regardless of what y is, and eventually get something that does not involve APPEND at all (except where it might occur in y), namely,
```
(CONS 1 (CONS 2 y)).
```
Of course, in general there may be more than one measured subset. For example, in LESSP,

**Definition**
```
(LESSP X Y)
  =
(IF (ZEROP Y)
    F
    (IF (ZEROP X)
        T
        (LESSP (SUB1 X) (SUB1 Y))))
```
there are two measured subsets, {X} and {Y}. Thus, we could open up (LESSP x 2) and eventually get (after simplification)
```
(OR (NOT (NUMBERP x))
    (EQUAL x 0)
    (EQUAL (SUB1 x) 0)).
```
That is, opening up (LESSP x 2) forces us to consider the cases: x is not a number, is 0, or is 1.

## C. DIVING IN TO SEE

The above two heuristics only scratch the surface and would not allow us to prove many theorems by themselves, because they attempt to answer the question "Should I open up $(f\ t_1\ \ldots\ t_n)$?" without considering the subterms of f's body. In general, the only way an intelligent decision can be made is to explore the body of f and see whether it has anything to contribute. This can be done by rewriting the body recursively, remembering that when a formal parameter $v_i$ of f is encountered it should be replaced by the already simplified $t_i$. The result will be a new term, val, and we must then decide whether we would prefer to keep $(f\ t_1\ \ldots\ t_n)$ or val.

One must be careful, when rewriting the body of f recursively, not to give similar consideration to the recursive calls of f, since that would lead to nonterminating recursion. We have found that a suitable rule to follow is to keep a list of the function names being tentatively opened up, and to refuse to consider opening recursive calls of those functions.

Note that because we rewrite the bodies of functions recursively before deciding whether to introduce those expression into the conjecture, we bring all our knowledge about type sets, previously proved results, and recursive functions to bear on the definition.

Sometimes the result of rewriting a body is just an instantiated copy of the body. For example, tentatively opening up (APPEND a b) may lead to

```
(IF (LISTP a)
    (CONS (CAR a) (APPEND (CDR a) b))
    b).
```

But, for example, if our type set or lemma reasoning could conclude that (LISTP a) were false, the result would be simply b. If a were of the form (CONS u v), then the result would be (CONS u (APPEND v b)), since axioms inform us of such things as (LISTP (CONS u v)) and (EQUAL (CAR (CONS u v)) u).

Sometimes the result is less trivial. For example, the recursive call in the definition of (QUOTIENT I J) is (QUOTIENT (DIFFERENCE I J) J). Had we previously proved the lemmas

```
(IMPLIES (NUMBERP I)
         (EQUAL (DIFFERENCE (PLUS J I) J) I))
```

and

```
(NOT (LESSP (PLUS I J) J)),
```

and if we knew that j were non-ZEROP and i were a number, then the result of tentatively opening up (QUOTIENT (PLUS j i) j) would be (ADD1 (QUOTIENT i j)).

Thus, on complicated function bodies, we might do a considerable amount of simplification before even beginning to decide whether we want to keep the expanded body. We have found that more economical methods of deciding whether to use a definition (such as preprocessing it, or exploring it without actually simplifying it) are inadequate because they do not take into account all that we know at the moment about the concepts involved. Since failure to open a definition at the right time will hide information from the proof and thus prevent a proof, it is crucial to bring one's full knowledge (both global and contextual) to the problem.

## VIII. USING DEFINITIONS

Thus, having obtained the simplified function body val we must decide whether it is "better" than the call it would replace, ( f $t_1$ ... $t_n$ ).

### 1. Keeping Nonrecursive Expansions

If val does not mention f as a function symbol, then we consider val to be better than ( f $t_1$ ... $t_n$ ). This happens frequently. For example, the function may be directed down a nonrecursive branch (as by the base case of an induction), or the recursive calls might be reduced to nonrecursive expressions by contextual information or lemmas. As an example of the latter, consider expanding (MEMBER a b) under the assumption that (MEMBER a (CDR b)) is non-F (as might be supplied by an induction hypothesis), where MEMBER is defined thus:

**Definition**
```
(MEMBER X L)
    =
(IF (LISTP L)
    (IF (EQUAL X (CAR L))
        T
        (MEMBER X (CDR L)))
    F).
```

In this case, type set reasoning will allow us to replace (MEMBER a (CDR b)) in the body by T. Then (IF (EQUAL a (CAR b)) T T) will reduce to T, and then (IF (LISTP b) T F) will reduce to (LISTP b) as the final result. As another example, consider (QUOTIENT I I). The recursive call is (QUOTIENT (DIFFERENCE I I) I), which, if we know the obvious facts about DIFFERENCE, will reduce to (QUOTIENT 0 I), which can be expanded again to eliminate QUOTIENT altogether by the previously mentioned heuristic involving explicit values in measured subsets.

### 2. Keeping Recursive Expansions

If the tentative result, val, involves recursive calls, then we are forced to choose between those calls and ( f $t_1$ ... $t_n$ ). We have found that it is usually a good idea to expand ( f $t_1$ ... $t_n$ ) (i.e., replace it by val) if each of the calls of f in val has one of three "good" properties compared to ( f $t_1$ ... $t_n$ ).

*a. No New Terms*

The first "good" property is simple and yet the most fundamental: if each of the arguments of a call of f in val already appears in the conjecture being proved, then the call is good. For example, if the conjecture already involves (CDR a) and b, then (APPEND a b) is best expressed in terms of (APPEND (CDR a) b). This heuristic has a strong normalizing effect on the conjecture; terms are expressed in common terms when possible. Of course, soon the conjecture "settles down" to those common terms. How is it that we ever open up anything again? The answer is induction.

For example, suppose we are proving some fact about (APPEND A B). We simplify it as much as possible. But if (CDR A) is not mentioned in the conjecture, we will not expand (APPEND A B) to introduce (APPEND (CDR A) B). But if induction supplies us with an induction hypothesis about (CDR A), it suddenly becomes a good idea to express (APPEND A B) in terms of (APPEND (CDR A) B).

This first "good" property accounts for the vast majority of openings performed by our program. As indicated, however, we have found two other useful properties.

*b. More Explicit Values As Arguments*

The second "good" property of a recursive call is that it contains more explicit values as arguments than ( f $t_1$ ... $t_n$ ). Introducing an explicit value where before we had an arbitrary term is a good idea, since we can usually further simplify matters.

*c. Less Complex Controllers*

The third "good" property of a recursive call in the tentative expansion is motivated by the example (LESSP I (ADD1 J)). We have found in our hand proofs that it is usually more convenient to express this as (LESSP (SUB1 I) J) — that is, to allow the recursive call to introduce (SUB1 I) where no (SUB1 I) appeared before, in order to get rid of (ADD1 J). As we will see in Chapter X, we can usually eliminate expressions like (SUB1 I) in other ways. Thus, the third "good" property of recursive calls is that the symbolic complexity of some measured subset is smaller than the complexity of that subset in the original ( f $t_1$ ... $t_n$ ) call. A heuristically adequate measure of symbolic complexity is the number of occurrences of function symbols. In computing the symbolic complexity of (IF x y z), however, we just take the maximum of the complexities of y and z, since after the IF is distributed on the far outside of the conjecture, only y and z will be arguments to the recursive call.

# IX

# Rewriting Terms and Simplifying Clauses

## A. REWRITING TERMS

The three preceding chapters have presented the major ideas involved in rewriting terms. In the next paragraph we explain the context in which we do the rewriting. Then, we describe how we put the various ideas together to rewrite terms.

When we rewrite a term, we do so in the context specified by two lists of assumptions. The first list, called the *type set alist*, is an alist of assumptions about the type sets of certain terms. The second list, called the *variable alist*, is an alist associating terms with variables and is used to avoid carrying out explicit substitutions. For example, when we explore the definition of APPEND while trying to expand (APPEND (CONS A B) C), we use the variable alist to remember that the first argument is (CONS A B) and the second is C, by associating (CONS A B) with the first formal parameter of APPEND, X, and associating C with the second formal parameter, Y. If the pair ⟨v, t⟩ is on the variable alist, we say v *is bound*, and that t *is the binding of* v. If v is bound to t, then v is assumed to be equal to t.

We proceed as follows to rewrite x under a type set alist and a variable alist.

### 1. Rewriting a Variable

If x is a variable, we ask whether x is bound. If so, let t be its binding. Because we will have previously rewritten t, we return t as the result of rewriting x.[17] If x is not bound, we return x.

If x is not a variable, we consider the cases on the form of x.

## 2. Rewriting Explicit Values

If x is an explicit value, we return x.

## 3. Rewriting IF-Expressions

If x is of the form (IF test left right), then we recursively rewrite test and obtain some new term, test'. If test' must be true or must be false under the assumptions in the type set alist, we recursively rewrite and return left or right, as appropriate. If test' can apparently be either true or false, we recursively rewrite left using the type set alist obtained by assuming test' true, and we recursively rewrite right using the type set alist obtained by assuming test' false. Suppose this produces left' and right'.

We then try to apply each of the following rewrite rules for IF-expressions to (IF test' left' right'). We return the result of the first applicable rule (if any), or else we return (IF test' left' right'):

(EQUAL (IF X Y Y) Y),

(EQUAL (IF X X F) X),

and

(EQUAL (IF X T F) X), applied only if X is Boolean.

## 4. Rewriting EQUAL-Expressions

If x is of the form (EQUAL t s), then we first obtain t' and s' by recursively rewriting t and s.

If t' and s' are identical, we return T.

If t' and s' could not possibly be equal, we return F. We know four ways to decide that two terms are definitely unequal: (a) their type sets do not intersect, (b) they are distinct explicit values, (c) one term is a bottom object and the other is a call of a shell constructor, or (d) one term is a call of a shell constructor and the other term occurs as a component (of the right type) in the first.

If none of the above rules apply, we next try rewriting (EQUAL t'

---

[17] Actually, t would have been rewritten under the assumptions available when it was bound to x. It is possible that by rewriting t again, under the current set of assumptions, we could further simplify it. In one of many heuristic compromises between power and efficiency, we do not resimplify t before returning it.

s') with each of the following rewrite rules:
>   (EQUAL (EQUAL X T) X), applied only if X is Boolean,
>   (EQUAL (EQUAL X (EQUAL Y Z))
>          (IF (EQUAL Y Z)
>              (EQUAL X T)
>              (EQUAL X F))),
>   (EQUAL (EQUAL X F) (IF X F T)).

Let val be the result of the first applicable rewrite, or (EQUAL t' s') if none applies. We return the result of rewriting val with all known rewrite lemmas.

To *rewrite a term* val *with lemmas,* we consider all known rewrite lemmas in the reverse order in which they were introduced and "apply" the first "applicable" lemma. If no lemma is applicable, we return val.

A rewrite lemma
>   (IMPLIES (AND $h_1$ ... $h_n$)
>            (EQUAL lhs rhs))

is applicable to val if val is an instance of lhs (under some substitution s on the variables of lhs), s does not violate the alphabetic ordering restriction if (EQUAL lhs rhs) is permutative, and we can establish the $h_i$.

To establish a hypothesis containing no free variables, we rewrite the hypothesis, using s as the alist specifying the values of variables. To establish a hypothesis containing free variables, we try to extend s so that under the extended substitution the hypothesis is one of the terms currently assumed true.

If the lemma is applicable, then we return the result of rewriting rhs (under the alist s plus any additional substitution pairs obtained in relieving hypotheses containing free variables).

Note that this scheme allows us to use lemmas that rewrite equalities. For an example of such a lemma, consider
>   (EQUAL (EQUAL (PLUS X Y) (PLUS X Z))
>          (EQUAL (FIX Y) (FIX Z))),

which allows us to "cancel" the identical first arguments of two equated PLUS-expressions.

### 5. Rewriting Recognizer Expressions

If x is of the form (r t), where r is a shell recognizer, we recursively rewrite t to obtain t'.

Then, if the type set of t' is {r}, we return T.
If the type set of t' does not include r as an element, we return F.
Otherwise, we rewrite ( r t' ) with lemmas (as above).

## 6. Rewriting Other Expressions

Otherwise, x is of the form ( f $t_1$ ... $t_n$ ). We first rewrite each $t_i$ to $t_i'$. If f is nonrecursive or a measured subset of the $t_i'$ are explicit values, we rewrite the body of f under a variable alist that associates each formal with the corresponding $t_i'$ and return the result. If f is on the list of function names being tentatively expanded, we return ( f $t_1'$ ... $t_n'$ ). Otherwise, we rewrite ( f $t_1'$ ... $t_n'$ ) with lemmas (as above).

If no rewrite rule is applicable, we determine whether f has a definition. If so, we add f to the list of functions being tentatively expanded, we rewrite the body of f under a variable alist associating the formals of f with the corresponding $t_i'$, and then we compare its rewritten body with ( f $t_1'$ ... $t_n'$ ) as in Chapter VIII to see which we should keep.

If f is not a defined function, we return ( f $t_1'$ ... $t_n'$ ).

## 7. Caveats

There are three discrepancies between the above description and our actual implementation.

First, as noted in Chapter VII, we distinguish between when we are concerned with maintaining the equality of the term being rewritten and when we are concerned only with maintaining its "truth value." In particular, we know that we are interested only in truth values when we are rewriting the atom of a literal in a clause, the test of an IF, or a hypothesis to a rewrite lemma. At all other times we must maintain strict equality (including, of course, when rewriting subterms of tests, etc.).

A second discrepancy is that we keep track of whether we are "hoping" the term we are rewriting will be rewritten to T or F (or be rewritten arbitrarily). In general, we do not care how the term rewrites. However, when we try to establish a hypothesis of a lemma, we do not waste time trying to rewrite the hypothesis to false. Thus, when we rewrite the hypotheses of lemmas, we actually rewrite the atom of the literal in the hypothesis. If the literal is positive, we "hope" the atom will be rewritten to T. If the literal is negative, we "hope" the atom will be rewritten to F. If we hope a term will be rewritten to T (or actually, non-F), we apply a rewrite rule only when the right-hand side of the conclusion is not F. The analogous statement holds when we hope the term will be rewritten to F. If we do not care how the term rewrites, we apply all rules as described.

Thus, if we were trying to establish that (VALUE x a) were true so that we could make a rewrite, we would not even consider applying a lemma with conclusion (NOT (VALUE x a)).

The third discrepancy is that before we return from rewriting, we ask whether the assumed type set of the answer is {T} or {F} (or, in the case where we are interested only in the "truth value" rather than the identity, we ask whether the type set does not include F), and if so, return instead T or F as appropriate.

## B. SIMPLIFYING CLAUSES

Given the ability to rewrite terms under a set of assumptions, it is easy to simplify clauses by sequentially rewriting the literals. The only interesting thing about simplifying clauses is how one can convert IF-expressions to clausal form "on-the-fly."

### 1. Converting IF-Expressions to Clausal Form

Suppose we are in the process of simplifying a clause
$$\{new_1 \ldots new_n \; old_1 \; old_2 \ldots old_k\},$$
by rewriting each of the literals in it. Suppose we have already rewritten those to the left of $old_1$ and are now ready to rewrite $old_1$ itself.

Then we assume all the literals except $old_1$ to be false. (Actually, if a literal has the form (NOT atm) we assume atm true.) Then we rewrite the atom of $old_1$ using the type set alist obtained from the foregoing assumptions and the empty variable alist. We thus obtain some value which we negate if $old_1$ was a negative literal. Let val be the result.

If val is T, we have established that the clause is true. If val is F, we delete $old_1$ from the clause and continue with $old_2$. Otherwise, we can replace $old_1$ by val and move on to $old_2$.

However, if val contains IF-expressions, we first split the clause into as many new clauses as necessary to remove all IFs from val. To do this, we first move all the IFs in val to the top, by repeatedly applying the rewrite rule

```
(EQUAL (f X1 ... (IF P Q R) ... Xn)
       (IF P
           (f X1 ... Q ... Xn)
           (f X1 ... R ... Xn))),
```

where f is any function symbol, to val until all the IFs are outside all other function symbols.

Then we repeatedly apply the rule

$$\{\text{new (IF p q r) old}\}$$
$$\leftrightarrow$$
$$\{\text{new (NOT p) q old}\} \land \{\text{new p r old}\}$$

until none of the IFs introduced by val remain.

This results in a conjunction of clauses which we continue to simplify recursively. We continue the rewriting process in each of these clauses, starting with $old_2$. The resulting set of clauses is returned as the value of simplifying our input clause.

Of course, if we changed anything (i.e., the resulting set of clauses is different from the singleton set containing our input clause), then each of the resulting clauses is poured over the waterfall again to be resimplified. Thus, we will get the opportunity to resimplify p, say, in the context of assuming the rewritten $old_2$ false.

## 2. Caveats

There are three discrepancies between this description and our actual implementation.

First, before we begin simplifying the literals of a clause, we check whether any literal has the form (NOT (EQUAL x t)), where x is a variable and t is a term not containing x as a variable. If so, we replace every occurrence of x in the clause with t and delete the literal.

The second discrepancy is that when we remove all of the IFs from val by splitting the clause into a set of clauses, we actually remove from the set any clause that is subsumed by any other clause.[18]

The third discrepancy is that we implement Robinson's "replacement principle" [49] here. In particular, if the splitting process pro-

---

[18] We use a subsumption algorithm of J. A. Robinson (private communication) combining backtracking with the delicate treatment of unification in [50].

## 126 / IX. REWRITING TERMS AND SIMPLIFYING CLAUSES

duces two clauses one of whose ground resolvents subsumes either of them, we throw out the subsumed clause(s) and keep the resolvent.

For example, the description above would transform the clause

```
{(IF P
    (IF Q R S)
    (IF Q R V))}
```

into the following four clauses

```
{(NOT P) (NOT Q) R}
{(NOT P) Q S}
{P (NOT Q) R}
{P Q V},
```

while our implementation would produce just three clauses:

```
{(NOT Q) R}
{(NOT P) Q S}
{P Q V}.
```

In particular, we must prove {(NOT Q ) R} given P and {(NOT Q) R} given (NOT P). Thus, we decide to prove {(NOT Q) R}.

### C. THE REVERSE EXAMPLE

Let us now look at an example theorem. We have picked a simple one so that we can explain each "move" carefully.

Consider the idea of reversing a list. The reverse of the empty list is the empty list. The reverse of the list $x_1$, $x_2$, ..., $x_n$ is $x_n$, ..., $x_2$, $x_1$. It can be recursively obtained by reversing $x_2$, ..., $x_n$ to get $x_n$, ..., $x_2$ and then adding $x_1$ to the right-hand end.

### Definition

```
(REVERSE X)
    =
(IF (LISTP X)
    (APPEND (REVERSE (CDR X))
            (CONS (CAR X) "NIL"))
    "NIL").
```

If X is a "proper list," then (REVERSE (REVERSE X)) is just X. A "proper list" is one that terminates in "NIL" rather than some other non-LISTP object. Its definition is

**Definition**

```
(PLISTP X)
   =
(IF (LISTP X)
    (PLISTP (CDR X))
    (EQUAL X "NIL")).
```

That is, if X is a LISTP, it must have a proper CDR, and if X is not a LISTP, it must be "NIL".

Let us prove

```
*RR    (IMPLIES (PLISTP X)
                (EQUAL (REVERSE (REVERSE X)) X)),
```

using only the axioms of our theory of Chapter III and the definitions of APPEND, REVERSE, and PLISTP.

## D. SIMPLIFICATION IN THE REVERSE EXAMPLE

A careful inspection of *RR (considered as a clause of two literals) will show that we can do nothing to simplify it further. In particular, we have no lemmas to apply to it, and none of the recursive functions will expand without violating our guidelines. Perhaps we can prove it by induction. We will discuss how induction analysis is handled in Chapter XV. At the moment, suffice it to say that we are led to the following three new conjectures to prove

```
*RR1   (IMPLIES (AND (NOT (LISTP X))
                     (PLISTP X))
                (EQUAL (REVERSE (REVERSE X)) X)),

*RR2   (IMPLIES (AND (LISTP X)
                     (NOT (PLISTP (CDR X)))
                     (PLISTP X))
                (EQUAL (REVERSE (REVERSE X)) X)),
```

and

```
*RR3   (IMPLIES
           (AND (LISTP X)
```

```
              (PLISTP X)
              (EQUAL (REVERSE (REVERSE (CDR X)))
                     (CDR X)))
       (EQUAL (REVERSE (REVERSE X)) X)).
```

(The conjunction of *RR2 and *RR3 is propositionally equivalent to the induction step.)

Let us now try to prove each of these by simplification. That means we try to rewrite each of the literals in the clausal form of the formulas, assuming the other literals to be false.

We begin with *RR1. The intuitive reason *RR1 is true is that if X is not a LISTP but is a proper list, it must be "NIL", and if it is "NIL", then we can confirm with computation that (REVERSE (REVERSE X)) is X. To show how our mechanical process follows this line of reasoning, we proceed to simplify each literal in turn.

*RR1 is the three-literal clause

```
       {(LISTP X)
        (NOT (PLISTP X))
        (EQUAL (REVERSE (REVERSE X)) X)}.
```

Assuming the second and third literals false, we first try to rewrite the first literal, (LISTP X). However, it cannot be simplified because the type set of X is UNIVERSE and we have no known rewrite lemmas for LISTP. Next, assuming the first and third literals false, we rewrite the atom of the second literal, (PLISTP X). When we assume (LISTP X) false, we note that X has type set UNIVERSE-{LISTP}. Thus, when we explore the definition of PLISTP while trying to expand (PLISTP X), we rewrite the test (LISTP X) to F and thus rewrite (PLISTP X) to (EQUAL X "NIL"). Thus, the intermediate clause obtained after rewriting the first two literals of *RR1 is

```
       {(LISTP X)
        (NOT (EQUAL X "NIL"))
        (EQUAL (REVERSE (REVERSE X)) X)}.
```

Finally, we explore the third literal, assuming (LISTP X) false and (EQUAL X "NIL") true. The inner REVERSE term in the third literal simplifies to "NIL" (because X is now known to have type set {LITATOM}) and so the outer REVERSE term opens up to "NIL". Thus, the value of the third literal is (EQUAL "NIL" X), which has type set {T} and hence is T. Having reduced a literal of *RR1 to T, we have proved it.

Goal *RR2 is also easy. Reasoning informally, we can see that if X is a LISTP and its CDR is not proper, then X cannot be a proper list. But

## D. SIMPLIFICATION IN THE REVERSE EXAMPLE / 129

this contradicts the third literal of *RR2. Our simplification heuristic proceeds as follows. The first literal in the clausal form of *RR2, (NOT (LISTP X)), cannot be simplified at all. The second, (PLISTP (CDR X)), is also in simplest form. Of course, we consider opening PLISTP up, but find that such a move would introduce (CDR (CDR X)), which is not currently involved in *RR2. But when we simplify the atom, (PLISTP X), of the third literal, (NOT (PLISTP X)), in the clausal representation of *RR2, we assume X has type set {LISTP} and (PLISTP (CDR X)) has type set {F} (i.e., is false). Thus, (PLISTP X) opens up to F (because its recursive call, (PLISTP (CDR X)), has been assumed false), and the third literal of *RR2 is thus rewritten to (NOT F) or T. Consequently, we are done with *RR2.

Goal *RR3 is the interesting one. The first hypothesis, (LISTP X), cannot be simplified. The second, (PLISTP X), can be replaced by (PLISTP (CDR X)), since when we tentatively open up PLISTP assuming (LISTP X) is true we obtain (PLISTP (CDR X)), and (CDR X) is mentioned in the conjecture already. The third hypothesis, (EQUAL (REVERSE (REVERSE (CDR X))) (CDR X)), cannot be simplified without introducing recursive calls (such as (REVERSE (CDR (CDR X)))) that violate our guidlines. Finally, in the conclusion, (EQUAL (REVERSE (REVERSE X)) X), we can open up (REVERSE X) to

(APPEND (REVERSE (CDR X))
        (CONS (CAR X) "NIL")),

since (CDR X) is already involved in the conjecture. None of our rules allow the resulting conjecture to be further simplifed and we are left with

*RR4

(IMPLIES
    (AND (LISTP X)
         (PLISTP (CDR X))
         (EQUAL (REVERSE (REVERSE (CDR X)))
                (CDR X)))
    (EQUAL (REVERSE (APPEND (REVERSE (CDR X))
                            (CONS (CAR X) "NIL")))
           X)).

To prove this, we must do something besides the kind of simplification we have been discussing. In fact, the proof will ultimately require many of our heuristics and we will advance the proof in each of the subsequent chapters on proof techniques.

# X

# Eliminating Destructors

## A. TRADING BAD TERMS FOR GOOD TERMS

A standard trick when trying to prove a theorem involving X − 1, where the variable X is known to be a number other than 0, is to replace X everywhere by Y + 1. This means that the formula now mentions Y and Y + 1 where before it mentioned X − 1 and X. This trading of SUB1 for ADD1 makes the relation between X − 1 and X more obvious: representing X as (ADD1 Y) makes it clear that X is a number, that it is not 0, and that (SUB1 X) (i.e., Y) literally occurs within its structure.

Consider another example. A formula about (QUOTIENT X Y) and (REMAINDER X Y) (i.e., the integer quotient and remainder of X divided by Y) can be reformulated by representing X as I + Y*J, where I is a number less than Y. This trades the terms (QUOTIENT X Y), (REMAINDER X Y), and X, for J, I, and I + Y*J. In particular, it eliminates QUOTIENT and REMAINDER at the expense of introducing PLUS and TIMES. This is usually a good trade because it makes clear that the quotient and remainder can have arbitrary values (within certain constraints). For example, we could induct on J, the number of times Y divides X. Furthermore, PLUS and TIMES are simpler functions than QUOTIENT and REMAINDER, and it happens that we know a lot of theorems about them. Once we have proved that any number can be represented as I + Y*J, where I is less than Y, we can deduce things about the quotient and remainder using our already established knowledge of addition and multiplication.

## A. TRADING BAD TERMS FOR GOOD TERMS / 131

What justifies this elimination of some terms in favor of others? Let us look at the elimination of SUB1 in favor of ADD1 more carefully and explain it in terms of axioms about arithmetic.

Suppose we have a formula of the form

*1      (p (SUB1 x) x),[19]

where x is a variable and (SUB1 x) actually occurs in the formula.

If x is not a number, or else is 0, then the (SUB1 x) term is degenerate and it is useful to consider that case separately. But if x is known to be a number and not 0, then we can soundly replace x by (ADD1 y), where y is a new variable known to be numeric. The result is that instead of proving *1 we try to prove both of the following:

*2      (IMPLIES (OR (NOT (NUMBERP x))
                     (EQUAL x 0))
                 (p (SUB1 x) x))

and

*3      (IMPLIES (NUMBERP y)
                 (p y (ADD1 y))).

Why is this move sound? That is, why does the conjunction of *2 and *3 suffice to prove *1? Clearly, if x is nonnumeric or is 0, then *2 implies *1. On the other hand, if x is numeric and non-0, then we can derive *1 from *3 as follows. Instantiate *3, replacing y by (SUB1 x), to obtain

         (IMPLIES (NUMBERP (SUB1 x))
                  (p (SUB1 x) (ADD1 (SUB1 x)))).

Then use the axiom

*A1     (IMPLIES (AND (NUMBERP X)
                      (NOT (EQUAL X 0)))
                 (EQUAL (ADD1 (SUB1 X))
                        X))

---

[19] We now loosen our notational conventions by permitting lower case words in function symbol positions to represent schemas. For example, we think of (p (SUB1 x) x) as standing for a term such as (EQUAL (PLUS V Y) (TIMES (SUB1 X) Z)). If (p (SUB1 x) x) was understood to denote the above example, then when we refer to (p A B) we would have in mind (EQUAL (PLUS V Y) (TIMES A Z)). It is possible to abuse notational conventions; however, since we are now engaged in explaining heuristics and assume that the reader is mathematically competent, we feel that our occasional use of this convention does not warrant the complexity that precision here would entail. In sections "precisely" describing how our theorem-prover works, we revert to using lower case words in function symbol positions to represent only function symbols, not schemas.

## 132 / X. ELIMINATING DESTRUCTORS

to rewrite (ADD1 (SUB1 x)) to x (since we are assuming x is a non-0 number). This produces

    (IMPLIES (NUMBERP (SUB1 x))
         (p (SUB1 x) x)).

Finally, use the axiom

*A2    (NUMBERP (SUB1 X))

to remove the hypothesis. The result is (p (SUB1 x) x), which is *1.

  Given that we believe it is easier to prove something about y and (ADD1 y) than about (SUB1 x) and x, the problem for a theorem-prover is to take note of *A1 and *A2 and generate *2 and *3 as goals when asked to prove *1. The process is exactly the reverse of the justification. Given *1, split it into two parts according to whether the hypotheses of *A1 hold. *2 is the case that they do not. In the case where they hold replace certain of the x's by (ADD1 (SUB1 x)), as allowed by *A1. Then generalize the result by replacing (SUB1 x) by y. The process of generalizing a conjecture is discussed in detail in Chapter XII. As we shall see then, when (SUB1 x) is generalized to some new variable y, it is usually a good idea to restrict the new variable to having some of the properties of (SUB1 x). In particular, we can take note of *A2 to restrict y to being numeric. The result is *3. The precise description of how we use *A1 and *A2 to generate *2 and *3 from *1 is given later in this chapter.

  The really difficult question is where we get the idea that REMAINDER and QUOTIENT are "bad" and PLUS and TIMES are "good." Merely knowing that we can eliminate REMAINDER and QUOTIENT because we have proved some facts analogous to *A1 and *A2 is not reason enough to eliminate them. For example, the following two axioms are exactly analogous to *A1 and *A2 above, and allow us to eliminate ADD1 in favor of SUB1:

    (IMPLIES (NUMBERP X)
         (EQUAL (SUB1 (ADD1 X)) X))

and

    (NUMBERP (ADD1 X)).

  In general, we let the user tell us which functions should be eliminated by allowing him to label certain theorems as "elimination" type theorems. For example, *A1 is an elimination theorem. As for *A2, it is to be explicitly labeled as a "generalization" lemma if it should be noted when (SUB1 x) is generalized.[20]

## B. THE FORM OF ELIMINATION LEMMAS

An elimination theorem must have the form

*ELIM   (IMPLIES hyp (EQUAL lhs var)),

where (1) var is a variable, (2) there is at least one proper subterm of lhs of the form (d $v_1$ ... $v_n$), where d is a function symbol and the $v_i$ are distinct variables and are the only variables in the theorem, and (3) var occurs in lhs only in such (d $v_1$ ... $v_n$). Examples of elimination lemmas are

**Axiom** SUB1.ELIM:
```
(IMPLIES (AND (NUMBERP X)
              (NOT (EQUAL X 0)))
         (EQUAL (ADD1 (SUB1 X))
                X)),
```

**Axiom** CAR/CDR.ELIM:
```
(IMPLIES (LISTP X)
         (EQUAL (CONS (CAR X) (CDR X))
                X)),
```

**Theorem** DIFFERENCE.ELIM:
```
(IMPLIES (AND (NUMBERP Y)
              (LESSEQP X Y))
         (EQUAL (PLUS X (DIFFERENCE Y X)) Y)).
```

We call the (d $v_1$ ... $v_n$) terms in the left-hand side of the conclusion "destructor terms." In the above examples, the destructor terms are (SUB1 X), (CAR X), (CDR X), and (DIFFERENCE Y X). The name "destructor" refers to the fact that these functions can be viewed as decomposing one of their arguments into its components in some representation.

---

[20] In the special case of shells, over which we have complete control, it is the implementation of the shell principle that declares how an axiom should be used. When the ADD1 shell is added, *A1 and *A2 are classed as "elimination" and "generalization" theorems.

## C. THE PRECISE USE OF ELIMINATION LEMMAS

An elimination lemma can be used to eliminate any instance of a destructor term by following the scenario sketched above for SUB1. However, we have found that it is usually a mistake to eliminate an instance in which the $v_i$ are bound to nonvariables, or when several of the $v_i$ are bound to the same variable. For example, eliminating (SUB1 (PLUS X Y)), while sound, usually produces something that is more general than the original conjecture and often a nontheorem, because the connection between the new variable introduced, and other uses of X and Y, is lost. Similarly, while (REMAINDER X X) could be eliminated as sketched, it would be a heuristic mistake (since (REMAINDER X X) is in fact 0).

Suppose we are trying to prove some conjecture p, involving a term (d $x_1$ ... $x_j$ ... $x_n$), where $x_1$, ..., $x_j$, ..., $x_n$ are distinct variables and (d $x_1$ ... $x_j$ ... $x_n$) is an instance of a destructor term (d $v_1$ ... $v_j$ ... $v_n$) in an elimination lemma:

*ELIM   (IMPLIES hyp (EQUAL lhs $v_j$)).

We will eliminate (d $x_1$ ... $x_j$ ... $x_n$) from p by using *ELIM to re-represent $x_j$. The result will be two new clauses, which are sufficient to establish p.

Let *ELIM' be the formula obtained by simultaneously replacing each $v_i$ in *ELIM by $x_i$:

*ELIM'  (IMPLIES hyp' (EQUAL lhs' $x_j$)).

Since *ELIM' is a theorem, our goal p is equivalent to the conjunction of

*ELIM1  (IMPLIES (NOT hyp') p)

and

*ELIM2  (IMPLIES (AND hyp' (EQUAL lhs' $x_j$))
              p).

*ELIM1 is one of the two formulas produced by eliminating (d $x_1$ ... $x_j$ ... $x_n$) from p. *ELIM1 is the degenerate case. We derive the second output formula from *ELIM2 as follows.

We generalize *ELIM2 in two steps. The first step is to add to *ELIM2 an additional hypothesis gen known to be a theorem:

*ELIM3  (IMPLIES (AND gen hyp' (EQUAL lhs' $x_j$))
              p).

Intuitively, gen is a theorem about the destructor terms in lhs' and serves to restrict the coming generalization. The selection of gen is described in Chapter XII. The second generalization step is to uniformly replace throughout *ELIM3 each of the destructor terms in lhs' by some distinct variable not already occuring in *ELIM3:

```
*ELIM4 (IMPLIES (AND gen' hyp'' (EQUAL lhs'' xⱼ))
               p').
```

Note that *ELIM4 is sufficient to prove *ELIM2: instantiate the new variables in *ELIM4 to produce *ELIM3 and then use the theorem gen to derive *ELIM2. (Actually, the generalization heuristic described in Chapter XII is used to produce *ELIM4 from *ELIM2 given the destructor terms in lhs.)

The final step in eliminating $(d\ x_1 \ldots x_j \ldots x_n)$ from p is to use the equality hypothesis (EQUAL lhs'' $x_j$) in *ELIM4 by uniformly replacing $x_j$ throughout *ELIM4 by lhs'' and deleting the hypothesis

```
*ELIM5 (IMPLIES (AND gen'' hyp''') p'').
```

*ELIM1 and *ELIM5 are the results of eliminating $(d\ x_1 \ldots x_j \ldots x_n)$ from p.

## D. A NONTRIVIAL EXAMPLE

This may seem like a very complicated way to replace X by (ADD1 Y). However, the idea of rerepresenting terms is important in many difficult proofs. Let us consider a nontrivial example. We will carry it out in the same step-by-step fashion we described above.

Suppose we have proved the following two theorems:

```
*T1     (IMPLIES (AND (NOT (ZEROP Y)) (NUMBERP X))
                (EQUAL (PLUS (REMAINDER X Y)
                            (TIMES Y
                                  (QUOTIENT X Y)))
                       X))
```

and

```
*T2     (EQUAL (LESSP (REMAINDER X Y) Y)
               (NOT (ZEROP Y))).
```

These are probably the two most important properties of REMAINDER and QUOTIENT. The first says that by adding the remainder to the product of the quotient and the divisor one obtains the dividend. The second says that the remainder is less than the divisor.[21] Our mechanical theorem-prover proves these two theorems by induction, using the recursive definitions of REMAINDER and QUOTIENT and previously proved theorems about PLUS, TIMES, LESSP, and DIFFERENCE. The theorem-prover's proof of *T1 is the last proof exhibited in Chapter XVI. Let us suppose *T1 and *T2 have been proved and that *T1 is available as an elimination theorem and *T2 as a generalization theorem (they are both useful as rewrite rules as well).

Let us now prove

```
(IMPLIES (DIVIDES A B)
         (DIVIDES A (TIMES B C))),
```

where the definition of (DIVIDES A B) is (EQUAL (REMAINDER B A) 0).

This is usually proved with the following argument. If (DIVIDES A B), then we can write B as (TIMES A J), making a factor of A in (TIMES B C) manifest. Of course, one has to consider the possibility that B cannot be represented as (TIMES A J). For example, what if B is nonnumeric or A is 0?

Let us now go through the mechanical proof and see how this reasoning is done formally and completely.

By opening up the definition of (DIVIDES A B) we obtain

*4
```
(IMPLIES (EQUAL (REMAINDER B A) 0)
         (EQUAL (REMAINDER (TIMES B C) A)
                0)).
```

Noting that (REMAINDER B A) is an instance of a destructor term in *T1 and that the arguments are all distinct variables, we can use *T1 to eliminate (REMAINDER B A). First we split *4 into two parts:

*5
```
(IMPLIES (AND (OR (NOT (NUMBERP B))
                  (ZEROP A))
              (EQUAL (REMAINDER B A) 0))
         (EQUAL (REMAINDER (TIMES B C) A) 0))
```

and

---

[21] This particular statement is stronger and more useful as a rewrite rule. It says that (REMAINDER X Y) is less than Y if and only if Y is non-ZEROP.

## D. A NONTRIVIAL EXAMPLE / 137

```
*6
(IMPLIES (AND (NUMBERP B)
              (NOT (ZEROP A))
              (EQUAL (PLUS (REMAINDER B A)
                           (TIMES A (QUOTIENT B A)))
                     B)
              (EQUAL (REMAINDER B A) 0))
         (EQUAL (REMAINDER (TIMES B C) A)
                0)).
```

Formula *5 corresponds to formula *ELIM1 in the schematic description of elimination above. We do not further manipulate it as part of eliminating (REMAINDER B A).

Formula *6 corresponds to formula *ELIM2 above, and we want to remove the destructor terms. First we generalize those terms away, replacing (REMAINDER B A) by the new variable I and (QUOTIENT B A) by the new variable J. The generalization heuristic will take note of *T2 as a generalization lemma and will also note that both REMAINDER and QUOTIENT are always numerically valued (by computing the type sets of the expressions generalized). The result of generalizing *6 is

```
(IMPLIES (AND (NUMBERP I)
              (NUMBERP J)
              (EQUAL (LESSP I A) (NOT (ZEROP A)))
              (NUMBERP B)
              (NOT (ZEROP A))
              (EQUAL (PLUS I (TIMES A J)) B)
              (EQUAL I 0))
         (EQUAL (REMAINDER (TIMES B C) A)
                0)).
```

(The first three hypotheses are the restrictions placed on the new variables by the generalization. The rest of the formula is just *6 with the destructor terms replaced by I and J.)

Now we use the hypothesis, (EQUAL (PLUS I (TIMES A J)) B), to replace B everywhere in the formula. Afterwards, we throw away the hypothesis. The result is

```
*7
(IMPLIES (AND (NUMBERP I)
              (NUMBERP J)
              (EQUAL (LESSP I A) (NOT (ZEROP A)))
              (NUMBERP (PLUS I (TIMES A J)))
```

## X. ELIMINATING DESTRUCTORS

```
              (NOT (ZEROP A))
              (EQUAL I 0))
         (EQUAL (REMAINDER (TIMES (PLUS I (TIMES A J))
                                  C)
                           A)
                0)).
```

This formula corresponds to *ELIM5 in the schematic presentation of elimination above. *5 and *7 together are the result of eliminating (REMAINDER B A) from *4.

At first sight things appear worse than they were. However, *5 is the degenerate case in which B is nonnumeric or A is ZEROP. Quoting from our sketch of the proof: "Of course, one has to consider the possibility that B cannot be represented as (TIMES A J). For example, what if B is nonnumeric or A is 0?" In fact, *5 can be simplified to true using only our definitions of TIMES and REMAINDER.

As for *7, it too can be simplified drastically. For example, since PLUS is always numeric, the (NUMBERP (PLUS I (TIMES A J))) term simplifies to T. But the really important observation is that our hypothesis (DIVIDES A B) has been transformed into (EQUAL I 0). Simplification will substitute 0 for I everywhere. After routine simplification the result is

```
(IMPLIES (AND (NUMBERP J)
              (NUMBERP A)
              (NOT (EQUAL A 0)))
         (EQUAL (REMAINDER (TIMES A (TIMES C J))
                           A)
                0)).
```

Once again quoting from our earlier sketch: "If (DIVIDES A B), then we can write B as (TIMES A J), making a factor of A in (TIMES B C) manifest." The actual process was: "If (EQUAL (REMAINDER B A) 0), then we can represent B as (PLUS I (TIMES A J)), where (EQUAL I 0)."

In the informal proof sketch, the word "manifest" assumed that the reader was familiar with the theorem (DIVIDES J (TIMES J I)), or, opening DIVIDES up, (EQUAL (REMAINDER (TIMES J I) J) 0). If we had previously proved this as a rewrite rule, then *7 would simplify to true (i.e., we would have gotten something out of the observation that indeed a factor of A was manifest). If we had not proved the above theorem then we would have to prove the simplified version of *7 by induction.

## E. MULTIPLE DESTRUCTORS AND INFINITE LOOPING

Two final aspects of destructor elimination should be mentioned. Occasionally one has the opportunity of eliminating several destructors. For example, SUB1, DIFFERENCE, and REMAINDER are all destructors and frequently occur in the same conjectures (since they are defined in terms of one another). We have found that when one has a choice of what terms to eliminate it is best to eliminate the simplest first. Thus we would eliminate SUB1 (with ADD1) before eliminating DIFFERENCE (with PLUS), and eliminate DIFFERENCE before we eliminated REMAINDER (with PLUS and TIMES). We define "simplest" here by considering the order in which the functions were introduced into the theory, since "subroutines" must be introduced before the functions using them.

When a conjecture contains several eliminable destructors, we eliminate the simplest first. Then we repeatedly eliminate destructor terms introduced by the previous elimination. The resulting set of clauses is returned to the top of the waterfall. For example, if a conjecture contained (SUB1 (SUB1 X)) and (REMAINDER U V), we would first eliminate (SUB1 X) by replacing X with (ADD1 J). This would transform (SUB1 (SUB1 X)) to (SUB1 J). Then we would eliminate (SUB1 J). Then we would return to simplification and not eliminate (REMAINDER U V) until after simplification. (The result in this case is just what we would have obtained by initially eliminating (SUB1 (SUB1 X)) by letting X be I + 2.)

In tandem with simplification, destructor elimination could cause infinite loops. An example is the term (LESSP (SUB1 I) I). Here we would "eliminate" (SUB1 I) by replacing it with (ADD1 J) to obtain (LESSP J (ADD1 J)). Then simplification would expand LESSP once to produce (LESSP (SUB1 J) J), and we would begin again. To avoid looping, we never eliminate a term that involves a variable that was introduced by a previous elimination pass unless there has been an intervening induction.

## F. WHEN ELIMINATION IS RISKY

As presented, and as implemented, the elimination of destructors can be heuristically risky, because it can result in a conjecture more general than the initial one. Elimination of destructors is always sound, because by construction, if *ELIM1 and *ELIM5 are theorems,

## 140 / X. ELIMINATING DESTRUCTORS

then p is a theorem. However, it is possible for p to be a theorem but for *ELIM5 to be a nontheorem. For example, if we transformed

```
(IMPLIES (AND (NUMBERP X)
              (NOT (EQUAL X 0)))
         (p (SUB1 X) X))
```

to

```
(p Y (ADD1 Y)),
```

instead of

```
(IMPLIES (NUMBERP Y)
         (p Y (ADD1 Y))),
```

the result might not be a theorem.

We now present sufficient conditions under which elimination is not risky. In the following proof we suppose that our elimination lemma has two destructor terms and that each destructor term has two arguments; a proof for the general case can be constructed in strict analogy with the following.

Suppose we are trying to prove

```
*GOAL   (p (d x y) (e x y) x y),
```

where x and y are distinct variables. Suppose further that we have an elimination lemma

```
*ELIM   (IMPLIES (hyp u v)
                 (EQUAL (lhs (d u v) (e u v) v)
                        u)).
```

Next suppose that we have the generalization lemma

```
*GEN    (g (d w z) (e w z) w z),
```

where w and z are distinct variables.

The results of using the elimination lemma *ELIM (together with the generalization lemma *GEN) upon *GOAL are the two formulas

```
*ELIM1  (IMPLIES (NOT (hyp x y))
                 (p (d x y) (e x y) x y))
```

and

```
*ELIM5  (IMPLIES (AND (hyp (lhs r s y) y)
                      (g r s (lhs r s y) y))
                 (p r s (lhs r s y) y)),
```

where r and s are new variables. Our no-risk insurance is: If *GOAL is a theorem, then *ELIM1 and *ELIM5 are theorems if

```
*COND  (IMPLIES (g r s (lhs r s y) y)
                (AND (EQUAL (d (lhs r s y) y) r)
                     (EQUAL (e (lhs r s y) y) s)))
```

is a theorem. *ELIM1 follows immediately from *GOAL. To prove *ELIM5, choose r, s, and y and assume the hypothesis (g r s (lhs r s y) y). Instantiate *GOAL to obtain

```
(p (d (lhs r s y) y)
   (e (lhs r s y) y)
   (lhs r s y)
   y).
```

From *COND, our assumptions, and the last formula, we conclude

```
(p r s (lhs r s y) y).
```

Q.E.D.

Given this theorem, we can check that the elimination of SUB1 by the introduction of ADD1 is not risky since we have

```
(IMPLIES (NUMBERP U)
         (EQUAL (SUB1 (ADD1 U))
                U)).
```

Similarly, we can check that the elimination of REMAINDER and QUOTIENT is not risky. The basic idea is that for any numeric U less than Y and any numeric V, U %  Y1*V has remainder U when divided by Y and quotient V when divided by Y.

Our implementation does not check that an elimination lemma is risk free, since it is not necessary to the soundness of the system. As users, we have never employed risky elimination theorems.

## G. DESTRUCTOR ELIMINATION IN THE REVERSE EXAMPLE

Let us now return to the proof of the theorem about (REVERSE (REVERSE X)). Recall that after induction we had three cases. Simplification reduced the first two to true and the third one to

```
*RR4
(IMPLIES
    (AND (LISTP X)
         (PLISTP (CDR X))
```

## 142 / X. ELIMINATING DESTRUCTORS

```
              (EQUAL (REVERSE (REVERSE (CDR X)))
                     (CDR X)))
      (EQUAL (REVERSE (APPEND (REVERSE (CDR X))
                              (CONS (CAR X) "NIL")))
             X)).
```

We can eliminate (CAR X) by using the elimination lemma for it added by our implementation of the shell principle:

**Axiom** CAR/CDR.ELIM:

```
      (IMPLIES (LISTP X)
               (EQUAL (CONS (CAR X) (CDR X))
                      X)).
```

In a conjecture of the form (p (CAR X) (CDR X) X), destructor elimination would produce two cases:

```
      (IMPLIES (NOT (LISTP X))
               (p (CAR X) (CDR X) X))
```

and

```
      (IMPLIES (LISTP (CONS A B))
               (p A B (CONS A B))).
```

Since our current conjecture, *RR4, has a (LISTP X) hypothesis, the first case produced by destructor elimination is trivial.[22] The second case, after simplification, is

```
*RR5
(IMPLIES (AND (PLISTP B)
              (EQUAL (REVERSE (REVERSE B)) B))
         (EQUAL (REVERSE (APPEND (REVERSE B)
                                 (CONS A "NIL")))
                (CONS A B))).
```

Note that in *RR5 the equality hypothesis about B can be used in the conclusion about (CONS A B) in a way that the equality hypothesis in *RR4 cannot. In particular, we see that there is an occurrence of B in the right-hand side of the conclusion, where none was obvious before.

We continue the proof of the REVERSE example in the next chapter.

The use of elimination in the REVERSE example illustrates an important aspect of destructor elimination not mentioned elsewhere.

---

[22] It is a propositional tautology and would not in fact be produced by the implementation for that reason.

Note that *RR5 is exactly what we would have had, had we initially used an induction step such as (IMPLIES (p B) (p (CONS A B))), instead of (IMPLIES (AND (LISTP X) (p (CDR X))) (p X)). Similarly, an elimination of (SUB1 X) would convert an induction step such as

```
(IMPLIES (AND (NUMBERP X)
              (NOT (EQUAL X 0))
              (p (SUB1 X)))
         (p X))
```

to the more conventional

```
(IMPLIES (AND (NUMBERP I)
              (p I))
         (p (ADD1 I))).
```

The latter induction is usually nicer for the same reasons that elimination of destructors is desirable at all.

One might ask why we did not do such an induction in the first place. The reason is that in general it is impossible because not all operations have inverses.

For example, consider the function DELETE,

**Definition**

```
(DELETE X Y)
   =
(IF (NLISTP Y)
    Y
    (IF (EQUAL X (CAR Y))
        (CDR Y)
        (CONS (CAR X) (DELETE X (CDR Y))))),
```

which returns the result of deleting the first occurence of X in Y. There is no function PUT.BACK such that (IMPLIES (PLISTP Y) (EQUAL (PUT.BACK X (DELETE X Y)) Y)). In the function DSORT, we use DELETE as the destructor function in the recursive call. In proving the correctness of DSORT in the theorem DSORT.SORT2, we perform an induction with the scheme

```
(AND (IMPLIES (NOT (LISTP X)) (p X))
     (IMPLIES (AND (LISTP X)
                   (p (DELETE (MAXIMUM X) X)))
              (p X))).
```

This induction is appropriate, and in the absence of an elimination lemma for DELETE, it is difficult to imagine the "constructive" (as opposed to "destructive") version of this induction.

Thus induction must, in general, operate by supplying hypotheses about the "destructors" actually employed by the recursive functions in the theorem, rather than conclusions about "constructors." Since some destructors are eliminable and the need to eliminate them arises outside of induction as well as inside it, we decided to formulate the destructor elimination heuristic in the general way described.

# XI

# Using Equalities

### A. USING AND THROWING AWAY EQUALITIES

When a formula has been maximally simplified and has had all the eligible destructor terms eliminated, we next try to use any equality hypotheses that may exist in the formula.

We try to use an equality hypothesis such as (EQUAL s' t') by substituting s' for t' elsewhere in the formula and deleting the equality. This procedure sometimes results in conjectures that can then be proved by simplification. For example, after the substitution it might become possible to open up a recursive function under our guidelines. But the motivation behind our handling of equality is far more heuristic: if a conjecture has not yielded to simplification and elimination of destructors, we will probably resort to induction to prove it. If so, it is best to clean up the conjecture as much as possible. Furthermore, if we have already applied induction, it is best to use the available induction hypotheses before resorting to induction again.

To explain our handling of equality we have to talk briefly about induction. The fundamental induction heuristic is to try to arrange things so that in an induction step of the form (IMPLIES (p t') (p t)),[23] simplification will be able to reduce the conclusion (p t) to

---

[23] Here and elsewhere in this chapter, we have ignored many of the details of induction, such as the case analysis responsible for making the induction hypotheses legal, so that we can concentrate on giving the reader our intuitions about the problems rather than the details.

some expression involving `t'`, so that the induction hypothesis can be used. Now suppose the theorem to be proved is of the form (`EQUAL s t`) (or merely concludes with such a literal). Then if the induction heuristic succeeds we will, after simplification, have a conjecture of the form

```
(IMPLIES (EQUAL s' t')
         (EQUAL s (h t'))).
```

In particular, `t'` will occur in the conclusion because we inducted in a way to make it do so.

If the implication above survives simplification we will probably have to prove it by induction. But proving something of the form (`IMPLIES p q`) by induction, where p is an equality, is often difficult. If we analyze the propositional calculus in the induction step for (`IMPLIES p q`),

```
(IMPLIES (IMPLIES p' q')
         (IMPLIES p q)),
```

we see that one of the cases is to prove (`IMPLIES p q`), given that `p'` does not hold. Since `p'` is an equality, its negation is not very strong, and since we have already failed to prove (`IMPLIES p q`), we do not have much with which to work.

From a purely heuristic standpoint it is better not to try to prove (`IMPLIES p q`) by induction but rather to try to use the hypothesis p and throw it away, before we go into induction again. Bledsoe noted the importance of "using up" a hypothesis and throwing it away in the "forcing principle" of [4]. In the resolution tradition of automatic theorem-proving, the idea of using up and throwing away hypotheses has received scant attention. One reason is that it is never actually necessary to throw away a clause in order to find a refutation of a set of clauses. However, when one is using a principle of induction, one has to select a particular formula upon which to do the induction. We believe that the "cleaning up" effect of fertilization, the simplifications performed by rewriting, and the generalizations to be described are essential to the performance of any theorem-proving system that handles proof by induction.

## B. CROSS-FERTILIZATION

In order to use (`EQUAL s' t'`) in

```
(IMPLIES (EQUAL s' t')
         (EQUAL s (h t'))),
```

we can substitute s' for t' in the other literals of the clause and delete the hypothesis. But rather than substitute for all occurrences of t' in (EQUAL s (h t')), we prefer to substitute just for those in (h t'). That is, if we have decided to use (EQUAL s' t') by substituting the left-hand side for the right, and one of the places into which we substitute is itself an equality, (EQUAL s (h t')), related to (EQUAL s' t') by induction, then we substitute only into the right-hand side. We call this "cross-fertilization."

To justify the plausibility of this heuristic, we must again discuss induction. If (EQUAL s t) is not proved by one induction, it is often because we managed to get t to simplify to t' but failed to get s to simplify to s', usually because s requires a different induction than t, e.g., induction on a different set of variables or with different hypotheses. By substituting s' for t' on t's side of the conjecture we can eliminate t from the problem and at the same time cast the problem entirely in terms of s and its descendants. In particular, by transforming

```
(IMPLIES (EQUAL s' t')
         (EQUAL s (h t')))
```

to

```
(EQUAL s (h s')),
```

we may be able to do an induction that lets both sides of the equality step through induction together.

## C. A SIMPLE EXAMPLE OF CROSS-FERTILIZATION

Let us now illustrate why we throw away the equality hypothesis after it has been used and how cross-fertilization often constructs theorems that go through induction cleanly. While we could cite complicated proofs in which cross-fertilization plays a role, we will deal here with a simple theorem, for pedagogical reasons. Let us prove that PLUS is commutative:

```
(EQUAL (PLUS X Y) (PLUS Y X)).
```

Note that the roles of X and Y are symmetric. Thus there is nothing to distinguish an induction on X from one on Y. An induction on X will let (PLUS X Y) step cleanly through the induction (in the sense that the (PLUS X Y) term in the induction conclusion will simplify to a term involving its counterpart in the induction hypothesis). Similarly, an induction on Y would "favor" (PLUS Y X). Neither induction is well suited to both terms. Let us do the induction on Y. Then the induction step is

```
(IMPLIES (AND (NOT (ZEROP Y))
              (EQUAL (PLUS X (SUB1 Y))
                     (PLUS (SUB1 Y) X)))
         (EQUAL (PLUS X Y) (PLUS Y X))).
```

After simplification and the elimination of (SUB1 Y) by replacing Y with (ADD1 Y), we get

```
(IMPLIES (EQUAL (PLUS X Y) (PLUS Y X))
         (EQUAL (PLUS X (ADD1 Y))
                (ADD1 (PLUS Y X)))).
```

Note that, as expected, the (PLUS Y X) term appears both in the hypothesis and the conclusion. Thus, we can use the equality by substituting (PLUS X Y) for (PLUS Y X) in the right-hand side of the conclusion and, after deleting the equality, get

```
(EQUAL (PLUS X (ADD1 Y)) (ADD1 (PLUS X Y))).
```

(Note that this equation is the "recursive" clause of a definition of PLUS that decomposes Y instead of X.) We have to prove this equation by induction, but now the theorem involves only (PLUS X Y) and its descendants. In particular, the commuted PLUS expression "favoring" Y has been eliminated (precisely because we inducted in the way we did and threw away the equality). Because all the PLUS terms descend from (PLUS X Y) there is a single, clear-cut induction we can perform to make them all step through induction, namely, induction on X. If we now perform an induction in which the induction step is

```
(IMPLIES (EQUAL (PLUS X (ADD1 Y))
                (ADD1 (PLUS X Y)))
         (EQUAL (PLUS (ADD1 X) (ADD1 Y))
                (ADD1 (PLUS (ADD1 X) Y)))),
```

the conclusion simplifies, first to

```
(EQUAL (ADD1 (PLUS X (ADD1 Y)))
       (ADD1 (ADD1 (PLUS X Y)))),
```

and then to

```
(EQUAL (PLUS X (ADD1 Y))
       (ADD1 (PLUS X Y))),
```

which is precisely our induction hypothesis.

Had we not performed the cross-fertilization, or had we done the substitution but not deleted the equality hypothesis, the conjecture would still have involved (PLUS Y X) terms and a subsequent induction on X would not have gone through (because our hypothesis would concern (PLUS Y X) and the conclusion (PLUS Y (ADD1 X))).

## D. THE PRECISE USE OF EQUALITIES

Here is the precise way we use equalities. Given a clause, we look for an equality hypothesis (EQUAL s' t') (i.e., a literal of the clause of the form (NOT (EQUAL s' t'))), where t' (or s') occurs in another literal of the clause and is not an explicit value template. Suppose we find such an (EQUAL s' t'), and that t' has the property described above. Then we will substitute s' for t' in a way described below.

If we are working on an induction step and there is an equality literal in the clause that mentions t' on the right-hand side, we decide to cross-fertilize (unless s' is an explicit value).

Then we consider substituting s' for t' in each literal of the clause, except the (EQUAL s' t') hypothesis. If we have not decided to cross-fertilize, we substitute s' uniformly for t' throughout the literal. But if we have decided to cross-fertilize, then (a) for an equality literal we substitute s' uniformly for t' only on the right side, (b) for a negative equality literal, we substitute s' uniformly for t' throughout the literal, and (c) for other literals we perform no substitution. Finally, if we are working on an induction step and s' was not an explicit value, we delete the (EQUAL s' t') hypothesis.

Of course, the resulting formula is then poured over the waterfall to be simplified again.

If there are multiple equality hypotheses that can be so used, we use only the first one we find. The others will be used on subsequent passes, after we have simplified the new formula and eliminated any destructor terms now eligible. By throwing away the equality hypothesis, we are generalizing the clause and are thus risking the adoption of a nontheorem as our goal. (For example, under the hypotheses elsewhere in the formula, the equality with which we substitute might be false. The input formula would thus be a theorem, but after eliminating the "reason" it was a theorem, it might not be.) As noted earlier, we prefer to do the safest things first, and both simplification and elimination of destructors are safe and can possibly simplify the formula before we have to resort to another substitution.

## E. CROSS-FERTILIZATION IN THE REVERSE EXAMPLE

Let us now return to the (REVERSE (REVERSE X)) example. Recall that after elimination of the (CAR X) and (CDR X) expressions we were left with

```
*RR5
(IMPLIES (AND (PLISTP B)
              (EQUAL (REVERSE (REVERSE B)) B))
         (EQUAL (REVERSE (APPEND (REVERSE B)
                                 (CONS A "NIL")))
                (CONS A B))).
```

Note that we can use the hypothesis (EQUAL (REVERSE (REVERSE B)) B) by substituting (REVERSE (REVERSE B)) for B in the conclusion. However, we will ignore the occurrences of B in the left-hand side of the conclusion, and will cross-fertilize for B in the right-hand side. Observe that we can cross-fertilize only because the elimination of destructors made it manifest that (CDR X) occurs in X when X is a list (i.e., B occurs in (CONS A B)). After deleting the equality hypothesis we are left with

```
*RR6
(IMPLIES (PLISTP B)
         (EQUAL (REVERSE (APPEND (REVERSE B)
                                 (CONS A "NIL")))
                (CONS A (REVERSE (REVERSE B))))).
```

We have thus managed to use our induction hypothesis, and we have produced a "balanced" equality with REVERSE and its descendants on both sides. In addition, since a subterm of s, namely, the innermost REVERSE expression, did reappear in s' because of our induction, we will be able to generalize it as described in the next chapter.

# XII

# Generalization

## A. A SIMPLE GENERALIZATION HEURISTIC

In proofs by induction, it is often easier to prove a theorem that is stronger than the one needed by some particular application. We have already seen several instances of this. For example, it arose in the TAUTOLOGYP proofs, where we proved

```
(IMPLIES (AND (NORMALIZED.IF.EXPRP X)
              (TAUTOLOGYP X A1))
         (VALUE X (APPEND A1 A2)))
```

rather than the more obvious

```
(IMPLIES (AND (NORMALIZED.IF.EXPRP X)
              (TAUTOLOGYP X "NIL"))
         (VALUE X A2)).
```

Another example arose in Chapter II when we proved

```
(EQUAL (MC.FLATTEN X ANS)
       (APPEND (FLATTEN X) ANS))
```

rather than the more obvious

```
(EQUAL (MC.FLATTEN X "NIL") (FLATTEN X)).
```

Of course, the reason that induction on stronger theorems is easier is that the main tools one has when proving a formula by induction are instances of the formula itself.

There has been some work on trying to generate mechanically the kind of generalizations performed in the MC.FLATTEN problem above [40, 2]. However, such major generalizations require "creative" insight into the problem.[24] We do not attempt to make such generalizations mechanically and leave such leaps to the user; he can always formulate and prove the desired theorem as a rewrite rule, so that instances of the theorem will be simplified to true and we will never have to guess the appropriate generalization.

However, if one inspects many inductive proofs, one discerns a common phenomenon involving generalization. Suppose we are trying to prove that some recursive function f always computes a value satisfying proposition p. That is, we are trying to prove (p (f $x_1$ ... $x_n$)). Suppose further that under some condition we know that (f $x_1$ ... $x_n$) computes its value by computing (f $x_1'$ ... $x_n'$) recursively, and then applying h. Thus, a reasonable induction step[25] to provide is

```
(IMPLIES (p (f x₁' ... xₙ'))
         (p (f x₁ ... xₙ))),
```

for then, when we simplify (f $x_1$ ... $x_n$) in the conclusion, we get

```
*1      (IMPLIES (p (f x₁' ... xₙ'))
                 (p (h (f x₁' ... xₙ')))),
```

where, as planned, we managed to cause (f $x_1'$ ... $x_n'$) to appear in both the hypothesis and the conclusion.

Now suppose that we are not able to prove this formula by the previously presented heuristics. Then we will probably have to appeal to induction again. We have found that in such a situation, it is fruitful to try instead to prove the more general

```
*2      (IMPLIES (p z) (p (h z))),
```

where z is a new variable. That is, if we were to prove *2, we could derive *1 from it by instantiating the z in *2 to be (f $x_1'$ ... $x_n'$). Another way to look at it is that by replacing (f $x_1'$ ... $x_n'$) in *1 by the new variable z, we can "guess" *2 as a possible way of proving *1.

We thus have another heuristic: if in the result of simplifying an in-

---

[24] That is, until someone produces a mechanization of the process, we say the process requires creativity.

[25] We will here ignore the finer points of induction, such as the case analysis, just as we did in the last chapter, and for the same reasons.

duction step we find subterms occuring in two or more literals, then we assume that those terms stepped through induction cleanly, have introduced their subsidiary functions and tests, and now merely represent "place holders" for arbitrary objects with the properties described by the hypotheses. Thus, those common subterms can be replaced by new variables.

If we were initially trying to prove an equality then it often happens that our hypothesis gets "used" (by fertilization) because one side of the equality comes through the induction cleanly. However, the other side of both the hypothesis and conclusion equalities may have common subterms as above. After cross-fertilizing, those common subterms are on opposite sides of the conclusion equality. Thus, we expand our heuristic to apply to common subterms on opposite sides of an equality literal.

## B. RESTRICTING GENERALIZATIONS

The heuristic of generalizing common subterms on either side of an equality or implication was first implemented in the theorem-prover described in [7]. We found it worked very well in simple examples. But it suffers from a tendency to produce overly general statements (i.e., nontheorems). In justifying the idea of generalizing the common subterm ( f $x_1'$ ... $x_n'$ ) , we said that perhaps it was a "place holder" for any object with the properties described by the induction hypotheses and case analysis.

But suppose that ( f $x_1'$ ... $x_n'$ ) has "intrinsic" properties that are relevant to the proof but not mentioned. For example, if ( f $x_1'$ ... $x_n'$ ) was known to be numeric, then it might be good to require that the new variable be numeric. Similarly, if we ever generalize ( SORT X ) , where SORT is a function that sorts the elements of its list argument into ascending order, it might be good to require that the new variable be an ordered list.

We do not know how to recognize mechanically what properties of the term being generalized to require of the new variable. So we permit the user to bring to our attention "generalization" lemmas that inform us of facts that are good to keep in mind when generalizing.

Suppose we have previously proved and noted as a generalization lemma a theorem such as

\*3    ( r ( f $t_1$ ... $t_n$ ) ) ,

where r is a schema. Suppose further that the term to be generalized, $(f\ x_1'\ \ldots\ x_n')$, is an instance of $(f\ t_1\ \ldots\ t_n)$.

The generalization of *1 above proceeds as follows. We notice that $(f\ x_1'\ \ldots\ x_n')$ is a common subterm and that it occurs in an instance of *3:

*3'     $(r\ (f\ x_1'\ \ldots\ x_n'))$.

Consider the conjecture obtained by adding *3' as a hypothesis to *1:

*4      (IMPLIES (AND $(r\ (f\ x_1'\ \ldots\ x_n'))$
                      $(p\ (f\ x_1'\ \ldots\ x_n')))$
                 $(p\ (h\ (f\ x_1'\ \ldots\ x_n'))))$.

Since *3' is a theorem, *1 is a theorem if *4 is a theorem. We then choose to try to prove *4 by trying to prove the more general conjecture obtained by replacing throughout *4 $(f\ x_1'\ \ldots\ x_n')$ with a new variable z:

         (IMPLIES (AND (r z) (p z))
                  (p (h z))).

If we know that the term being generalized always returns an object of a single shell type (that is, if we can observe that the type set of the term is a singleton), then we have found it useful to restrict the new variable to being in that class. For example, if $(f\ x_1'\ \ldots\ x_n')$ is numeric, then it is both sound and generally useful to add as an additional hypothesis (NUMBERP z). However, if the type set of $(f\ x_1'\ \ldots\ x_n')$ contains more than one type, the corresponding restriction would be a disjunction (e.g., z would be required to satisfy either NUMBERP or LISTP). We have found, empirically, that such weak constraints usually contribute little and split the theorem into several parts. In this case we simply ignore the type set information we have about $(f\ x_1'\ \ldots\ x_n')$.

## C. EXAMPLES OF GENERALIZATIONS

We will describe the whole generalization procedure precisely in a moment. But first we look more closely at two example generalizations that were mentioned earlier. The first was in the MC.FLATTEN proof in Chapter II. There, in proving

         (EQUAL (MC.FLATTEN X ANS)
                (APPEND (FLATTEN X) ANS)),

we inducted on X, simplified, eliminated the destructors CAR and CDR, performed two cross-fertilizations, and were left with

*5    (EQUAL (APPEND (FLATTEN Z)
                     (APPEND (FLATTEN V) ANS))
             (APPEND (APPEND (FLATTEN Z)
                             (FLATTEN V)) ANS)).

Let us now proceed to justify and carry out the generalization step done automatically in Chapter II. Note that (FLATTEN Z) and (FLATTEN V) occur on both sides of the equality. Assuming that FLATTEN has played its role and that the conjecture is true because of the relationships introduced by expanding functions and using the induction hypotheses, we now replace the two FLATTEN expressions by Y and A, respectively. However, since we know that FLATTEN always returns a list, we restrict the two variables to be LISTPs. The result is

*6    (IMPLIES (AND (LISTP Y)
                   (LISTP A))
              (EQUAL (APPEND Y (APPEND A ANS))
                     (APPEND (APPEND Y A) ANS))).

This is just a weak version of the associativity of APPEND and can now be proved by induction on the new variable, Y. In fact, *5 does not yield to induction (without a generalization somewhere along the line) because induction on the variables inside the FLATTEN expressions will simply produce deeper and deeper nests of APPENDs. It is thus crucial to guess that FLATTEN has played its role and to eliminate it.

We discussed a second example of generalization in connection with the elimination of destructors in Chapter X. When we generalize destructor terms, we do not bother to look for common subterms: we know exactly which terms we want to replace with variables. However, since we are performing a generalization, we do take advantage of our type set knowledge and available generalization lemmas.

For example, if, for any reason, we generalize (REMAINDER B A) by replacing it with I (as we did in Chapter X), then we add the additional restriction (NUMBERP I) because we know from its type set that REMAINDER always returns a number. Furthermore, if we know

            (EQUAL (LESSP (REMAINDER X Y) Y)
                   (NOT (ZEROP Y)))

as a generalization lemma, we add the additional restriction

            (EQUAL (LESSP I A) (NOT (ZEROP A))).

Thus, the result of generalizing the REMAINDER expression in

        (p (REMAINDER B A) A B)

is

        (IMPLIES (AND (NUMBERP I)
                      (EQUAL (LESSP I A)
                             (NOT (ZEROP A))))
                 (p I A B)).

This would be split into two conjectures when simplified, and we would consider the case that A was ZEROP on one branch and that A was non-ZEROP and I less than A on the other one.

## D. THE PRECISE STATEMENT OF THE GENERALIZATION HEURISTIC

There are two parts to the generalization heuristic: the choice of terms to generalize and the actual generalization and use of known facts about the terms. The elimination of destructors relies only upon the second phase of the generalization heuristic.

We say a term is *generalizable* unless it is a variable, an explicit value template, or its function symbol is EQUAL or a destructor. We first collect all the generalizable terms t such that either t occurs in two or more literals or there occurs a literal with atom (EQUAL x y) and t occurs in both x and y. After collecting such common terms, we delete from further consideration any term that properly contains another that we are considering. The remaining terms collected will be generalized away. By working our way outwards from the minimal common subterms in single generalization steps we are able to use our generalization lemmas to catch relations between the new and old variables that might be lost if we replaced the largest common subterms in one pass.

We have found it inappropriate to generalize explicit values (such as 5) or even terms that are explicit value templates because they contain too much information. We do not generalize destructor terms; if the destructor elimination heuristic has not eliminated them, they should probably not be eliminated. Note that after generalizing the minimal terms, it may be possible to eliminate destructor terms that could not have been eliminated earlier.

Having obtained a set of minimal common subterms (or having been supplied them by the destructor elimination heuristic) we then proceed to carry out the actual generalization. For each term t being generalized we search through all known generalization lemmas, looking for all lemmas that contain a subterm of which t is an instance under some substitution s. Whenever we find such a lemma, we instantiate it with s and add it as a hypothesis to the formula we are generalizing (i.e., we add the negation of the instantiated lemma as a literal to the clause being generalized). In addition, we obtain the type set of t (in the context of assuming all the literals in the clause false), and if that set contains only one type, r, we add ( r t ) as a hypothesis.

When we have so considered each of the terms being generalized we will have produced a new, expanded formula equivalent to the original one. We then generalize this expanded formula by uniformly substituting distinct new variables (i.e., ones not occurring in the formula) for each of the terms being generalized. This more general formula is then poured over the waterfall again (or, in the case of a generalization for the elimination of destructors, given back to that heuristic).

We have found it inappropriate to use a generalization lemma to restrict the generalization of a term if the generalization of the added hypothesis still mentions the function symbol of the term we were generalizing. Thus, a lemma such as

```
(EQUAL (FLATTEN (GOPHER X))
       (FLATTEN X))
```

might be a good generalization lemma when (GOPHER x) is generalized to Z (because it effectively adds the restriction that Z has the same fringe as x). But that lemma is not a good lemma to use when (FLATTEN x) is generalized to Z. Such a use of the lemma would add the hypothesis that (EQUAL (FLATTEN (GOPHER x)) Z), so we would not only fail to eliminate FLATTEN but would actually complicate matters by transforming (FLATTEN x) to (FLATTEN (GOPHER x)).

## E. GENERALIZATION IN THE REVERSE EXAMPLE

We now return to our ongoing proof that when a proper list is reversed twice the result is the list itself.

Recall that after cross-fertilizing we had

```
*RR6
(IMPLIES (PLISTP B)
         (EQUAL (REVERSE (APPEND (REVERSE B)
                                 (CONS A "NIL")))
                (CONS A (REVERSE (REVERSE B))))).
```

Note that (REVERSE B) occurs on both sides of the equality. It came through the induction cleanly and was involved in the left-hand side of our induction hypothesis and the left-hand side of our simplified induction conclusion. After using the hypothesis we find it on opposite sides of the equality. We thus generalize it to Z. Since we have no generalization lemmas about REVERSE and since (REVERSE B) is not always in a single shell class, we do not restrict the generalization. We thus obtain

```
*RR7
(IMPLIES (PLISTP B)
         (EQUAL (REVERSE (APPEND Z (CONS A "NIL")))
                (CONS A (REVERSE Z)))).
```

We can interpret the conclusion of this new conjecture in the following way. Suppose we have any list Z and we insert A as its last element. Then reversing the resulting list is equivalent to reversing Z and adding A as the first element. Note, however, that the hypothesis of *RR7 is not relevant to the truth of the conjecture.

# XIII

# Eliminating Irrelevance

## A. TWO SIMPLE CHECKS FOR IRRELEVANCE

Eliminating irrelevant hypotheses in a formula before trying to prove it by induction is just another way of obtaining a stronger conjecture to prove. Furthermore, eliminating irrelevant terms from a clause simplifies the task of finding an appropriate induction. In general, recognizing that a hypothesis is irrelevant to the truth of a formula takes a deep understanding of the problem at hand. However, there are simple cases where it is clear that a hypothesis is irrelevant.

The most obvious place to look for irrelevance is in hypotheses that are completely disconnected from the rest of the formula. Thus, the first thing we do when looking for irrelevance is to partition the literals of the clause according to shared variables, putting two literals in the same partition if they share variables. Then we try to decide which partitions are probably falsifiable and we delete any such partition from the clause. If a partition is falsifiable, then the result of deleting the partition from the clause is a theorem if and only if the original clause is a theorem. We have two simple heuristics for deciding that a partition is "probably" falsifiable. But even when these heuristics fail us (by guessing incorrectly that a partition is falsifiable), we are assured that if the result of deleting the partition (or indeed any set) from the clause is a theorem, then the original clause is a theorem.

The first heuristic tests whether the partition mentions any recursive function. If not, then it is composed entirely of EQUAL, shell recognizers, constructors, bottom objects, etc. But if such a partition were always true we should have proved this clause with simplification. Since we did not, we can reasonably assume the partition can be falsified. We give an example of such a partition in a moment.

The second heuristic tests whether the partition contains exactly one term and that term has the form ( f $v_1$ . . . $v_n$ ), where f is a recursive function and the $v_i$ are distinct variables. The only way such a partition can be a theorem is if the function f always returns true. But to have survived simplification it would have to appear sometimes to return F or else we would have simplified the literal to true by type set considerations. While it is certainly easy to write recursive functions that always return T but for which we compute the type set {T F}, we have never had occasion to use them. Thus, we feel justified in assuming we could find values for the $v_i$ that would falsify the literal, since the $v_i$ are completely unconstrained. An analogous treatment eliminates any singleton partition whose member is a term of the form ( NOT ( f $v_1$ . . . $v_n$ ) ).

## B. THE REASON FOR ELIMINATING ISOLATED HYPOTHESES

One might wonder why we bother to advocate any mechanical irrelevance-checking, given that our two ideas are so trivial. Generally, isolated hypotheses do not prevent the correct induction from succeeding, since they survive the induction step untouched. However, they do obscure the choice of the correct induction. In particular, irrelevant recursive hypotheses compete with the other terms in the conjecture for our attention when we are trying to choose the correct induction. Without recognizing that they are irrelevant, we may decide to induct "for" them rather than for the important terms.

The elimination of irrelevance is valuable as well in detecting that we are trying to prove a nontheorem. It is frequently the case, when our current conjecture is in fact not a theorem, that we are led to consider demonstrably falsifiable goals.

For example, suppose we tried to prove ( EQUAL ( REVERSE ( REVERSE X ) ) X ) without bothering to require that X be a proper list. The mechanical proof attempt goes as follows:

```
Give the conjecture the name *1.
We will try to prove it by induction.  There is
```

## B. THE REASON FOR ELIMINATING ISOLATED HYPOTHESES / 161

only one plausible induction. We will induct according to the following scheme:

```
(AND (IMPLIES (NOT (LISTP X)) (p X))
     (IMPLIES (AND (LISTP X) (p (CDR X)))
              (p X))).
```

The inequality CDR.LESSP establishes that the measure (COUNT X) decreases according to the well-founded relation LESSP in the induction step of the scheme. The above induction scheme leads to two new goals:

*Case 1.* (IMPLIES (NOT (LISTP X))
          (EQUAL (REVERSE (REVERSE X)) X)).

This simplifies, opening up the definition of REVERSE, to the goal:

```
(IMPLIES (NOT (LISTP X))
         (EQUAL "NIL" X)),
```

which has two irrelevant terms in it. By eliminating these terms we get:

    F.

Why say more?

```
*********************************************************
***                                                   ***
***              F A I L E D !                        ***
***                                                   ***
*********************************************************
```

CPU time (devoted to theorem-proving): .465 seconds

Note that in the base case we ended up having to prove that if X is not a list, then it is "NIL". Since this partition contains no recursive function we should have been able to simplify it to true were it a theorem. Since we could not simplify it, we eliminated those two literals, leaving us, in this case, with the empty clause. In general, one cannot conclude anything about the input conjecture if our heuristics lead to the empty clause. In particular, one cannot conclude that the input conjecture was not a theorem, since we might have generalized it in one of many different ways, and succeeded in generalizing it "too much." Nevertheless, it is usually worthwhile for the user to construct a counterexample to the conjecture that led to the empty clause and

## C. ELIMINATION OF IRRELEVANCE IN THE REVERSE EXAMPLE

see whether that falsifies the input. In the above example, the user need only identify a nonlist other than "NIL".

The example REVERSE proof we have been conducting illustrates how irrelevance can crop up in the proof of a well-stated theorem. Recall that after generalizing (REVERSE B) to Z we were left with

```
*RR7
(IMPLIES (PLISTP B)
         (EQUAL (REVERSE (APPEND Z (CONS A "NIL")))
                (CONS A (REVERSE Z)))).
```

The hypothesis (PLISTP B) was important at the beginning of the proof. Indeed, the above attempt at proving the theorem without the proper list hypothesis led us to a counterexample. (The reader should recall that our proofs of the first two induction cases, *RR1 and *RR2, of the correct statement of the theorem, *RR, in Chapter IX were based largely on reasoning about PLISTP.) However, after the generalization step, (PLISTP B) is disconnected from the rest of the conjecture.

If *RR7 were true because (PLISTP B) were always false, then PLISTP would have to return F on every input, since there are no restrictions on B. Because the type set of (PLISTP B) is {T F}, we can reasonably suppose (PLISTP B) is sometimes T and hence is truly irrelevant to *RR7. We produce

```
*RR8    (EQUAL (REVERSE (APPEND Z (CONS A "NIL")))
               (CONS A (REVERSE Z))).
```

Our heuristics have thus led us to a natural lemma about APPEND and REVERSE. We must prove it by induction.

# XIV

## Induction and the Analysis of Recursive Definitions

If the steps described above do not prove the conjecture, we will have reduced it to a conjunction of formulas, each of which is as simple as we can make it. Having nothing else left in our arsenal, we must prove each of these formulas by induction. Therefore, we choose one of them, formulate an induction scheme that seems appropriate for it, and then apply the scheme to the formula to obtain a new set of formulas to prove. We use the methods we have already discussed (and additional inductions, of course) to prove each of the formulas produced by the induction scheme.

Thus, the only outstanding question is the major one: how do we invent an induction scheme appropriate for a formula?

The answer lies in the similarity of recursion and induction. The reader may have noticed the similarity between our statements of the induction principle and the definition principle.

Very roughly speaking, the induction principle lets us prove (p $x_1$ ... $x_n$), where the $x_i$ are distinct variables, assuming (p $y_1$ ... $y_n$), provided that for some measure m and well-founded relation r, it is the case that (m $y_1$ ... $y_n$) is r-smaller than (m $x_1$ ... $x_n$), under the assumptions of the case analysis.

Very roughly speaking, the definition principle lets us define (f $x_1$ ... $x_n$), where the $x_i$ are distinct variables, in terms of (f $y_1$ ... $y_n$), provided that for some measure m and well-founded relation r, it is the case that (m $y_1$ ... $y_n$) is r-smaller than (m $x_1$ ... $x_n$), under the assumptions of the case analysis in the body of the function.

We exploit this similarity (or, rather, contrived this similarity) to determine a reasonable induction to perform when confronted with a conjecture involving recursive functions. In particular, suppose we were trying to prove a conjecture that contained some call of a recursive function f, and suppose that we knew that in its recursion, f drove down some measure of certain of its arguments. Then if the call of f in question contains variables in those argument positions, that call "suggests" a plausible induction, namely, under the conditions that ensure that the measure is driven down, assume the instances of the conjecture obtained by replacing those variables by what they will be when that call opens up. Such an induction is sound because we know that the measure of the indicated arguments is being decreased. Such an induction is heuristically plausible because it gives us induction assumptions about terms that would occur in the induction conclusion were we only to open up the appropriate call.

Thus, to set up an induction for a conjecture we must have a good understanding of how the functions in it recurse. In particular, we want to be able to look at a term in the formula and see immediately the inductions it suggests. In the remainder of this chapter we discuss how we analyze each recursive function when it is first introduced to check the requirements of the definition principle and to note the inductions the function suggests. In the next chapter we will discuss how we refine these suggested inductions to form one appropriate for the formula as a whole.

Even considered in isolation from the proof-time induction analysis, the definition-time analysis of functions is complicated. Two distinct problems are intertwined in the actual processing: we must establish that the function is well defined and we must analyze its recursion to determine all the possible inductions it suggests. We do these two tasks simultaneously. However, we start out by describing roughly how we prove that functions are well defined, and then we describe the kind of information we want to have by the time we actually try formulating inductions. Once both tasks have been sketched and the important heuristic components identified, we give a more precise statement of the actual process.

## A. SATISFYING THE PRINCIPLE OF DEFINITION

The definition principle puts several restrictions on the form of definitions. Most of these are trivial syntactic requirements. However, the requirement of a well-founded relation and measure justifying the definition is nontrivial. If ( f $x_1$ ... $x_n$) is defined to be some term, body, then the following must hold (according to Chapter III):

(d) there is a well-founded relation, denoted by a function symbol r, and a function symbol m of n arguments, such that for each occurrence of a subterm of the form (f $y_1$ ... $y_n$) in body and the f-free terms $t_1$, ..., $t_k$ governing it, it is a theorem that

```
(IMPLIES (AND t₁ ... tₖ)
         (r (m y₁ ... yₙ) (m x₁ ... xₙ))).
```

Note that we must establish that the tests governing each recursive call imply that some measure (fixed for all recursive calls) of the arguments is getting smaller according to a well-founded relation (fixed for all recursive calls). But we are not permitted to use the tests that involve the function symbol f, which has not yet been admitted into the theory. We describe below how we find a measure that is decreasing in each recursive call.

## 1. Machines

We start by flattening the nested structure of the body into a table that enumerates the branches through the definition that lead to recursions, and lists the governing tests and the recursive calls on each branch. We call this table the "machine" of the function definition and it is useful both in applying the definition principle and in formulating the case analysis of an induction. To build a machine for a function body we walk through the body collecting the tests governing the current expression (in the spirit of TAUTOLOGYP). We stop as soon as the current expression is either not an IF, or is an IF but a recursive call occurs in the test, or the set of recursive calls on one branch is a nonempty subset of those on the other. Each time we stop, we add an entry to the emerging table if the current expression contains a recursive call. The entry contains the tests collected and all the recursive calls in the current expression.

For example, the machine for Peter's version of Ackermann's function [46],

**Definition**

```
(ACK M N)
   =
(IF (ZEROP M)
    (ADD1 N)
    (IF (ZEROP N)
        (ACK (SUB1 M) 1)
        (ACK (SUB1 M) (ACK M (SUB1 N))))),
```

is

| case | tests | recursive calls |
|---|---|---|
| (1) | (AND (NOT (ZEROP M)) (ZEROP N)) | (ACK (SUB1 M) 1) |
| (2) | (AND (NOT (ZEROP M)) (NOT (ZEROP N))) | (ACK (SUB1 M) (ACK M (SUB1 N))) (ACK M (SUB1 N)) |

For a second example, consider the function that returns T if X is a subtree of Y and F otherwise:

**Definition**

```
(OCCUR X Y)
 =
(IF (EQUAL X Y)
    T
    (IF (LISTP Y)
        (IF (OCCUR X (CAR Y))
            T
            (OCCUR X (CDR Y)))
        F)).
```

The machine for OCCUR is

| case | tests | recursive calls |
|---|---|---|
| (1) | (AND (NOT (EQUAL X Y)) (LISTP Y)) | (OCCUR X (CAR Y)) (OCCUR X (CDR Y)) |

To satisfy restriction (d), given the machine for a definition, it suffices to find a measure and well-founded relation such that for each case in the machine and for each recursive call in the case, we can prove that the tests in the case imply that the measure of the arguments in the recursive call is smaller than the measure of the formals.

## 2. The Form of Induction Lemmas

To select an appropriate measure and well-founded relation, we rely upon axioms and previously proved theorems labeled by the user with the hint "induction." An induction theorem points out that some operation drives some measure down according to some well-founded relation. The general form of an induction lemma is

(IMPLIES h
       (r (m $y_1$ ... $y_j$) (m $x_1$ ... $x_j$))),

where the $x_i$ are distinct variables, all the variables in h occur in the conclusion, r and m are function symbols, and r is known to be well founded.[26]

Note that our definition of an induction lemma requires that the same explicit measure function m be the outermost function symbol of both arguments to the well-founded relation in the conclusion. Thus, certain useful rewrite lemmas, such as one with the conclusion (LESSP (DIFFERENCE I N) I), would not be allowed to double as induction lemmas. Instead, slightly reformulated versions would have to be proved for use as induction lemmas. However, note that if y is a numerically valued term, then (LESSP y x) implies (LESSP (COUNT y) (COUNT x)), because (COUNT y) is y and (COUNT x) is greater than or equal to x. Our implementation thus contains the following feature: if the conclusion of a lemma has the form (LESSP y x), where y has type set {NUMBERP}, then if the lemma is to be processed as an induction lemma, the conclusion is treated as though it were (LESSP (COUNT y) (COUNT x)).

Note that if one of the hypotheses is of the form (NOT (EQUAL (m $y_1$ ... $y_j$) (m $x_1$ ... $x_j$))), the lemma tells us that the measure is either decreasing or stays fixed. The knowledge that a measure does not increase—even though it may not decrease—is important when trying to establish that a lexicographic measure is decreasing.

Here are three examples of induction lemmas:

       (IMPLIES (NOT (ZEROP X))
             (LESSP (COUNT (SUB1 X)) (COUNT X))),

       (IMPLIES (LISTP X)
             (LESSP (COUNT (CDR X)) (COUNT X))),

and

       (IMPLIES (LESSP X Y)
             (LESSP (DIFFERENCE Y (ADD1 X))
                  (DIFFERENCE Y X))).

---

[26] In our current implementation, r must be LESSP. The theorem-prover constructs lexicographic combinations automatically, as will be described. The only other function we have been tempted to assume to be well founded is Gentzen's [19] well-founded relation on $\epsilon_0$.

The first two are intrinsic to the axiomatization of the ADD1 and CONS shells. The third one is the theorem COUNTING.UP.BY.1, which informs us that the measure DIFFERENCE of X and Y decreases if X is replaced by (ADD1 X) and Y is held constant, provided X is less than Y. Other examples of induction theorems can be found in Appendix A.

The basic idea behind the use of induction theorems is simple: consider the machine for some function f whose formal parameters are $x_1$, ..., $x_n$, and consider a particular recursive call (f $y_1$ ... $y_n$) in some case of the machine. We look for an induction lemma with a conclusion that says that some measure of our $y_i$ is smaller than the same measure of our $x_i$. If we find such a lemma we then try to prove that the tests in the machine imply the hypothesis of the induction lemma.

Thus, the theorems we try to prove while applying the principle of definition are usually quite simple since they involve showing only that the tests in the machine imply the hypothesis of the induction lemma, and not the usually more difficult fact that the measure is decreasing according to a well-founded relation. Of course, that more difficult fact was established once and for all when the induction lemma was proved.

## 3. A Simple Example

Now let us consider an example of using induction lemmas. Suppose PRED and FN have been previously defined. Consider the function WHILELOOP:

**Definition**
```
(WHILELOOP I MAX X)
  =
(IF (LESSP I MAX)
    (IF (PRED I X)
        T
        (WHILELOOP (ADD1 I) MAX (FN X)))
    F).
```

We have contrived this function to have several interesting properties and will discuss it many times in this and the next chapter. WHILE-LOOP is the obvious recursive expression of a simple loop that counts I up to MAX by 1. If it reaches or exceeds MAX, it returns F. If ever (PRED I X) holds, it stops and returns T. Otherwise, WHILELOOP resets X to (FN X) and iterates.

## A. SATISFYING THE PRINCIPLE OF DEFINITION

The machine for WHILELOOP is

| case | tests | recursive calls |
|---|---|---|
| (1) | (AND (LESSP I MAX)<br>(NOT (PRED I X))) | (WHILELOOP (ADD1 I)<br>MAX<br>(FN X)) |

Note that the recursive call changes I to (ADD1 I) and holds MAX fixed. (It also happens to change X to (FN X), but that is not important at the moment.) Recall the example induction lemma COUNTING.UP.BY.1 about DIFFERENCE (now instantiated with the variables used in WHILELOOP in one of nine possible ways):

(IMPLIES (LESSP I MAX)
    (LESSP (DIFFERENCE MAX (ADD1 I))
       (DIFFERENCE MAX I))).

The conclusion of this lemma informs us that the DIFFERENCE between MAX and I may be decreasing in WHILELOOP. To guarantee it, we must prove only that the tests in the machine for this case imply the hypothesis of the induction lemma. That is, we must prove

(IMPLIES (AND (LESSP I MAX)
       (NOT (PRED I MAX)))
  (LESSP I MAX)).

This is trivial.

Thus, to satisfy requirement (d) of the definition principle for WHILELOOP, we let the well-founded relation r be LESSP, and we let m be M, where

(M I MAX X) = (DIFFERENCE MAX I).

Thus, WHILELOOP is accepted under the definition principle as a well-defined function (and would be accepted by our implementation if it were cognizant of the induction lemma used).

### 4. Lexicographic Measures and Relations

Let us illustrate one further aspect of finding a measure and well-founded relation that explain a function. Consider Ackermann's function as defined above.

Note that the first argument, M, sometimes goes down in recursion and sometimes stays fixed. The second argument, N, sometimes goes up (quite rapidly) and sometimes goes down. The function is well defined because a lexicographic measure is going down. That is, in each recursive call (ACK m n), the pair ⟨m,n⟩ is lexicographically smaller than the pair ⟨M, N⟩. To say it another way, the ordinal ω*m + n is less than ω*M + N.

We can say this in our theory easily since we have ordered pairs.[27] In particular, the measure (CONS (COUNT M) (COUNT N)) gets LEX-smaller on every recursive call, where LEX is the lexicographic relation induced by LESSP and LESSP:

**Definition**

```
(LEX P1 P2)
  =
(IF (LESSP (CAR P1) (CAR P2))
    T
    (IF (EQUAL (CAR P1) (CAR P2))
        (LESSP (CDR P1) (CDR P2))
        F)).
```

If one knows enough to try lexicographic measures, the justification of ACK can be derived from the previously mentioned induction lemma intrinsic to the ADD1 shell:

```
(IMPLIES (NOT (ZEROP X))
         (LESSP (COUNT (SUB1 X)) (COUNT X))).
```

It is easy, and useful, to be able to formulate such lexicographic measures and well-founded relations from arbitrarily many other measures and well-founded relations.

For example, consider the function GCD:

**Definition**

```
(GCD X Y)
  =
(IF (ZEROP X)
    (FIX Y)
    (IF (ZEROP Y)
        X
        (IF (LESSP X Y)
            (GCD X (DIFFERENCE Y X))
            (GCD (DIFFERENCE X Y) Y)))).
```

---

[27] Indeed, it is quite easy to write functions for doing ordinal arithmetic.

One explanation of why GCD is well defined is that the measure (CONS (COUNT X) (COUNT Y)) decreases according to the relation LEX above.[28]

## B. INDUCTION SCHEMES SUGGESTED BY RECURSIVE FUNCTIONS

We will be more precise later about how we use a function's machine and known induction theorems to find a measure and well-founded relation justifying the function's definition. We delay that discussion because we want to get more from a recursive definition than the assurance that it is well defined. We want to be able to look at a call of a recursive function and see inductions that are appropriate. So we will now turn our attention to induction.

To apply the principle of induction to a conjecture we must invent a case analysis, together with some substitutions, a variable n-tuple, a measure, and a well-founded relation. We must prove the conjecture under each of the cases. In all but the base case we may assume instances of the conjecture obtained by uniformly replacing some of its variables by terms. However, we must also establish that in each such case the substitutions decrease the measure of the variable n-tuple according to the well-founded relation. Inventing an induction scheme that actually makes it possible to prove a given conjecture, while satisfying all the foregoing constraints, is sometimes quite difficult.

But the recursive definitions of functions in the conjecture "suggest" appropriate inductions. In particular, suppose the conjecture mentions the call ( f $x_1$ ... $x_n$) of a recursively defined function, suppose further that when we defined f we found a measured subset of the function's arguments, a measure on that subset, and well-founded relation justifying the definition of f, and finally suppose that all the $x_i$ in those measured positions are distinct variables. Then the sense in which ( f $x_1$ ... $x_n$) suggests an induction is as follows. By opening up ( f $x_1$ ... $x_n$) we could reexpress the conjecture in

---

[28] When we, as users of our program, introduced the definition of GCD, we had previously proved as an induction lemma that (DIFFERENCE X Y) is smaller than X when X and Y are non-ZEROP. We proved this so that we could introduce REMAINDER and QUOTIENT, both of which recurse by subtracting the second argument from the first. It had not occurred to us that GCD could be explained lexicographically until the definitional mechanism used the measure and relation above to explain it.

terms of the recursive calls ( f $y_1$ ... $y_n$ ) in the definition of f. But, we can obtain inductive hypotheses about ( f $y_1$ ... $y_n$ ) —at least for those $y_i$ in the measured subset—by using the substitutions that replace the measured $x_i$ by $y_i$ and replace other variables arbitrarily. By using a case analysis similar to the one in the definition of f, we know that our previously discovered measure on the subset decreases under these substitutions. Thus, it is easy to justify the induction scheme.

**1. Why We Use Measures and Relations**

The key to the above induction heuristic is that under certain conditions recursive definitions allow us to rewrite some function calls in the induction conclusion to terms involving instances of those calls—namely, the recursive calls in the body of the definitions—and that those instances can also be provided in inductive hypotheses. This reasoning can be expressed in a way in which no explicit mention is made of measures and well-founded relations: having accepted a function definition we can induct according to it (e.g., the "subgoal induction" scheme of Morris and Wegbreit [42].)

Let us consider a simple example. Suppose we wished to prove a conjecture of the form (p (WHILELOOP I MAX X)). One appropriate induction for it is closely analogous to the definition of WHILELOOP. The induction step provides the induction hypothesis (p (WHILE-LOOP (ADD1 I) MAX (FN X))), under the conditions (LESSP I MAX) and (NOT (PRED I X)). Two base cases are necessary, namely, those obtained by negating the conditions defining the induction step. The induction scheme can be justified by any measure and well-founded relation justifying the definition of WHILELOOP.

More generally, if f is a recursive function, then the term (f A B C) suggests an induction scheme which, under exactly the case analysis in the body of f, supplies induction hypotheses obtained by instantiating A, B, and C exactly as they are instantiated in the recursive calls in (f A B C). The suggested induction is justified by the same measure and well-founded relation justifying the definition of f. Thus, at first sight there is no apparent need for us to consider measures and well-founded relations explicitly for induction.

However, we want the terms in the conjecture being proved to suggest inductions to perform. For example, suppose the term (f A B term), where term is not a variable, is mentioned in the conjecture. If f is a well-defined recursive function, but no measure analysis is available, no legal induction is suggested by (f A B term). In particular, we are not free to obtain our induction hypotheses by instantitiating A, B, and term as they are instantiated by recursive calls in f, because term is not a variable, and we have no reason to believe that the measure justifying f decreases under a substitution instantiating only A and B.

But (WHILELOOP I MAX term) nevertheless suggests an induction, namely, the one in which we have the induction assumption for (ADD1 I) and MAX in the case (LESSP I MAX). We can look at (WHILELOOP I MAX term) and see an induction because we know that only the first two arguments of WHILELOOP are critical to the justification of the definition of WHILELOOP.

Because we want terms in the conjecture to suggest inductions, it is to our advantage to analyze each recursive definition carefully, so as to discover all the measures and well-founded relations that explain it.

## 2. Measured Subsets

The key to seeing an induction in (WHILELOOP I MAX term) is knowing that I and MAX, and just those two, are crucial in the measure justifying the definition. Our definition principle requires that the measure m justifying a definition of a function f, with formal parameters $x_1, \ldots, x_n$, be a function of n arguments. But it is often the case that there exists a measure m' and a subset $s = \{x'_1, \ldots, x'_j\}$ of $\{x_1, \ldots, x_n\}$ such that the requirements of the definition principle are met when (m $x_1 \ldots x_n$) is defined to be (m' $x'_1 \ldots x'_j$). If such an m' and s exist, we call s a *measured subset* for f.

Thus, a precise statement of the fact that only I and MAX are crucial to the measure justifying WHILELOOP is that {I MAX} is a measured subset of WHILELOOP. The only measured subset for ACK is {M N}. The fact that we explained ACK lexicographically, piecing together measures on M and measures on N, is not relevant here.

The ability to spot when a term suggests an induction is stregthened if we know all the possible justifications of the function's definition. Consider WHILELOOP. It is possible, for example, that (FN X) drives X down if (PRED I X) is false. If that were known, then {X} would also be a measured subset of WHILELOOP and we would be able to spot a potentially useful induction in a term such as (WHILELOOP $term_1$ $term_2$ X).

It is not unusual for a function to have more than one measured subset. Consider the definition of DIFFERENCE:

## Definition

```
(DIFFERENCE I J)
  =
(IF (ZEROP I)
    0
    (IF (ZEROP J)
        I
        (DIFFERENCE (SUB1 I) (SUB1 J)))).
```

Note that there are two explanations of this definition: the COUNT of the first argument decreases in every call, and the COUNT of the second argument decreases in every call.

Consequently, both {X} and {Y} are measured subsets of DIFFER-ENCE and both (DIFFERENCE X term) and (DIFFERENCE term Y) suggest legal inductions.

When we enforce the requirements of the principle of definition, we consider all the different combinations of arguments (that is, all the subsets of the formals), measured in all possible ways by known measures and lexicographic combinations of known measures. We even consider the possibility that unchanged arguments are contributing to the justification; for example, {I MAX} is a measured subset for WHILELOOP, even though MAX is unchanged in the recursion. Should we find any subset, measure, and well-founded relation decreasing in every recursive call, then we accept the definition under our principle of definition. To satisfy our need to spot inductions we find every nontrivial explanation.[29]

## 3. Specifying an Induction Scheme

It is not enough to know that some given measure of some subset of a function's arguments is decreased in recursion. That is enough to let us spot that an induction is suggested, but not to tell us what the induction is. We would also like to know, for each explanation found, the case analysis for the induction and the inductive instances we can (and supposedly should) assume for each case.

Recall that when we induct we are free to pick an arbitrary case analysis (together with a well-founded relation, measure, and n-tuple of variables). If we can show that a certain instantiation decreases the

---

[29] We do not bother to look for measures on supersets of identified measured subsets. In addition, we do not consider a lexicographic combination of two measures if either measure alone is always decreasing.

measure of our n-tuple of variables under the hypotheses of the case analysis, then we can legally assume that instance of the conjecture being proved. Thus, a typical inductive step is an implication in which we have the conditions defining the case and a set of instances of the conjecture as our hypotheses, and the conjecture itself as our conclusion.

It should be fairly clear that the case analysis for any induction for a function should be somewhat along the lines of the machine for the function. To illustrate this, consider Ackermann's function again and recall its machine:

| case | tests | recursive calls |
|------|-------|-----------------|
| (1) | (AND (NOT (ZEROP M)) (ZEROP N)) | (ACK (SUB1 M) 1) |
| (2) | (AND (NOT (ZEROP M)) (NOT (ZEROP N))) | (ACK (SUB1 M) (ACK M (SUB1 N))) (ACK M (SUB1 N)) |

If we were trying to prove a proposition about (ACK M N) we would clearly want to have two separate induction steps. One, governed by the tests in case (1), would supply as an inductive assumption the conjecture with M replaced by (SUB1 M) and N replaced by 1, because in that case the term (ACK M N) in the conclusion will open up to (ACK (SUB1 M) 1). The second inductive step, governed by the tests in case (2), would supply two inductive assumptions: one in which M is replaced by (SUB1 M) and N is replaced by (ACK M (SUB1 N)), and one in which M is replaced by M, and N is replaced by (SUB1 N). This choice of instances is motivated by the observation that the (ACK M N) term in the conclusion will open up to involve both of the two ACK terms we will obtain in the hypothesis under these instantiations. The base case of our induction would be obtained by negating the disjunction of the tests in cases (1) and (2).

So in general the machine for a function suggests the case analysis and the induction hypotheses. We now describe the selection of the case analysis and induction hypotheses more carefully.

*a. Defining the Cases*

How shall we define the cases of the induction argument? As we noted, we could in principle use the tests in the machine. We know that if we are free to instantiate all the arguments in any measured subset, then the associated measure will be decreased under the case analysis of the machine. This will give us n + 1 cases in an induction argument about a machine with n entries. The extra case is the base case obtained by negating the tests leading to induction steps.

We have found that using the machine in this way, while ideal if we are considering only a single term in the conjecture, leads to trouble if there are interesting interactions between the various recursive functions involved in the formula being proved.

i. Irrelevant Tests and Weak Base Cases

Let us look at another example. Consider the definition of LESSP:

**Definition**

```
(LESSP X Y)
   =
(IF (ZEROP Y)
    F
    (IF (ZEROP X)
        T
        (LESSP (SUB1 X) (SUB1 Y)))).
```

The machine for LESSP is

| case | tests | recursive calls |
|---|---|---|
| (1) | (AND (NOT (ZEROP Y)) (NOT (ZEROP X))) | (LESSP (SUB1 X) (SUB1 Y)) |

{X} and {Y} are both measured subsets of LESSP.

Consider the term (LESSP term I), where term is a nonvariable. This term suggests an induction on I, in which we have one induction step, providing one inductive hypothesis, namely, that obtained by replacing I by (SUB1 I). But what exactly should be the conditions governing the step? Note that two tests govern the recursion, (NOT (ZEROP I)) and (NOT (ZEROP term)). The question is, do we want these two tests to define our induction step as well?

While the second test is important in making LESSP compute the desired value, it is completely irrelevant to the recursion on I. Consider the consequences of using both tests to define our induction step. We would be obliged to prove the weak base case in which I was assumed to be non-ZEROP but term was assumed to be ZEROP. While this may be a natural base case for (LESSP term I), it may not be a natural base case for other expressions in the conjecture. For example, suppose (PLUS I J) were also involved in the conjecture. Then, in

## B. INDUCTION SCHEMES SUGGESTED BY RECURSIVE FUNCTIONS / 177

the case in which I is assumed non-ZEROP, we have to use recursion to determine the value of (PLUS I J) and thus would naturally want the case to be an induction step, not a base case.

For example, consider proving the theorem

(NOT (LESSP (PLUS I J) I)).

If we do not treat the second test in the machine for LESSP as irrelevant, the (LESSP term I) subterm suggests an induction in which the induction step is governed by (NOT (ZEROP I)) and (NOT (ZEROP (PLUS I J))). Thus, we are obliged to prove the base case

(IMPLIES (AND (NOT (ZEROP I))
              (ZEROP (PLUS I J)))
         (NOT (LESSP (PLUS I J) I))),

which reduces to

(IMPLIES (ZEROP (PLUS I J)) (ZEROP I)).

The proof of the above formula requires expanding the definition of PLUS, which we would do by performing a second induction on I.

On the other hand, if we recognize that the second test in the machine for LESSP is irrelevant here, the (LESSP term I) subterm suggests an induction in which the induction step is governed by (NOT (ZEROP I)), giving rise to the "natural" base case (ZEROP I). Both the induction step and the base case of the suggested induction simplify to true.

Recognizing certain tests to be irrelevant has another important effect on the analysis of appropriate inductions. Note that in (NOT (LESSP (PLUS I J) I)) there are two suggested inductions, the one suggested by (LESSP term I) and the one suggested by (PLUS I J). If we recognize the second test in the machine for LESSP to be irrelevant to an induction on the second argument, then the induction suggested by (LESSP term I) is precisely the induction suggested by (PLUS I J). If, on the other hand, we use the irrelevant test to define the induction step, the (LESSP term I) induction is different from the (PLUS I J) induction and we must choose between them.

Eliminating irrelevant tests from induction schemes thus has two advantageous effects: we eliminate weak base cases and we can arrange for different terms to suggest identical inductions. The latter effect is important because it permits us to focus our attention on a smaller number of candidate inductions, each "satisfying" a larger number of terms in the conjecture.

In the foregoing example, it is easy to spot that the test (NOT

(ZEROP X)) is irrelevant to the measured subset {Y}. But in general, the situation is more subtle. For example, consider the function MEMBER:

**Definition**
```
(MEMBER X Y)
    =
(IF (LISTP Y)
    (IF (EQUAL X (CAR Y))
        T
        (MEMBER X (CDR Y)))
    F).
```

MEMBER returns T if X is an element of the list Y, and F otherwise. The machine for the function is

| cases | tests | recursive calls |
|---|---|---|
| (1) | (AND (LISTP Y) (NOT(EQUAL X (CAR Y)))) | (MEMBER X (CDR Y)) |

Note that the second test, while important in making MEMBER compute the desired value, is completely irrelevant to the justification of the recursion. Recognizing the irrelevancy of the second test in MEMBER permits up to spot that (MEMBER A X), (APPEND X Y), and (REVERSE X) all suggest similiar inductions.

ii. Weeding Out Irrelevant Tests

So how do we weed out the irrelevant tests in a function definition? The answer is fairly simple: we consider the induction lemmas that explain recursive calls, and we use the hypotheses of the lemmas, rather than the tests in the machine, to determine the cases.

That is, if a machine tests $t_1, \ldots,$ and $t_j$ and then recurses with $(f\ y_1 \ldots y_n)$ and we know the induction lemma

(IMPLIES h
    (r (m $y_1$ ... $y_n$) (m $x_1$ ... $x_n$))))

and can establish

(IMPLIES (AND $t_1$ ... $t_j$) h),

then we know m goes down in that recursive call. Regardless of what terms actually govern the recursion, h is sufficient to guarantee that m decreases. In the case analysis for the induction, we will use h rather than the $t_i$, as the test for this case of the induction.

iii. An Example of Weeding Out Irrelevant Tests

For example, in processing MEMBER at definition time, we try {Y} as a possible measured subset. We see that in the only recursion, Y is being changed to (CDR Y). In our list of known induction lemmas we find the following lemma added by our implementation of the shell principle:

```
(IMPLIES (LISTP X)
         (LESSP (COUNT (CDR X)) (COUNT X))).
```

Instantitating the lemma by matching the (CDR X) in its conclusion with the (CDR Y) in the recursive call, we learn that the COUNT of Y decreases if the tests in the machine imply the instantiated hypothesis of the lemma, (LISTP Y). We thus try to prove the trivial implication

```
(IMPLIES (AND (LISTP Y)
              (NOT (EQUAL X (CAR Y))))
         (LISTP Y)).
```

Since this succeeds (and there is only one recursive call), we have positively identified {Y} as a measured subset. For the purposes of induction according to this measured subset, we associate with {Y} a revised version of the machine for MEMBER:

| case | tests | recursive calls |
|------|-------|-----------------|
| (1)  | (LISTP Y) | (MEMBER X (CDR Y)) |

where the test is that from the induction lemma, not the function body.

Note that this use of the hypotheses of induction lemmas is most effective when the user states induction lemmas with the cleanest hypotheses available.

*b. Selecting the Induction Hypotheses*

The second part of specifying an induction scheme is to decide, for each case in the revised machine for a measured subset, what the legal and supposedly appropriate induction instances are.

From the point of view of any given term ($f\ x_1\ \ldots\ x_n$) in a conjecture to be proved by induction, we know exactly which instances we would like in a given case of the analysis: those instances ($f\ y_1\ \ldots\ y_n$) that arise when the recursive definition of f is expanded under that case. If some subset of the $x_i$ consists of distinct variables, and that same subset is measured, then we know that instantiating those $x_i$ in accordance with the recursion is sound. But what of the other $x_i$? Clearly, if $x_i$ is not in the measured subset, then we can ignore it if we wish (from the soundness point of view), but (from the heuristic point of view) we should do our best to get induction hypotheses in which the terms occupying $x_i$'s position are those that arise when ( f $x_1$ ... $x_n$) is expanded. The induction principle allows us to substitute arbitrarily for variables not in the measured subset.

Thus, in the ideal case, when all the $x_i$ are distinct variables, we will substitute as directed by the recursion regardless of which measure we use to justify the induction. At definition time we cannot anticipate which argument positions might be occupied by nonvariables and which by variables, so we just note the "ideal" hypotheses in which we substitute for all the positions. We can represent a hypothesis as a substitution scheme on the formals of the function—a substitution that, for each recursive call, maps each formal to the corresponding term in the recursive call. At induction time we substitute the actual values of the formals into the substitution scheme to obtain the substitution used to produce an induction hypothesis. At that time we must throw out any substitution pair that would have us substitute for a nonvariable. Of course, not any pair can be thrown out if the resulting induction is to be sound. In the next chapter, we specify which pairs may be deleted.

## C. THE DETAILS OF THE DEFINITION-TIME ANALYSIS

We now summarize what we observe about a function's definition when it is introduced.

We first compute the machine for the function. Then, we consider all possible subsets of the formal parameters. For each subset of cardinality j we produce the j-tuples containing exactly the variables of the subset but in all possible permutations. For each such variable j-tuple we look at every known j-ary measure and well-founded relation in an induction lemma and note, for every recursive call in the machine, whether the measure goes down or does not increase on the j-tuple. When we find an induction lemma whose conclusion suggests that a measure decreases or is at least nonincreasing, we use our simplification heuristics to try to prove that the tests in the machine governing that recursion imply the hypothesis of the lemma.

In searching for an appropriate induction lemma, we elaborate the

## C. THE DETAILS OF THE DEFINITION-TIME ANALYSIS / 181

set of user supplied lemmas by exploiting the fact that LESSP is transitive. This is useful when the recursion involves nests of function symbols. For example, consider the definition of HALF:

**Definition**

```
(HALF I)
    =
(IF (ZEROP I)
    0
    (IF (ZEROP (SUB1 I))
        0
        (ADD1 (HALF (SUB1 (SUB1 I)))))))
```

Note that in the recursive call, I is replaced by (SUB1 (SUB1 I)). We can use the following induction lemma to establish that a measure of the variable 1-tuple ⟨I⟩ is decreasing:

**Theorem** SUB1.LESSP:

```
(IMPLIES (NOT (ZEROP X))
         (LESSP (COUNT (SUB1 X)) (COUNT X))).
```

We use the lemma as follows. We match (SUB1 (SUB1 I)) with the (SUB1 X) in the conclusion of the lemma. Instantiating the lemma with the resulting substitution we get

```
(IMPLIES (NOT (ZEROP (SUB1 I)))
         (LESSP (COUNT (SUB1 (SUB1 I)))
                (COUNT (SUB1 I)))).
```

Had the right-hand side of the instantiated conclusion been (COUNT I) (i.e., the measure applied to our variable 1-tuple), we would be done. Since it is (COUNT (SUB1 I)) instead, we recursively try to establish that (COUNT (SUB1 I)) is no bigger than (COUNT I) according to the same measure and LESSP. Thus, we again appeal to SUB1.LESSP, matching (SUB1 I) with (SUB1 X), and obtain

```
(IMPLIES (NOT (ZEROP I))
         (LESSP (COUNT (SUB1 I)) (COUNT I))).
```

This time the right-hand side is (COUNT I), so, provided the hypotheses hold, we have a transitivity argument that (COUNT (SUB1 (SUB1 I))) is smaller than (COUNT I). To complete the argument we must show that the tests in the machine governing the recursion imply the conjunction of the tests we have accumulated: (NOT (ZEROP (SUB1 I))) and (NOT (ZEROP I)). That is, (COUNT (SUB1 (SUB1 I))) is smaller than (COUNT I) if I is a number, not 0, and not 1.

## XIV. INDUCTION AND THE ANALYSIS OF RECURSIVE DEFINITIONS

In some cases we might find more than one way to establish that (SUB1 I) is no bigger than I. For example, we might know the induction lemma

```
(IMPLIES (NOT (EQUAL (COUNT (SUB1 X))
                     (COUNT X)))
         (LESSP (COUNT (SUB1 X)) (COUNT X))).
```

Recall that such a lemma means that (COUNT (SUB1 X)) is less than or equal to (COUNT X). Thus, if our implementation were aware of the above lemma, HALF would be an acceptable definition even if it only tested (ZEROP I). That is, the above lemma allows the implementation to deduce that (COUNT (SUB1 (SUB1 I))) is smaller than (COUNT I) if I is a number and not 0.

The foregoing procedure for unraveling nests of function symbols such as (SUB1 (SUB1 I)) is not limited to induction lemmas about COUNT and LESSP. The procedure may be generalized and applied to any measure and any well-founded relation that is known to be transitive. In fact, any well-founded relation may be extended to a transitive well-founded relation by taking its transitive closure.

Any time we find that a measure goes down or is nonincreasing on a particular call, we remember the variable j-tuple, measure, well-founded relation, recursive call, and the hypotheses of the induction lemma(s) used.

Having observed the behavior of every known measure and well-founded relation on every variable j-tuple and recursive call, we then try to find a way to piece this information together to find a lexicographic relation on m-tuples of the variable j-tuples. We proceed to find all possible lexicographic combinations justifying the definition. Of course, we do not construct trivial combinations in which, say, the first component by itself is always decreasing.

We now illustrate how we piece measures together lexicographically. Suppose we have found that some measure $m_1$ on the variable j-tuple $\langle x_1, \ldots, x_j \rangle$ gets $r_1$-smaller on some recursive calls and is merely nonincreasing on others. Also, suppose that on these other recursive calls we have found that another measure, $m_2$, on the variable k-tuple $\langle y_1, \ldots, y_k \rangle$ gets $r_2$-smaller. Then we construct a measure that produces pairs of the form (CONS ($m_1 x_1 \ldots x_j$) ($m_2 y_1 \ldots y_k$)) and we form the lexicographic well-founded relation induced by $r_1$ and $r_2$. Either $r_1$ or $r_2$ may itself be lexicographic in nature.

If we ever find a measure and relation that is always decreasing, we can easily use the induction lemma hypotheses to reconstruct a revised version of the machine with the lemma hypotheses, rather than the tests from the function body, defining the cases. We conjoin all the hypotheses involved in explaining all the recursions in a given case and use that conjunction as the tests for the case. In addition, we convert each recursive call in each case into a substitution by pairing each formal with the term occupying the corresponding argument position in the recursive call.

When we have finished, we will have found all the measured subsets we could identify. For each measured subset we will have noted the measure and well-founded relation associated with it, and the revised machine. We call this information associated with a measured subset an *induction template* because it describes one of the inductions suggested by the function.

As an example of an induction template, let us consider the function WHILELOOP: it has a measured subset {I MAX}, the measure is (DIFFERENCE MAX I), and the measure decreases according to the well-founded relation LESSP. The case analysis contains just one case, namely, that to prove (p I MAX X) one may inductively assume (p (ADD1 I) MAX (FN X)) under the condition that (LESSP I MAX). This induction step is encoded as the substitution that replaces I by (ADD1 I), MAX by MAX, and X by (FN X). Note that the definition of the case does not mention (PRED I X). In general, a case may have more than one such substitution (e.g., FLATTEN, ACK, and NORMALIZE), there may be more than one case (e.g., ACK and NORMALIZE), and there may be more than one induction template (e.g., LESSP and DIFFERENCE).

## D. RECURSION IN THE REVERSE EXAMPLE

Recall that we are in the process of proving the theorem
(IMPLIES (PLISTP X)
         (EQUAL (REVERSE (REVERSE X))
                X)).

There are three recursive functions involved in the proof of this theorem, PLISTP, REVERSE, and APPEND (which, recall, is a "subroutine" of REVERSE).

Given the single induction lemma
(IMPLIES (LISTP X)
         (LESSP (COUNT (CDR X)) (COUNT X))),

we make the following straightforward observations about those three functions.

(PLISTP X) has one measured subset, {X}, which is measured with COUNT and is getting LESSP-smaller. The appropriate case analysis contains one case, namely, that to prove (p X) one may inductively assume (p (CDR X)) under the condition (LISTP X).

(REVERSE X) has exactly the same analysis as PLISTP.

(APPEND X Y) also has exactly the same analysis as PLISTP, except we note that to prove (p X Y) one should assume (p (CDR X) Y) under the condition (LISTP X). In particular, we note that in this, the only induction case, APPEND holds its second argument, Y, fixed while changing X to (CDR X). Although Y is not essential to the measure (it is not in the measured subset), our inductions for APPEND should try to keep the second argument fixed.

# XV

# Formulating an Induction Scheme for a Conjecture

We now return to the problem of setting up an induction for a particular formula. The basic idea is to look at the terms in the formula and to note the inductions suggested by each of them. Then, after expanding each scheme to account for as many terms as possible, we pick the "best" according to various heuristics.

## A. COLLECTING THE INDUCTION CANDIDATES

Suppose that the term ( f $t_1$ ... $t_n$) occurs in the formula and we know an induction template for the recursive function f. Then there are two obvious questions: does the template apply (e.g., are enough of the $t_i$ terms distinct variables to permit a sound induction) and what is the induction suggested?

### 1. Deciding That a Template Applies

If the $t_i$ occupying the argument positions measured by the induction template we have in mind are all distinct variables, then we can use the template to construct a sound induction scheme. And in general, if any of the measured $t_i$ is not a variable, or is the same variable as another measured $t_i$, we cannot. However, there is an important exception to the rule. If $t_j$ is a nonvariable term in a measured position, and if the jth argument of f is never changed in any of the substitutions in the cases (i.e., in any of the recursive calls of f), then we can get induction hypotheses that force the measure down by not changing any of the variables occurring in $t_j$. (The second argument of WHILELOOP is an example of a $t_j$ in a measured position not changed in any substitution; the second argument of APPEND is not.)

We now make this precise. Let the *changeables* be those $t_i$ that occur in measured positions that are also sometimes changed in the recursion. Let the *unchangeables* be the variables occurring in those $t_i$ occupying measured positions that never change in the recursion. A template *applies* to a term if the changeables are all distinct variables and none of them is among the unchangeables.

For example, the template of WHILELOOP applies to (WHILELOOP X Y (H Z)) because X is the only changeable (and is a variable) and Y is the only unchangeable (and is different from X). The template does not apply to (WHILELOOP (G I) MAX X), because (G I) is a changeable that is not a variable. The template does not apply to (WHILE-LOOP I (G I) X), because I is both a changeable and an unchangeable. We justify the definition of "applies" after explaining how it is used.

## 2. Obtaining the Induction Scheme Suggested

Given a template that applies to a term, we can produce an induction scheme. An induction scheme is a case analysis together with a set of substitutions giving the induction hypotheses for each case, exactly analogous to the revised machine in the template. The only problem is that the template is expressed in terms of the formals of the function. We must instantiate it with the actual arguments of the term in the formula.

For example, the template for WHILELOOP (which was defined using the formals I, MAX, and X) tells us that under the condition (LESSP I MAX) we should assume the formula with I replaced by (ADD1 I), MAX replaced by MAX, and X replaced by (FN X). Consider the term (WHILELOOP K (PLUS X Y) V). K is the only changeable and X and Y are both unchangeables. Thus, the template applies. The induction suggested by (WHILELOOP K (PLUS X Y) V) is to assume the formula with K replaced by (ADD1 K), X and Y unchanged, and V replaced by (FN V), under the condition (LESSP K (PLUS X Y)). This is an appropriate induction because when (WHILELOOP K (PLUS X Y) V), in the induction conclusion, is opened up under the case (LESSP K (PLUS X Y)), the resulting WHILELOOP expression is identical to that provided in the induction hypothesis. But note that in instantiating the

## A. COLLECTING THE INDUCTION CANDIDATES / 187

substitution scheme we had to throw out the pair that would have instructed us to replace (PLUS X Y) by (PLUS X Y). In general, when we instantiate the substitution schemes in the template with the actual arguments, we must then go through the substitutions and delete certain illegal substitution pairs. There are three kinds of pairs that must be deleted (and can be deleted without affecting the soundness of the resulting induction scheme, provided the induction template indeed applies to the term in question).

As noted above, we must throw out any pair that would have us substitute for a nonvariable.

The second kind of pair we must throw out is any that requires that we substitute for an unchangeable. For example, in (WHILELOOP K (G X) X), we know we must keep X unchanged because it occurs in the second argument. However, it also occurs in the third argument, and the substitution scheme in the template would have us replace X by (FN X). This would violate the need to keep X fixed. Hence we must throw that pair out.

The third criterion for deleting a pair applies to an ambiguous "substitution," one that would have us substitute twice for the same variable. For example, the term (WHILELOOP I MAX I) would suggest we substitute (ADD1 I) for I (because it occurs in the first argument) and (FN I) for I (because it occurs in the last argument). If we find such duplicate pairs we must keep the one that substitutes for a measured variable (if either does). If neither variable is measured, we can choose arbitrarily which to delete. (Given that the template applies, we know that both are not measured since we have already determined that no two changeables are identical.)

Thus, we obtain the induction scheme, if any, suggested by a template and a term, in two steps. We first decide whether the template applies. If so, we then instantiate the case analysis and substitution schemes in the template and delete any illegal pairs from the resulting "substitutions." Both steps are trivial.

We now give some example induction schemes suggested by certain terms. The term (WHILELOOP K (PLUS X Y) K) suggests the scheme in which, under the condition (LESSP K (PLUS X Y)), we provide an inductive assumption for K replaced by (ADD1 K), and X and Y unchanged.

The term (LESSP I (G I J)) suggests the induction scheme in which, under the condition (NOT (ZEROP I)), we provide an inductive assumption for I replaced by (SUB1 I).

Finally, (LESSP I J) suggests two schemes since there are two templates that apply. The first provides an inductive assumption for I replaced by (SUB1 I) and J replaced by (SUB1 J), under the condition (NOT (ZEROP I)). The second is the symmetric one for J.

## 3. Proof that the Inductions Suggested Are Sound

We claim that if given a template and a term ( f $t_1$ ... $t_k$) to which the template applies, the above process produces a case analysis and some substitutions satisfying the requirements of the induction principle.

Recall that we must exhibit some conditions $q_i$, and some substitutions $s_{i,j}$, together with a well-founded relation r, variable n-tuple $x_1$, ..., $x_n$, and measure m such that, for each i and j

```
*goal   (IMPLIES q_i
                (r (m x_1 ... x_n)/s_{i,j} (m x_1 ... x_n)))).
```

Let s be the substitution mapping the formals of f to the corresponding $t_i$. Let $v_1$, ..., $v_a$ be the formals of f in the measured subset of the template. Suppose the condition defining the ith case in the template is $q_i'$. Then the $q_i$ we use in the induction principle is $q_i'$/s, as described. The $s_{i,j}$ are the substitutions obtained by instantiating the substitution schemes of the template with s and deleting illegal pairs, as described.

The proof that the $s_{i,j}$ are indeed substitutions (as opposed to arbitrary sets of pairs) follows trivially from the definition of "applies" and the first and third deletion criteria. The hard part still remains: what r, $x_1$, ..., $x_n$, and m do we use?

The well-founded relation r is that of the induction template. The $x_i$ are the union of the changeables and the unchangeables defined above (in any order). Observe that the $x_i$ include every variable mentioned in any term in a measured position of ( f $t_1$ ... $t_k$).

We now construct a measure function m satisfying *goal. Let m' be the measure function associated with the induction template. Consider the ith case and the i,jth recursive call in the template (i.e., the jth call in the ith case). By construction, m' applied to the terms $y_1$, ..., $y_a$ occupying the measured positions in the i,jth recursive call is r-smaller than m' applied to the measured subset itself:

```
        (IMPLIES q_i'
                (r (m' y_1 ... y_a) (m' v_1 ... v_a))).
```

Thus, the above theorem holds if we instantiate it with s:

```
*thm     (IMPLIES qᵢ'/s
                  (r (m' y₁ ... yₐ)/s
                     (m' v₁ ... vₐ)/s)).
```

We define m, the required measure function, with the equation

```
(m x₁ ... xₙ)
    =
(m' v₁ ... vₐ)/s.
```

That such a nonrecursive definition is well formed (i.e., that the body mentions no variable except the $x_i$) follows from the definition of the $x_i$ and that the $v_i$ are the measured subset.

We now show that *goal holds under the foregoing definitions. Note that the hypothesis $q_i'$/s of *thm, is by definition, $q_i$, and that the right-hand side of the conclusion, (m' $v_1$ ... $v_a$)/s, is equal to (m $x_1$ ... $x_n$) by the definition of m. Hence, *goal is equivalent to *thm if (m' $y_1$ ... $y_a$)/s is equal to (m $x_1$ ... $x_n$)/$s_{i,j}$, or equivalently, if (m' $y_1$/s ... $y_a$/s) is (m' ($v_1$/s)/$s_{i,j}$ ... ($v_a$/s)/$s_{i,j}$). In fact, it is the case that for each c from 1 to a, $y_c$/s is ($v_c$/s)/$s_{i,j}$. The reasoning is as follows: $y_c$/s is the term in the cth measured position of the i,jth recursive call introduced by expanding (f $t_1$ ... $t_n$). But ($v_c$/s) is the $t_i$ in the cth measured position. Let us call it $t_c$. Either $t_c$ is one of the changeables or else it is a term all whose variables are unchangeables. If it is a changeable, $s_{i,j}$ maps it to $y_c$/s by construction of the induction template substitution schemes and the instantiation and pair-deletion process. If $t_c$ is a term all whose variables are unchangeables, then $s_{i,j}$ will not effect it, by the second pair-deletion criterion. Thus, $t_c$/$s_{i,j}$ is $t_c$. But by definition of an unchangeable, $t_c$ is $y_c$/s.    Q.E.D.

Thus, the simple definition of "applies" and the instantiation and deletion process produce sound induction schemes.

## B. THE HEURISTIC MANIPULATION OF INDUCTION SCHEMES

Once we have noted each induction scheme (i.e., case analysis and substitutions) suggested by any term in the formula, we begin the mainly heuristic phase of manipulating induction schemes and choosing the best one we can formulate. We have discovered many heuristics for obtaining new schemes from old ones and for choosing among the various candidate inductions.

To facilitate the discussion of these heuristics, we associate with each scheme

the term for which it was produced,

the set of all variables that are substituted for (whether measured or unmeasured), henceforth called the *changing variables*,

the set consisting of all the unchangeables (as defined previously) and all the variables occurring in arguments to the term that are changed by no substitution (that is, the variables that must stay fixed for the induction to be sound plus those we are keeping fixed because the function merely "wants" them that way), henceforth called the *unchanging* variables, and

a *score* that reflects our estimation of how good a match this induction scheme is for the term for which it accounts.

Initially, the score is the rational quotient of the number of formals for which we substituted, divided by the number of formals of the function symbol of the term that suggested the induction template. Thus, initially, the score reflects how many of the arguments of the function we expect to "pick up" by our induction hypotheses when the function opens up.

## 1. Subsumption of Induction Schemes

Our first step in cleaning up the set of possible induction schemes is to throw away any "subsumed" by another.

Suppose that some scheme $s_1$ has only the single case that says, "Under the assumption (LISTP X) assume the conjecture with X replaced with (CDR X)." Further, suppose that scheme $s_2$ has a richer case structure, but one of its cases is, "Under the assumption (LISTP X) and (LISTP (CDR X)) assume the conjecture with X replaced with (CDR (CDR X)))." Then from a heuristic point of view, $s_2$ "subsumes" $s_1$ because the $s_2$ case above will essentially give the term by which $s_1$ was suggested permission to recurse twice, once to (CDR X) and then again to (CDR (CDR X)). We throw out the $s_1$ induction in favor of the $s_2$ induction.

In order for $s_2$ to *subsume* $s_1$ we require three things. First, the changing variables of $s_1$ must be a subset of those of $s_2$. Second, the unchanging variables of $s_1$ must be a subset of those of $s_2$. Finally we require that every case of $s_1$ be subsumed by a case of $s_2$, in the sense that the tests of the $s_1$ case are a subset of those of the $s_2$ case, and that for every substitution in the $s_1$ case there be a substitution in the $s_2$ case such that for each component in the $s_1$ substitution, there is a component in the $s_2$ substitution with the same variable and a term that mentions the term of the $s_1$ component. Further, no $s_2$ case is permitted to subsume two $s_1$ cases, and no $s_2$ substitution is permitted to subsume two $s_1$ substitutions.

When we throw out $s_1$ because $s_2$ subsumes it, we add $s_1$'s score to that of $s_2$, and we consider that $s_2$ accounts for all the terms for which it used to account plus all those for which $s_1$ was accounting.

Note that if two induction schemes are identical, then one subsumes the other. Hence, if several terms in the formula suggests the same induction, as we saw happen with (PLISTP X) and (REVERSE X), we will have eliminated duplicate suggestions after the subsumption check. We will have noted also that the suggested induction accounts for several terms.

As an example of an induction scheme being subsumed by a nonidentical one, consider the theorem that (HALF I) is less than or equal to I:

(NOT (LESSP I (HALF I))).

The LESSP-expression suggests inducting on I, assuming the theorem for (SUB1 I) under the case (NOT (ZEROP I)). The HALF-expression suggests induction on I, assuming the theorem for (SUB1 (SUB1 I)) under the case (NOT (ZEROP I)) and (NOT (ZEROP (SUB1 I))). The induction for LESSP is subsumed by the induction for HALF. Note that the induction we choose is equivalent to assuming the conjecture for I and proving it for I + 2 in the induction step, handling as base cases the possibilities that I is nonnumeric, 0, or 1. This induction accounts for the LESSP term because the LESSP term in the induction conclusion can be expanded twice to transform its argument from I + 2 to I.

At the conclusion of this chapter we present several thoroughly described example inductions involving subsumption and our other heuristics.

## 2. Merging of Induction Schemes

To see the need for the next heuristic, let us look at an example. Suppose we are trying to prove the transitivity of LESSP

(IMPLIES (AND (LESSP X Y)
              (LESSP Y Z))
         (LESSP X Z)).

## XV. FORMULATING AN INDUCTION SCHEME FOR A CONJECTURE

Since LESSP changes both of its arguments, and a measure of either is sufficient to justify the definition, we start out with six suggested schemes. They are (grouping them by two's in correspondence with the LESSP terms above)

Under the condition (NOT (ZEROP X)), assume the conjecture with X replaced with (SUB1 X) and Y with (SUB1 Y).

Under the condition (NOT (ZEROP Y)), assume the conjecture with X replaced with (SUB1 X) and Y with (SUB1 Y).

Under the condition (NOT (ZEROP Y)), assume the conjecture with Y replaced with (SUB1 Y) and Z with (SUB1 Z).

Under the condition (NOT (ZEROP Z)), assume the conjecture with Y replaced with (SUB1 Y) and Z with (SUB1 Z).

Under the condition (NOT (ZEROP X)), assume the conjecture with X replaced with (SUB1 X) and Z with (SUB1 Z).

Under the condition (NOT (ZEROP Z)), assume the conjecture with X replaced with (SUB1 X) and Z with (SUB1 Z).

Suppose we were to induct according to any one of these—for example, the first. Then our hypothesis would be obtained by replacing X and Y by (SUB1 X) and (SUB1 Y), respectively. But consider the term (LESSP X Z) in the conclusion. Note that its counterpart in the induction hypothesis (i.e., the term obtained by instantiating it as above) is (LESSP (SUB1 X) Z). If we open up the (LESSP X Z) in the conclusion we will obtain (LESSP (SUB1 X) (SUB1 Z)), which is not in the hypothesis because in the hypothesis we did not replace Z by (SUB1 Z). But if we do not open it up, and leave it (LESSP X Z), it is not in the hypothesis because in the hypothesis we did replace X by (SUB1 X).

Clearly, if the changing variables of two schemes overlap, then doing either induction by itself will throw the terms associated with the other induction "out of sync" by replacing some but not all the variables they need to reoccur. That is the heuristic motivation of "merging."

We *merge* scheme $s_1$ into $s_2$ provided their changing variables have a nonempty intersection, the unchanging variables of each has an empty intersection with the changing variables of the other, and we can merge each case of $s_1$ into a case of $s_2$ in the following sense: for every substitution in the $s_1$ case we can find a substitution in the $s_2$

## B. THE HEURISTIC MANIPULATION OF INDUCTION SCHEMES / 193

case such that the two substitutions substitute for some common variable, the two substitutions substitute identically on all common variables, and there is at least one common variable v such that the term replacing it is not v itself. Further, no two $s_1$ cases may merge into the same $s_2$ case, and no two $s_1$ substitutions may merge into the same $s_2$ substitution.

The *result of such a merge* is like $s_2$, except that for each case in $s_2$ that absorbed an $s_1$ case we add the tests from the $s_1$ case to those of the $s_2$ case, and we extend each substitution in the $s_2$ case by adding any new pairs from an $s_1$ substitution with which it overlapped (if any). In addition, we merge each $s_1$ case into as many other $s_2$ cases as we can. We add the changing variables of $s_1$ to the changing variables of $s_2$, and we add the unchanging variables of $s_1$ to the unchanging variables of $s_2$. The new scheme's score is the sum of the scores of $s_1$ and $s_2$, and the new scheme accounts for all the terms of the two older schemes. We throw out $s_1$ and retain the modified $s_2$.

To see how merging works, we return to the example of the transitivity of LESSP. The six schemes noted above merge into one:

Under the conditions (NOT (ZEROP X)), (NOT (ZEROP Y)), and (NOT (ZEROP Z)), assume the conjecture with X replaced with (SUB1 X), Y with (SUB1 Y), and Z with (SUB1 Z).

If we let (p X Y Z) be the statement of the transitivity of LESSP, then the above induction scheme produces three base cases, one each for X, Y, or Z being ZEROP. The induction step is

```
(IMPLIES (AND (NOT (ZEROP X))
              (NOT (ZEROP Y))
              (NOT (ZEROP Z))
              (p (SUB1 X) (SUB1 Y) (SUB1 Z)))
         (p X Y Z)).
```

Note that if the LESSP expressions in the conclusion (p X Y Z) are expanded under the conditions given, the result is precisely the induction hypothesis.

The result of merging is sound for the same reasons that $s_2$ was sound. For example, if we merged the last five LESSP induction schemes above into the first scheme, then the new scheme is sound because the first induction was sound. In particular, we still replace X with (SUB1 X) under the hypothesis that (NOT (ZEROP X)) is true, so that the same measure and well-founded relation explain the new induction scheme.

In general, merging adds additional tests to the induction cases and

## 194 / XV. FORMULATING AN INDUCTION SCHEME FOR A CONJECTURE

instantiates some variables not involved in the justification. The additional tests in the induction cases will cause additional base cases to be generated (when we negate the conditions of the induction steps to obtain the base cases), but that is a heuristic problem, not a logical one.

We feel it is reasonable to add the extra tests (for example, the tests on (NOT (ZEROP Y)) and (NOT (ZEROP Z)) above) to the induction steps since those steps introduce instances (involving (SUB1 Y) and (SUB1 Z) in our example) that generally do not make sense without such tests.

### 3. Flawed Induction Schemes

Once we have done all the possible subsumptions and merges we analyze the schemes to decide whether some of the recursions in the conjecture are getting in the way of others. Let us consider the associativity of APPEND:

```
(EQUAL (APPEND (APPEND A B) C)
       (APPEND A (APPEND B C))).
```

Recall that (APPEND X Y) recursively changes X and leaves Y fixed. We thus observe three suggested induction schemes, namely, one for each of the terms above in which APPEND has a variable in its first argument. Two of the suggested inductions merge and we must thus consider the two alternative schemes:

Under the condition (LISTP A), assume the conjecture with A replaced with (CDR A) and B and C unchanged.

Under the condition (LISTP B), assume the conjecture with B replaced with (CDR B) and C unchanged.

Now consider what would happen if we inducted according to the second scheme. The induction step would be

```
(IMPLIES (AND (LISTP B)
              (EQUAL (APPEND (APPEND A (CDR B)) C)
                     (APPEND A (APPEND (CDR B) C))))
         (EQUAL (APPEND (APPEND A B) C)
                (APPEND A (APPEND B C)))).
```

The term (APPEND B C) in the conclusion would open up to involve (APPEND (CDR B) C), as planned. However, the term (APPEND A B) in the conclusion, whether we open it or not, will not look like its

### B. THE HEURISTIC MANIPULATION OF INDUCTION SCHEMES / 195

counterpart in the hypothesis because there we replaced B with ( CDR B ) in an argument of APPEND that does not change.

On the other hand, if we induct according to the first scheme, then all the terms mentioning A in the conclusion will recurse once to give rise to the terms mentioning A in the hypothesis (since A is never used by any term other than the ones the scheme accounts for), and furthermore, by holding all the other variables fixed, the induction on A guarantees that terms mentioning only those variables will reappear in the hypothesis simply because they will not change. Thus, the first scheme is virtually perfect while the second is worthless. In this case we say that the first scheme is "unflawed."

We consider any scheme, $s_1$, *flawed* if there is another scheme $s_2$ such that some changing or unchanging variable of $s_2$ is an "induction" variable of $s_1$. We say that v is an *induction* variable of $s_1$ provided there is a term t for which $s_1$ accounts and a template temp for the function symbol of t such that v is a changeable with respect to t and temp.

The example just given illustrates why no induction variable of $s_1$ should be an unchanging variable of $s_2$. We consider $s_1$ flawed if one of its induction variables is a changing variable of $s_2$ because the two schemes must disagree on some variables since they did not merge. We do not care whether $s_1$'s unchanging variables overlap other schemes since $s_1$ may account for the terms suggesting those schemes by not touching those variables.

If some scheme is unflawed, we throw out all flawed schemes. If all schemes are flawed, we simply throw out none and proceed (but usually with some trepidation).[30]

### 4. Tie-Breaking Rules

It is surprising how often, in well-stated theorems, only one suggested induction scheme survives the foregoing heuristics. The most common exception arises when the conjecture is symmetric in several variables, in which case it usually does not matter which induction is chosen. However, if after all the above we are still left with more than one candidate scheme, we consider the one with the highest score as

---

[30] We have seen a few examples where the merging heuristic flaws the "right" induction scheme, by introducing into it a variable over which there is competition. As users we tend to rephrase the theorem to avoid the problem. See the footnote about ID in Chapter XIX.

being the most likely to succeed (if there is exactly one with the highest score).

If at this point two or more schemes are tied with the highest score, we have one useful tie-breaking rule. It is most likely that the terms we induct "for" will occur in both the hypothesis and the conclusion after simplification. Therefore, these terms are most likely to be eliminated by fertilization and generalization. Consequently, by inducting on the "nastiest" functions in the conjecture we may well eliminate them now and have something better to work on next time. We use a simple measure of "nastiness": a function is nasty if it is not primitive-recursive (in that its arguments are either held fixed or decremented by shell accessors).

Thus, to break ties we choose the scheme that is credited with the largest number of terms that are not primitive-recursive. If we still have a tie, we choose arbitrarily. It is interesting to note that in proving the several hundred theorems in Appendix A, we choose arbitrarily only in symmetric cases.

### 5. Superimposing the Machine

Finally, having chosen a scheme, we perform one last operation on it. If the scheme is credited with exactly one term, then it just represents the induction for that term. Consequently, we can ensure that the term in the induction conclusion opens up to precisely the terms in the induction hypothesis (at least on the measured subset) by superimposing on the case structure of the scheme the case structure of the machine for that function. Since the two case structures have exactly the same number of cases (we originally created the case analysis for the scheme by replacing the tests in the case analysis of the machine), this amounts to going over the scheme's cases and adding in the tests from the corresponding cases of the instantiated machine.

If, on the other hand, we have merged other schemes into the winner, then the case structure of any particular machine probably will not account for all the terms credited to the scheme. (Recall the discussion of irrelevant tests and weak base cases in Chapter XIV.) If we have merged, we simply check that each of the cases of the scheme is different (in the sense that each set of tests is different), and if we find any two that are identical we add the substitutions of one case into those of the other and throw out the first case. This is sound since the new case has the same tests, and all the induction hypotheses drive the same measure down under those tests. It is heuristically necessary since, if we were to leave the two identical cases, with two different induction hypotheses, then when the two conclusions opened up there would be nothing to force them down the branch of the machine handled by their respective hypothesis (since the tests are equivalent). Thus, we cover our bets by providing one induction step with all the relevant hypotheses.

### 6. Producing the Induction Formula

Having finally obtained the $x_i$, $q_i$, and $s_{i,j}$ from the selected induction scheme for the clause in question, we apply the induction principle to the term corresponding to the disjunction of its literals, producing $k + 1$ terms. We then convert those terms into a propositionally equivalent set of IF-free clauses after expanding all calls of AND, OR, NOT, and IMPLIES. The resulting set of clauses is then poured over the waterfall.

## C. EXAMPLES OF INDUCTION

We now consider three example inductions. There are many other induction examples elsewhere in this book.

The first example comes from the proof that a fast string searching algorithm is correct. The ideas behind the algorithm and the proofs are discussed in Chapter XVIII. The intuitive statement of one of the main lemmas is that if PAT is a nonempty character sequence, if I is a legal character position in another sequence STR, and if the right end of the leftmost occurrence of PAT in STR is beyond position I, then it is also beyond position (PLUS I (DELTA1 (NTHCHAR I STR) PAT)). Stated formally this is

```
(IMPLIES (AND (LISTP PAT)
              (LESSP I (LENGTH STR))
              (LESSP I
                     (PLUS (LENGTH PAT)
                           (STRPOS PAT STR))))
         (LESSP (PLUS I (DELTA1 (NTHCHAR I STR) PAT))
                (PLUS (LENGTH PAT)
                      (STRPOS PAT STR)))).
```

## XV. FORMULATING AN INDUCTION SCHEME FOR A CONJECTURE

This simplifies, by opening up the nonrecursive functions, to

```
(IMPLIES
  (AND (LISTP PAT)
       (LESSP I (LENGTH STR))
       (LESSP I
              (PLUS (LENGTH PAT) (STRPOS PAT STR))))
  (LESSP (PLUS I
               (STRPOS (CONS (CAR (NTH STR I)) "NIL")
                       (REVERSE PAT)))
         (PLUS (LENGTH PAT) (STRPOS PAT STR)))).
```

In order to understand how we set up an induction for this formula, it is not necessary to understand what these functions do, only how they do it. Their definitions are in Appendix A. However, below we give a brief sketch of the recursive structure of the functions:

| | |
|---|---|
| (LESSP X Y) | decrements X and Y with SUB1, under non-ZEROP tests, |
| (LENGTH X) | decrements X with CDR under a (LISTP X) test, |
| (PLUS X Y) | decrements X with SUB1 and leaves Y unchanged, under a non-ZEROP test on X, |
| (STRPOS X Y) | decrements Y with CDR and leaves X unchanged, under (among other things) a (LISTP Y) test, |
| (REVERSE X) | decrements X with CDR under a (LISTP X) test, |
| (NTH X Y) | decrements X with CDR and Y with SUB1, under a non-ZEROP test on Y. |

When presented with the above simplified formula, our induction heuristics lead to the following analysis (the following is the output generated by the implementation):

```
We will appeal to induction. The recursive terms in
the conjecture suggest ten inductions. They merge
into two likely candidate inductions. However, only
one is unflawed. We will induct according to the
following scheme:
```

## C. EXAMPLES OF INDUCTION / 199

```
(AND (IMPLIES (NOT (LISTP STR))
              (p I STR PAT))
     (IMPLIES (NOT (NUMBERP I))
              (p I STR PAT))
     (IMPLIES (EQUAL I 0) (p I STR PAT))
     (IMPLIES (AND (LISTP STR)
                   (NUMBERP I)
                   (NOT (EQUAL I 0))
                   (p (SUB1 I) (CDR STR) PAT))
              (p I STR PAT))).
```

The inequality SUB1.LESSP establishes that the measure (COUNT I) decreases according to the well-founded relation LESSP in the induction step of the scheme. Note, however, the inductive instance chosen for STR. The above induction scheme produces six new conjectures.

The ten plausible inductions should be evident. All of them involve instantiating either STR with (CDR STR), I with (SUB1 I), or PAT with (CDR PAT). But the I and STR inductions are linked by the (NTH STR I) term and they all merge into one scheme. The three schemes involving PAT are all identical and merge. Thus, we are left with two schemes: induction on I and STR together, or on PAT by itself. But we can see that the PAT induction is flawed because the (STRPOS PAT STR) terms mention PAT in an unchanging argument position. Hence, the heuristics have left us with only one induction: simultaneous induction on I (decomposed by SUB1s) and STR (decomposed by CDRs). We thus have three base cases, one for STR being non-LISTP, one for I being nonnumeric, and one for I being 0. The single induction step supposes STR is a list and I is a positive integer, and inductively assumes the theorem for (SUB1 I) and (CDR STR). It should be noted that no function in the theorem actually recurses in exactly this way (in particular, NTH, the function that linked the I and STR inductions, does not bother to test STR before applying CDR because NTH happens to be controlled entirely by its numeric argument).

This reduction in the number of plausible inductions, from ten to two to one in this case, is the rule rather than the exception. Out of about 400 inductions involved in our standard library of proofs, we have found that about 90% of the time the field of possible schemes is narrowed to one by subsumption, merging, and the elimination of flawed schemes.

## XV. FORMULATING AN INDUCTION SCHEME FOR A CONJECTURE

About half the time the remaining 10% are symmetric in several variables and there is no "right" induction. In the remaining 5% of the cases, our tie-breaking rules apply. To illustrate these (as well as the contributions of our other heuristics), we consider two proofs from Chapter IV.

To illustrate how scores occasionally enter, consider the theorem TAUTOLOGYP.IS.SOUND. The opening "move" in this proof is as follows:

**Theorem** TAUTOLOGYP.IS.SOUND:
```
    (IMPLIES (AND (NORMALIZED.IF.EXPRP X)
                  (TAUTOLOGYP X A1))
             (VALUE X (APPEND A1 A2)))
```
Give the conjecture the name *1.

Let us appeal to the induction principle. The
recursive terms in the conjecture suggest four
inductions. They merge into three likely candidate
inductions, none of which is unflawed. However, one
is more likely than the others. We will induct
according to the following scheme:
```
    (AND (IMPLIES (NOT (IF.EXPRP X))
                  (p X A1 A2))
         (IMPLIES (AND (IF.EXPRP X)
                       (p (RIGHT.BRANCH X)
                          (CONS (CONS (TEST X) F) A1)
                          A2)
                       (p (LEFT.BRANCH X)
                          (CONS (CONS (TEST X) T) A1)
                          A2)
                       (p (RIGHT.BRANCH X) A1 A2)
                       (p (LEFT.BRANCH X) A1 A2))
                  (p X A1 A2))).
```
The inequalities LEFT.BRANCH.LESSP and RIGHT.BRANCH.-
LESSP establish that the measure (COUNT x)
decreases according to the well-founded relation
LESSP in the induction step of the scheme. Note,
however, the inductive instances chosen for A1. The
above induction scheme generates the following 26 new
conjectures.

The four initial schemes were those suggested by NORMA-LIZED.IF.EXPRP, TAUTOLOGYP, VALUE, and APPEND. The first is subsumed by the second because NORMALIZED.IF.EXPRP and TAUTOLOGYP both recurse on the LEFT.BRANCH and RIGHT.BRANCH of IF-expressions and nowhere else. (In describing its proofs, our theorem-proving program does not distinguish subsumption from merging.) The VALUE induction is different because it recurses on the TEST of IF-expressions as well. Of course, the APPEND expression is altogether different, since it recurses down the CDRs of its first argument. After the subsumption we are left with three possibilities and all are flawed. The two possible induction schemes on X are flawed by one another, and the scheme on A1 is flawed by the TAUTOLOGYP induction. But since we have two "votes" in favor of the TAUTOLOGYP induction while we have only one in favor of each of the other two, we go with that induction. (Note that had the votes been tied we would have chosen this induction anyway, because it accounts for the TAUTOLOGYP-expression, and TAUTOLOGYP is the only nonprimitive-recursive function in the conjecture.) Since the induction scheme accounts for two terms, we do not superimpose the machine for TAUTOLOGYP but since the revised machine for TAUTOLOGYP has several different cases, all governed by the same test, (IF.EXPRP X), we "cover our bets" by grouping them all together in a single case with four different induction hypotheses.

It happens that as often as not, scores are tied (because there are just two obvious inductions that "perfectly" satisfy their terms but do not merge). This is where our final tie-breaking rule enters.

Consider the statement that NORMALIZE preserves the VALUE of its argument:

**Theorem** NORMALIZE.IS.SOUND:

    (EQUAL (VALUE (NORMALIZE X) A)
       (VALUE X A))

Give the conjecture the name *1.

Perhaps we can prove it by induction. The recursive terms in the conjecture suggest two inductions, neither of which is unflawed, and both of which appear equally likely. So we will choose the one that will probably lead to eliminating the nastiest expression. We will induct according to the following scheme:

## XV. FORMULATING AN INDUCTION SCHEME FOR A CONJECTURE

```
(AND
 (IMPLIES (NOT (IF.EXPRP X)) (p X A))
 (IMPLIES
    (AND (IF.EXPRP X)(IF.EXPRP (TEST X))
        (p (CONS.IF (TEST (TEST X))
                    (CONS.IF (LEFT.BRANCH (TEST X))
                             (LEFT.BRANCH X)
                             (RIGHT.BRANCH X))
                    (CONS.IF (RIGHT.BRANCH (TEST X))
                             (LEFT.BRANCH X)
                             (RIGHT.BRANCH X)))
           A))
    (p X A))
 (IMPLIES (AND (IF.EXPRP X)
               (NOT (IF.EXPRP (TEST X)))
               (p (RIGHT.BRANCH X) A)
               (p (LEFT.BRANCH X) A))
          (p X A))).
```

The inequalities IF.COMPLEXITY.GOES.DOWN1, IF.-COMPLEXITY.GOES.DOWN2, IF.COMPLEXITY.STAYS.EVEN and IF.DEPTH.GOES.DOWN establish that the measure:

(CONS (IF.COMPLEXITY X) (IF.DEPTH X))

decreases according to the well-founded lexicographic relation induced by LESSP and LESSP in each induction step of the scheme. The above induction scheme generates three new conjectures.

Here we have only two choices: induction on X as suggested by NORMALIZE or induction on X as suggested by VALUE. They do not merge and each flaws the other. We thus choose to eliminate NORMALIZE because it is not primitive-recursive.

### D. THE ENTIRE REVERSE EXAMPLE

Recall the example theorem that has been used to tie together all the foregoing proof techniques:

**Theorem** REVERSE.REVERSE:

```
(IMPLIES (PLISTP X)
         (EQUAL (REVERSE (REVERSE X)) X)).
```

We reproduce below the theorem-prover's output in response to the four user commands to define APPEND, REVERSE, and PLISTP and to prove the above theorem. In addition to recalling the definition-time analysis of the recursive functions and the previous steps in the proof, the reader should note the induction used on the original theorem, the induction used on the subgoal generated (formula *1.1 below), and the proof of each of the formulas produced by the second induction.

**Definition**

```
(APPEND X Y)
    =
(IF (LISTP X)
    (CONS (CAR X) (APPEND (CDR X) Y))
    Y).
```

The lemma CDR.LESSP informs us that (COUNT X) goes down according to the well-founded relation LESSP in each recursive call. Hence, APPEND is accepted under the definition principle. Observe that:

```
(OR (LISTP (APPEND X Y))
    (EQUAL (APPEND X Y) Y))
```

is a theorem.

CPU time (devoted to theorem-proving): .281 seconds

**Definition**

```
(REVERSE X)
    =
(IF (LISTP X)
    (APPEND (REVERSE (CDR X))
            (CONS (CAR X) "NIL"))
    "NIL").
```

The lemma CDR.LESSP can be used to show that (COUNT X) goes down according to the well-founded relation LESSP in each recursive call. Hence, REVERSE is accepted under the definition principle. Observe that:

```
(OR (LITATOM (REVERSE X))
    (LISTP (REVERSE X)))
```

is a theorem.

CPU time (devoted to theorem-proving): .34 seconds

## Definition

```
(PLISTP X)
   =
(IF (LISTP X)
    (PLISTP (CDR X))
    (EQUAL X "NIL")).
```

The lemma CDR.LESSP informs us that (COUNT X) goes down according to the well-founded relation LESSP in each recursive call. Hence, PLISTP is accepted under the definition principle. From the definition we can conclude that:

```
(OR (EQUAL F (PLISTP X))
    (EQUAL T (PLISTP X)))
```

is a theorem.

CPU time (devoted to theorem-proving): .317 seconds

## Theorem REVERSE.REVERSE:

```
(IMPLIES (PLISTP X)
         (EQUAL (REVERSE (REVERSE X)) X)).
```

Name the conjecture *1.

Perhaps we can prove it by induction. There are two plausible inductions. However, they merge into one likely candidate induction. We will induct according to the following scheme:

```
(AND (IMPLIES (NOT (LISTP X)) (p X))
     (IMPLIES (AND (LISTP X) (p (CDR X)))
              (p X))).
```

The inequality CDR.LESSP establishes that the measure (COUNT X) decreases according to the well-founded relation LESSP in the induction step of the scheme. The above induction scheme leads to the following three new conjectures:

*Case 1.*
```
(IMPLIES (AND (NOT (LISTP X)) (PLISTP X))
         (EQUAL (REVERSE (REVERSE X)) X)),
```

## D. THE ENTIRE REVERSE EXAMPLE / 205

which we simplify, expanding the definitions of
PLISTP and REVERSE, to:

    (TRUE).

*Case 2.*
(IMPLIES (AND (LISTP X)
            (NOT (PLISTP (CDR X)))
            (PLISTP X))
       (EQUAL (REVERSE (REVERSE X)) X)),

which we simplify, unfolding PLISTP, to:

    (TRUE).

*Case 3.*
(IMPLIES (AND (LISTP X)
            (EQUAL (REVERSE (REVERSE (CDR X)))
                (CDR X))
            (PLISTP X))
       (EQUAL (REVERSE (REVERSE X)) X)),

which we simplify, expanding PLISTP and REVERSE, to:

(IMPLIES
    (AND (LISTP X)
        (EQUAL (REVERSE (REVERSE (CDR X)))
            (CDR X))
        (PLISTP (CDR X)))
    (EQUAL (REVERSE (APPEND (REVERSE (CDR X))
                        (CONS (CAR X) "NIL")))
       X)).

Appealing to the lemma CAR/CDR.ELIM, we now replace X
by (CONS A B) to eliminate (CDR X) and (CAR X). This
generates:

(IMPLIES
 (AND (LISTP (CONS A B))
     (EQUAL (REVERSE (REVERSE B)) B)
     (PLISTP B))
 (EQUAL
    (REVERSE (APPEND (REVERSE B) (CONS A "NIL")))
    (CONS A B))),

which further simplifies, obviously, to the
conjecture:

## XV. FORMULATING AN INDUCTION SCHEME FOR A CONJECTURE

```
(IMPLIES
 (AND (EQUAL (REVERSE (REVERSE B)) B)
      (PLISTP B))
 (EQUAL
    (REVERSE (APPEND (REVERSE B) (CONS A "NIL")))
    (CONS A B))).
```

We now use the above equality hypothesis by cross-fertilizing (REVERSE (REVERSE B)) for B and throwing away the equality. This produces:

```
(IMPLIES
 (PLISTP B)
 (EQUAL
    (REVERSE (APPEND (REVERSE B) (CONS A "NIL")))
    (CONS A (REVERSE (REVERSE B))))),
```

which we generalize by replacing (REVERSE B) by Z. The result is the conjecture:

```
(IMPLIES (PLISTP B)
         (EQUAL (REVERSE (APPEND Z (CONS A "NIL")))
                (CONS A (REVERSE Z)))),
```

which has an irrelevant term in it. By eliminating this term we get:

```
        (EQUAL (REVERSE (APPEND Z (CONS A "NIL")))
               (CONS A (REVERSE Z))),
```

which we will finally name *1.1.

Let us appeal to the induction principle. There are two plausible inductions. However, they merge into one likely candidate induction. We will induct according to the following scheme:

```
        (AND (IMPLIES (NOT (LISTP Z)) (p Z A))
             (IMPLIES (AND (LISTP Z) (p (CDR Z) A))
                      (p Z A))).
```

The inequality CDR.LESSP establishes that the measure (COUNT Z) decreases according to the well-founded relation LESSP in the induction step of the scheme. The above induction scheme produces the following two new goals:

*Case 1.*
```
(IMPLIES
      (NOT (LISTP Z))
      (EQUAL (REVERSE (APPEND Z (CONS A "NIL")))
             (CONS A (REVERSE Z)))),
```
which simplifies, applying the lemmas CAR.CONS and CDR.CONS, and expanding the definitions of APPEND and REVERSE, to:

```
      (TRUE).
```

*Case 2.*
```
(IMPLIES
 (AND
  (LISTP Z)
  (EQUAL
        (REVERSE (APPEND (CDR Z) (CONS A "NIL")))
        (CONS A (REVERSE (CDR Z)))))
 (EQUAL (REVERSE (APPEND Z (CONS A "NIL")))
        (CONS A (REVERSE Z)))),
```
which simplifies, using the lemmas CAR.CONS and CDR.CONS, and expanding the functions APPEND and REVERSE, to:

```
(IMPLIES
 (AND
  (LISTP Z)
  (EQUAL
        (REVERSE (APPEND (CDR Z) (CONS A "NIL")))
        (CONS A (REVERSE (CDR Z)))))
 (EQUAL
  (APPEND
        (REVERSE (APPEND (CDR Z) (CONS A "NIL")))
        (CONS (CAR Z) "NIL"))
  (CONS A
        (APPEND (REVERSE (CDR Z))
                (CONS (CAR Z) "NIL"))))).
```
Appealing to the lemma CAR/CDR.ELIM, we now replace Z by (CONS B X) to eliminate (CDR Z) and (CAR Z). This generates:

```
(IMPLIES
 (AND (LISTP (CONS B X))
      (EQUAL (REVERSE (APPEND X (CONS A "NIL")))
             (CONS A (REVERSE X))))
 (EQUAL
    (APPEND (REVERSE (APPEND X (CONS A "NIL")))
            (CONS B "NIL"))
    (CONS A
          (APPEND (REVERSE X)
                  (CONS B "NIL"))))),
```

which we further simplify, trivially, to:

```
(IMPLIES
 (EQUAL (REVERSE (APPEND X (CONS A "NIL")))
        (CONS A (REVERSE X)))
 (EQUAL
    (APPEND (REVERSE (APPEND X (CONS A "NIL")))
            (CONS B "NIL"))
    (CONS A
          (APPEND (REVERSE X)
                  (CONS B "NIL"))))).
```

We use the above equality hypothesis by cross-fertilizing (CONS A (REVERSE X)) for:

```
(REVERSE (APPEND X (CONS A "NIL")))
```

and throwing away the equality. This generates:

```
(EQUAL (APPEND (CONS A (REVERSE X))
               (CONS B "NIL"))
       (CONS A
             (APPEND (REVERSE X)
                     (CONS B "NIL")))).
```

However this simplifies further, applying the lemmas CDR.CONS and CAR.CONS, and unfolding the definition of APPEND, to:

```
(TRUE).
```

That finishes the proof of *1.1, which, consequently, also finishes the proof of *1. Q.E.D.

CPU time (devoted to theorem-proving): 5.037 seconds

# XVI

## Illustrations of Our Techniques via Elementary Number Theory

In this chapter we illustrate the foregoing proof techniques on several theorems in elementary number theory. Assuming only the axioms, lemmas, and definitions in Chapter III, our theorem-prover establishes a sequence of definitions and theorems leading to the basic theorem: if Y is nonzero and X is a number, then X = R + Q*Y, where R is the remainder of X by Y and Q is the quotient of X by Y.

The following points are worth noting.

Although we explicitly request the theorem-prover to prove a few simple arithmetic theorems, many other such propositions are raised as subgoals by our heuristics and proved inductively. The reader should note carefully each of the subgoals proved by induction. For example, in the proof of the associativity of multiplication (ASSOCIATIVITY.OF.TIMES), the theorem-prover generates as a subgoal the statement that multiplication distributes over addition.

The induction lemma RECURSION.BY.DIFFERENCE is proved in order to admit the definitions of REMAINDER and QUOTIENT.

The remainder of this chapter was produced by our automatic theorem-prover in response to the user commands supplying function definitions and statements of theorems to prove.

### A. PLUS.RIGHT.ID

**Theorem**  PLUX.RIGHT.ID (rewrite):
    (EQUAL (PLUS X 0) (FIX X)).

This simplifies, unfolding the functions FIX and PLUS, to the conjecture:

> (IMPLIES (NUMBERP X)
>          (EQUAL (PLUS X 0) X)),

which we will name *1.

Let us appeal to the induction principle. There is only one plausible induction. We will induct according to the following scheme:

> (AND (IMPLIES (EQUAL X 0) (p X))
>      (IMPLIES (NOT (NUMBERP X)) (p X))
>      (IMPLIES (AND (NOT (EQUAL X 0))
>                    (NUMBERP X)
>                    (p (SUB1 X)))
>               (p X))).

The inequality SUB1.LESSP establishes that the measure (COUNT X) decreases according to the well-founded relation LESSP in the induction step of the scheme. The above induction scheme generates the following three new conjectures:

*Case 1.* (IMPLIES (AND (EQUAL X 0) (NUMBERP X))
                   (EQUAL (PLUS X 0) X)).

This simplifies, expanding PLUS, to:

> (TRUE).

*Case 2.* (IMPLIES (AND (NOT (EQUAL X 0))
                        (NOT (NUMBERP (SUB1 X)))
                        (NUMBERP X))
                   (EQUAL (PLUS X 0) X)).

Of course, this simplifies, clearly, to:

> (TRUE).

*Case 3.* (IMPLIES (AND (NOT (EQUAL X 0))
                        (EQUAL (PLUS (SUB1 X) 0) (SUB1 X))
                        (NUMBERP X))
                   (EQUAL (PLUS X 0) X)),

which we simplify, opening up the function PLUS, to:

```
(IMPLIES (AND (NOT (EQUAL X 0))
              (EQUAL (PLUS (SUB1 X) 0) (SUB1 X))
              (NUMBERP X))
         (EQUAL (ADD1 (PLUS (SUB1 X) 0)) X)).
```

Appealing to the lemma SUB1.ELIM, we now replace X by (ADD1 Z) to eliminate (SUB1 X). We use the type restriction lemma noted when SUB1 was introduced to restrict the new variable. We must thus prove:

```
(IMPLIES (AND (NUMBERP Z)
              (NOT (EQUAL (ADD1 Z) 0))
              (EQUAL (PLUS Z 0) Z)
              (NUMBERP (ADD1 Z)))
         (EQUAL (ADD1 (PLUS Z 0)) (ADD1 Z))).
```

This further simplifies, rewriting with ADD1.EQUAL, to:

```
(TRUE).
```

That finishes the proof of *1. Q.E.D.

CPU time (devoted to theorem-proving): 1.425 seconds

## B. COMMUTATIVITY2.OF.PLUS

**Theorem** COMMUTATIVITY2.OF.PLUS (rewrite):

```
(EQUAL (PLUS X (PLUS Y Z))
       (PLUS Y (PLUS X Z))).
```

Call the conjecture *1.

Perhaps we can prove it by induction. There are four plausible inductions. They merge into two likely candidate inductions, both of which are unflawed, and both of which appear equally likely. So we will choose arbitrarily. We will induct according to the following scheme:

```
(AND (IMPLIES (NOT (NUMBERP Y)) (p X Y Z))
     (IMPLIES (EQUAL Y 0) (p X Y Z))
     (IMPLIES (AND (NUMBERP Y)
                   (NOT (EQUAL Y 0))
                   (p X (SUB1 Y) Z))
              (p X Y Z))).
```

## 212 / XVI. ILLUSTRATIONS OF OUR TECHNIQUES VIA ELEMENTARY NUMBER THEORY

The inequality SUB1.LESSP establishes that the measure (COUNT Y) decreases according to the well-founded relation LESSP in the induction step of the scheme. The above induction scheme generates the following three new conjectures:

*Case 1.* (IMPLIES (NOT (NUMBERP Y))
        (EQUAL (PLUS X (PLUS Y Z))
              (PLUS Y (PLUS X Z)))),

which simplifies, applying PLUS.RIGHT.ID, and opening up PLUS, to:

      (IMPLIES (AND (NOT (NUMBERP Y))
              (NOT (NUMBERP Z))
              (NUMBERP X))
        (EQUAL X (PLUS X Z))).

Eliminate the irrelevant term. This produces:

      (IMPLIES (AND (NOT (NUMBERP Z)) (NUMBERP X))
        (EQUAL X (PLUS X Z))),

which we will name *1.1.

*Case 2.* (IMPLIES (EQUAL Y 0)
        (EQUAL (PLUS X (PLUS Y Z))
              (PLUS Y (PLUS X Z)))),

which we simplify, applying the lemma PLUS.RIGHT.ID, and opening up PLUS, to the conjecture:

      (IMPLIES (AND (NOT (NUMBERP Z)) (NUMBERP X))
        (EQUAL X (PLUS X Z))).

Name the above subgoal *1.2.

*Case 3.*
(IMPLIES (AND (NUMBERP Y)
          (NOT (EQUAL Y 0))
          (EQUAL (PLUS X (PLUS (SUB1 Y) Z))
               (PLUS (SUB1 Y) (PLUS X Z))))
      (EQUAL (PLUS X (PLUS Y Z))
          (PLUS Y (PLUS X Z)))),

which simplifies, expanding the function PLUS, to the new conjecture:

```
(IMPLIES (AND (NUMBERP Y)
              (NOT (EQUAL Y 0))
              (EQUAL (PLUS X (PLUS (SUB1 Y) Z))
                     (PLUS (SUB1 Y) (PLUS X Z))))
         (EQUAL (PLUS X (ADD1 (PLUS (SUB1 Y) Z)))
                (ADD1 (PLUS (SUB1 Y) (PLUS X Z))))).
```

Appealing to the lemma SUB1.ELIM, we now replace Y by (ADD1 V) to eliminate (SUB1 Y). We use the type restriction lemma noted when SUB1 was introduced to constrain the new variable. We thus obtain:

```
(IMPLIES (AND (NUMBERP V)
              (NUMBERP (ADD1 V))
              (NOT (EQUAL (ADD1 V) 0))
              (EQUAL (PLUS X (PLUS V Z))
                     (PLUS V (PLUS X Z))))
         (EQUAL (PLUS X (ADD1 (PLUS V Z)))
                (ADD1 (PLUS V (PLUS X Z))))).
```

Of course, this simplifies further, trivially, to:

```
(IMPLIES (AND (NUMBERP V)
              (EQUAL (PLUS X (PLUS V Z))
                     (PLUS V (PLUS X Z))))
         (EQUAL (PLUS X (ADD1 (PLUS V Z)))
                (ADD1 (PLUS V (PLUS X Z))))).
```

We use the above equality hypothesis by cross-fertilizing (PLUS X (PLUS V Z)) for (PLUS V (PLUS X Z)) and throwing away the equality. This generates the new conjecture:

```
(IMPLIES (NUMBERP V)
         (EQUAL (PLUS X (ADD1 (PLUS V Z)))
                (ADD1 (PLUS X (PLUS V Z))))),
```

which we generalize by replacing (PLUS V Z) by Y. We restrict the new variable by appealing to the type restriction lemma noted when PLUS was introduced. This generates:

```
(IMPLIES (AND (NUMBERP Y) (NUMBERP V))
         (EQUAL (PLUS X (ADD1 Y))
                (ADD1 (PLUS X Y)))).
```

Eliminate the irrelevant term. We would thus like to prove:
```
(IMPLIES (NUMBERP Y)
         (EQUAL (PLUS X (ADD1 Y))
                (ADD1 (PLUS X Y)))).
```
Finally name the above subgoal *1.3.

Let us appeal to the induction principle. There are two plausible inductions. However, they merge into one likely candidate induction. We will induct according to the following scheme:
```
(AND (IMPLIES (NOT (NUMBERP X)) (p X Y))
     (IMPLIES (EQUAL X 0) (p X Y))
     (IMPLIES (AND (NUMBERP X)
                   (NOT (EQUAL X 0))
                   (p (SUB1 X) Y))
              (p X Y))).
```
The inequality SUB1.LESSP establishes that the measure (COUNT X) decreases according to the well-founded relation LESSP in the induction step of the scheme. The above induction scheme generates three new formulas:

*Case 1.*
```
(IMPLIES (AND (NOT (NUMBERP X)) (NUMBERP Y))
         (EQUAL (PLUS X (ADD1 Y))
                (ADD1 (PLUS X Y)))),
```
which we simplify, unfolding PLUS, to:
```
(TRUE).
```

*Case 2.*
```
(IMPLIES (AND (EQUAL X 0) (NUMBERP Y))
         (EQUAL (PLUS X (ADD1 Y))
                (ADD1 (PLUS X Y)))).
```
This simplifies, expanding the definition of PLUS, to:
```
(TRUE).
```

*Case 3.*
```
(IMPLIES (AND (NUMBERP X)
              (NOT (EQUAL X 0))
              (EQUAL (PLUS (SUB1 X) (ADD1 Y))
                     (ADD1 (PLUS (SUB1 X) Y)))
              (NUMBERP Y))
         (EQUAL (PLUS X (ADD1 Y))
                (ADD1 (PLUS X Y)))),
```

which simplifies, appealing to the lemma ADD1.EQUAL, and unfolding PLUS, to:
>        (TRUE).
That finishes the proof of *1.3.

So let us turn our attention to:
>        (IMPLIES (AND (NOT (NUMBERP Z)) (NUMBERP X))
>                 (EQUAL X (PLUS X Z))),

which is formula *1.2 above. But this conjecture is subsumed by formula *1.1 above.

So we now return to:
>        (IMPLIES (AND (NOT (NUMBERP Z)) (NUMBERP X))
>                 (EQUAL X (PLUS X Z))),

named *1.1 above. Perhaps we can prove it by induction. There is only one plausible induction. We will induct according to the following scheme:
>        (AND (IMPLIES (EQUAL X 0) (p X Z))
>             (IMPLIES (NOT (NUMBERP X)) (p X Z))
>             (IMPLIES (AND (NOT (EQUAL X 0))
>                           (NUMBERP X)
>                           (p (SUB1 X) Z))
>                      (p X Z))).

The inequality SUB1.LESSP establishes that the measure (COUNT X) decreases according to the well-founded relation LESSP in the induction step of the scheme. The above induction scheme produces the following three new conjectures:

*Case 1.* (IMPLIES (AND (EQUAL X 0)
>                        (NOT (NUMBERP Z))
>                        (NUMBERP X))
>                   (EQUAL X (PLUS X Z))).

This simplifies, unfolding the definition of PLUS, to:
>        (TRUE).

*Case 2.* (IMPLIES (AND (NOT (EQUAL X 0))
>                        (NOT (NUMBERP (SUB1 X)))
>                        (NOT (NUMBERP Z))
>                        (NUMBERP X))
>                   (EQUAL X (PLUS X Z))),

which we simplify, trivially, to:

(TRUE).

*Case 3.* (IMPLIES (AND (NOT (EQUAL X 0))
                       (EQUAL (SUB1 X)
                              (PLUS (SUB1 X) Z))
                       (NOT (NUMBERP Z))
                       (NUMBERP X))
                  (EQUAL X (PLUS X Z))),

which we simplify, unfolding the function PLUS, to:

(IMPLIES (AND (NOT (EQUAL X 0))
              (EQUAL (SUB1 X)
                     (PLUS (SUB1 X) Z))
              (NOT (NUMBERP Z))
              (NUMBERP X))
         (EQUAL X (ADD1 (PLUS (SUB1 X) Z)))).

Applying the lemma SUB1.ELIM, replace X by (ADD1 V) to eliminate (SUB1 X). We use the type restriction lemma noted when SUB1 was introduced to constrain the new variable. This generates:

(IMPLIES (AND (NUMBERP V)
              (NOT (EQUAL (ADD1 V) 0))
              (EQUAL V (PLUS V Z))
              (NOT (NUMBERP Z))
              (NUMBERP (ADD1 V)))
         (EQUAL (ADD1 V) (ADD1 (PLUS V Z)))),

which we further simplify, using the lemma ADD1.EQUAL, to:

(TRUE).

That finishes the proof of *1.1, which finishes the proof of *1. Q.E.D.

CPU time (devoted to theorem-proving): 5.861 seconds

## C. COMMUTATIVITY.OF.PLUS

**Theorem** COMMUTATIVITY.OF.PLUS (rewrite):

(EQUAL (PLUS X Y) (PLUS Y X)).

Call the conjecture *1.
Let us appeal to the induction principle. Two in-

ductions are suggested by terms in the conjecture, neither of which is unflawed, and both of which appear equally likely. So we will choose arbitrarily. We will induct according to the following scheme:
```
(AND (IMPLIES (EQUAL X 0) (p X Y))
     (IMPLIES (NOT (NUMBERP X)) (p X Y))
     (IMPLIES (AND (NOT (EQUAL X 0))
                   (NUMBERP X)
                   (p (SUB1 X) Y))
              (p X Y))).
```
The inequality SUB1.LESSP establishes that the measure (COUNT X) decreases according to the well-founded relation LESSP in the induction step of the scheme. The above induction scheme generates three new formulas:

*Case 1.* (IMPLIES (EQUAL X 0)
                    (EQUAL (PLUS X Y) (PLUS Y X))).

This simplifies, appealing to the lemma PLUS.RIGHT.ID, and opening up the definition of PLUS, to:
      (TRUE).

*Case 2.* (IMPLIES (NOT (NUMBERP X))
                    (EQUAL (PLUS X Y) (PLUS Y X))).

This simplifies, unfolding the function PLUS, to:
```
(IMPLIES (AND (NOT (NUMBERP X)) (NUMBERP Y))
         (EQUAL Y (PLUS Y X))).
```
Call the above conjecture *1.1.

*Case 3.* (IMPLIES (AND (NOT (EQUAL X 0))
                         (NUMBERP X)
                         (EQUAL (PLUS (SUB1 X) Y)
                                (PLUS Y (SUB1 X))))
                    (EQUAL (PLUS X Y) (PLUS Y X))),

which we simplify, expanding the function PLUS, to:
```
(IMPLIES (AND (NOT (EQUAL X 0))
              (NUMBERP X)
              (EQUAL (PLUS (SUB1 X) Y)
                     (PLUS Y (SUB1 X))))
         (EQUAL (ADD1 (PLUS (SUB1 X) Y))
                (PLUS Y X))).
```

Appealing to the lemma SUB1.ELIM, we now replace X by
(ADD1 Z) to eliminate (SUB1 X). We employ the type
restriction lemma noted when SUB1 was introduced to
restrict the new variable. We thus obtain:

```
(IMPLIES (AND (NUMBERP Z)
              (NOT (EQUAL (ADD1 Z) 0))
              (NUMBERP (ADD1 Z))
              (EQUAL (PLUS Z Y) (PLUS Y Z)))
         (EQUAL (ADD1 (PLUS Z Y))
                (PLUS Y (ADD1 Z)))).
```

Of course, this simplifies further, clearly, to:

```
(IMPLIES (AND (NUMBERP Z)
              (EQUAL (PLUS Z Y) (PLUS Y Z)))
         (EQUAL (ADD1 (PLUS Z Y))
                (PLUS Y (ADD1 Z)))).
```

We use the above equality hypothesis by
cross-fertilizing (PLUS Y Z) for (PLUS Z Y) and
throwing away the equality. We would thus like to
prove:

```
(IMPLIES (NUMBERP Z)
         (EQUAL (ADD1 (PLUS Y Z))
                (PLUS Y (ADD1 Z)))),
```

which we will name *1.2.

We will try to prove it by induction. Two
inductions are suggested by terms in the conjecture.
However, they merge into one likely candidate
induction. We will induct according to the following
scheme:

```
(AND (IMPLIES (NOT (NUMBERP Y)) (p Y Z))
     (IMPLIES (EQUAL Y 0) (p Y Z))
     (IMPLIES (AND (NUMBERP Y)
                   (NOT (EQUAL Y 0))
                   (p (SUB1 Y) Z))
              (p Y Z))).
```

The inequality SUB1.LESSP establishes that the
measure (COUNT Y) decreases according to the
well-founded relation LESSP in the induction step of

the scheme. The above induction scheme generates
three new conjectures:

*Case 1.* (IMPLIES (AND (NOT (NUMBERP Y)) (NUMBERP Z))
              (EQUAL (ADD1 (PLUS Y Z))
                     (PLUS Y (ADD1 Z)))),

which we simplify, expanding the definition of PLUS,
to:
    (TRUE).

*Case 2.* (IMPLIES (AND (EQUAL Y 0) (NUMBERP Z))
              (EQUAL (ADD1 (PLUS Y Z))
                     (PLUS Y (ADD1 Z)))).

This simplifies, expanding the definition of PLUS,
to:
    (TRUE).

*Case 3.* (IMPLIES (AND (NUMBERP Y)
                   (NOT (EQUAL Y 0))
                   (EQUAL (ADD1 (PLUS (SUB1 Y) Z))
                          (PLUS (SUB1 Y) (ADD1 Z)))
                   (NUMBERP Z))
              (EQUAL (ADD1 (PLUS Y Z))
                     (PLUS Y (ADD1 Z)))).

This simplifies, using the lemma ADD1.EQUAL, and
opening up the function PLUS, to:
    (TRUE).

That finishes the proof of *1.2.

So next consider
        (IMPLIES (AND (NOT (NUMBERP X)) (NUMBERP Y))
              (EQUAL Y (PLUS Y X))),

which we named *1.1 above. Let us appeal to the
induction principle. There is only one suggested
induction. We will induct according to the following
scheme:
        (AND (IMPLIES (EQUAL Y 0) (p Y X))
             (IMPLIES (NOT (NUMBERP Y)) (p Y X))
             (IMPLIES (AND (NOT (EQUAL Y 0))
                           (NUMBERP Y)
                           (p (SUB1 Y) X))
                      (p Y X))).

The inequality SUB1.LESSP establishes that the measure (COUNT Y) decreases according to the well-founded relation LESSP in the induction step of the scheme. The above induction scheme produces three new goals:

*Case 1.* (IMPLIES (AND (EQUAL Y 0)
              (NOT (NUMBERP X))
              (NUMBERP Y))
       (EQUAL Y (PLUS Y X))).

This simplifies, unfolding the definition of PLUS, to:

    (TRUE).

*Case 2.* (IMPLIES (AND (NOT (EQUAL Y 0))
              (NOT (NUMBERP (SUB1 Y)))
              (NOT (NUMBERP X))
              (NUMBERP Y))
       (EQUAL Y (PLUS Y X))).

Of course, this simplifies, clearly, to:

    (TRUE).

*Case 3.* (IMPLIES (AND (NOT (EQUAL Y 0))
              (EQUAL (SUB1 Y)
                  (PLUS (SUB1 Y) X))
            (NOT (NUMBERP X))
            (NUMBERP Y))
      (EQUAL Y (PLUS Y X))),

which we simplify, expanding the function PLUS, to the new conjecture:

    (IMPLIES (AND (NOT (EQUAL Y 0))
              (EQUAL (SUB1 Y)
                  (PLUS (SUB1 Y) X))
            (NOT (NUMBERP X))
            (NUMBERP Y))
       (EQUAL Y (ADD1 (PLUS (SUB1 Y) X)))).

Appealing to the lemma SUB1.ELIM, replace Y by (ADD1 Z) to eliminate (SUB1 Y). We employ the type restriction lemma noted when SUB1 was introduced to restrict the new variable. This produces:

```
(IMPLIES (AND (NUMBERP Z)
              (NOT (EQUAL (ADD1 Z) 0))
              (EQUAL Z (PLUS Z X))
              (NOT (NUMBERP X))
              (NUMBERP (ADD1 Z)))
         (EQUAL (ADD1 Z) (ADD1 (PLUS Z X)))).
```

However this simplifies further, applying the lemma ADD1.EQUAL, to:

```
(TRUE).
```

That finishes the proof of *1.1, which finishes the proof of *1. Q.E.D.

CPU time (devoted to theorem-proving): 4.002 seconds


## D. ASSOCIATIVITY.OF.PLUS

**Theorem** ASSOCIATIVITY.OF.PLUS (rewrite):

```
(EQUAL (PLUS (PLUS X Y) Z)
       (PLUS X (PLUS Y Z))).
```

This formula simplifies, appealing to the lemmas COMMUTATIVITY.OF.PLUS and COMMUTATIVITY2.OF.PLUS, to:

```
(TRUE).
```

Q.E.D.

CPU time (devoted to theorem-proving): .229 seconds


## E. TIMES

### Definition

```
(TIMES I J)
   =
(IF (ZEROP I)
    0
    (PLUS J (TIMES (SUB1 I) J))).
```

The lemma SUB1.LESSP establishes that (COUNT I) goes down according to the well-founded relation LESSP in each recursive call. Hence, TIMES is

accepted under the principle of definition. Note that
(NUMBERP (TIMES I J)) is a theorem.

CPU time (devoted to theorem-proving): .611 seconds

## F. TIMES.ZERO

**Theorem** TIMES.ZERO (rewrite):

(EQUAL (TIMES X 0) 0).

Name the conjecture *1.

Let us appeal to the induction principle. There is
only one suggested induction. We will induct
according to the following scheme:

(AND (IMPLIES (EQUAL X 0) (p X))
     (IMPLIES (NOT (NUMBERP X)) (p X))
     (IMPLIES (AND (NOT (EQUAL X 0))
                   (NUMBERP X)
                   (p (SUB1 X)))
              (p X))).

The inequality SUB1.LESSP establishes that the
measure (COUNT X) decreases according to the
well-founded relation LESSP in the induction step of
the scheme. The above induction scheme produces three
new conjectures:

*Case 1.* (IMPLIES (EQUAL X 0)
                 (EQUAL (TIMES X 0) 0)),

which we simplify, expanding the definition of TIMES,
to:

(TRUE).

*Case 2.* (IMPLIES (NOT (NUMBERP X))
                 (EQUAL (TIMES X 0) 0)).

This simplifies, expanding the definition of TIMES,
to:

(TRUE).

*Case 3.* (IMPLIES (AND (NOT (EQUAL X 0))
                      (NUMBERP X)
                      (EQUAL (TIMES (SUB1 X) 0) 0))
                 (EQUAL (TIMES X 0) 0)).

This simplifies, unfolding PLUS and TIMES, to:

(TRUE).

That finishes the proof of *1.  Q.E.D.

CPU time (devoted to theorem-proving): .543 seconds


## G. TIMES.ADD1

**Theorem** TIMES.ADD1 (rewrite):

```
(EQUAL (TIMES X (ADD1 Y))
       (IF (NUMBERP Y)
           (PLUS X (TIMES X Y))
           (FIX X))).
```

This formula simplifies, using the lemma
SUB1.TYPE.RESTRICTION, and unfolding the definitions
of FIX and TIMES, to two new conjectures:

*Case 1.* (IMPLIES (NUMBERP Y)
                  (EQUAL (TIMES X (ADD1 Y))
                         (PLUS X (TIMES X Y)))),

which we will name *1.

*Case 2.* (IMPLIES (AND (NOT (NUMBERP Y)) (NUMBERP X))
                  (EQUAL (TIMES X 1) X)).

Eliminate the irrelevant term. The result is the new
formula:

```
(IMPLIES (NUMBERP X)
         (EQUAL (TIMES X 1) X)),
```

which we would usually push and work on later by
induction. But since we have already pushed one goal
split off of the original input we will disregard all
that we have previously done, give the name *1 to the
original input, and work on it.

So now let's consider:

```
(EQUAL (TIMES X (ADD1 Y))
       (IF (NUMBERP Y)
           (PLUS X (TIMES X Y))
           (FIX X))).
```

We named this *1. We will appeal to induction. Three
inductions are suggested by terms in the conjecture.
However, they merge into one likely candidate

induction. We will induct according to the following scheme:

```
(AND (IMPLIES (NOT (NUMBERP X)) (p X Y))
     (IMPLIES (EQUAL X 0) (p X Y))
     (IMPLIES (AND (NUMBERP X)
                   (NOT (EQUAL X 0))
                   (p (SUB1 X) Y))
              (p X Y))).
```

The inequality SUB1.LESSP establishes that the measure (COUNT X) decreases according to the well-founded relation LESSP in the induction step of the scheme. The above induction scheme leads to the following three new conjectures:

*Case 1.*
```
(IMPLIES (NOT (NUMBERP X))
         (EQUAL (TIMES X (ADD1 Y))
                (IF (NUMBERP Y)
                    (PLUS X (TIMES X Y))
                    (FIX X)))).
```

This simplifies, applying PLUS.RIGHT.ID, and unfolding the definitions of TIMES and FIX, to:

(TRUE).

*Case 2.*
```
(IMPLIES (EQUAL X 0)
         (EQUAL (TIMES X (ADD1 Y))
                (IF (NUMBERP Y)
                    (PLUS X (TIMES X Y))
                    (FIX X)))),
```

which simplifies, expanding the definitions of TIMES, PLUS and FIX, to:

(TRUE).

*Case 3.*
```
(IMPLIES (AND (NUMBERP X)
              (NOT (EQUAL X 0))
              (EQUAL (TIMES (SUB1 X) (ADD1 Y))
                     (IF (NUMBERP Y)
                         (PLUS (SUB1 X)
                               (TIMES (SUB1 X) Y))
                         (FIX (SUB1 X)))))
```

## G. TIMES.ADD1 / 225

```
(EQUAL (TIMES X (ADD1 Y))
       (IF (NUMBERP Y)
           (PLUS X (TIMES X Y))
           (FIX X)))),
```

which simplifies, applying SUB1.ADD1 and SUB1.TYPE.-
RESTRICTION, and opening up the definitions of FIX,
PLUS and TIMES, to the following two new goals:

*Case 1.*
```
(IMPLIES (AND (NUMBERP X)
              (NOT (EQUAL X 0))
              (NUMBERP Y)
              (EQUAL (TIMES (SUB1 X) (ADD1 Y))
                     (PLUS (SUB1 X)
                           (TIMES (SUB1 X) Y))))
         (EQUAL (ADD1 (PLUS Y (TIMES (SUB1 X)
                                     (ADD1 Y))))
                (PLUS X
                      (PLUS Y (TIMES (SUB1 X) Y))))).
```

But this simplifies again, applying ADD1.EQUAL, and
opening up PLUS, to:

```
(IMPLIES (AND (NUMBERP X)
              (NOT (EQUAL X 0))
              (NUMBERP Y)
              (EQUAL (TIMES (SUB1 X) (ADD1 Y))
                     (PLUS (SUB1 X)
                           (TIMES (SUB1 X) Y))))
         (EQUAL (PLUS Y (TIMES (SUB1 X) (ADD1 Y)))
                (PLUS (SUB1 X)
                      (PLUS Y (TIMES (SUB1 X) Y))))).
```

But this again simplifies, applying the lemma
COMMUTATIVITY2.OF.PLUS, to:

```
(IMPLIES (AND (NUMBERP X)
              (NOT (EQUAL X 0))
              (NUMBERP Y)
              (EQUAL (TIMES (SUB1 X) (ADD1 Y))
                     (PLUS (SUB1 X)
                           (TIMES (SUB1 X) Y))))
         (EQUAL (PLUS Y (TIMES (SUB1 X) (ADD1 Y)))
                (PLUS Y
                      (PLUS (SUB1 X)
                            (TIMES (SUB1 X) Y))))).
```

## 226 / XVI. ILLUSTRATIONS OF OUR TECHNIQUES VIA ELEMENTARY NUMBER THEORY

Applying the lemma SUB1.ELIM, replace X by (ADD1 Z) to eliminate (SUB1 X). We employ the type restriction lemma noted when SUB1 was introduced to restrict the new variable. This produces:

(IMPLIES (AND (NUMBERP Z)
              (NUMBERP (ADD1 Z))
              (NOT (EQUAL (ADD1 Z) 0))
              (NUMBERP Y)
              (EQUAL (TIMES Z (ADD1 Y))
                     (PLUS Z (TIMES Z Y))))
         (EQUAL (PLUS Y (TIMES Z (ADD1 Y)))
                (PLUS Y (PLUS Z (TIMES Z Y))))).

This simplifies further, clearly, to the goal:

(IMPLIES (AND (NUMBERP Z)
              (NUMBERP Y)
              (EQUAL (TIMES Z (ADD1 Y))
                     (PLUS Z (TIMES Z Y))))
         (EQUAL (PLUS Y (TIMES Z (ADD1 Y)))
                (PLUS Y (PLUS Z (TIMES Z Y))))).

We now use the above equality hypothesis by cross-fertilizing (TIMES Z (ADD1 Y)) for (PLUS Z (TIMES Z Y)) and throwing away the equality. We must thus prove:

        (IMPLIES (AND (NUMBERP Z) (NUMBERP Y))
                 (EQUAL (PLUS Y (TIMES Z (ADD1 Y)))
                        (PLUS Y (TIMES Z (ADD1 Y))))).

Of course, this finally simplifies, obviously, to:

        (TRUE).

*Case 2.* (IMPLIES (AND (NUMBERP X)
                      (NOT (EQUAL X 0))
                      (NOT (NUMBERP Y))
                      (EQUAL (TIMES (SUB1 X) 1)
                             (SUB1 X)))
                 (EQUAL (TIMES X 1) X)).

This simplifies again, rewriting with SUB1.ADD1, and unfolding the functions PLUS and TIMES, to:

```
(IMPLIES (AND (NUMBERP X)
              (NOT (EQUAL X 0))
              (NOT (NUMBERP Y))
              (EQUAL (TIMES (SUB1 X) 1)
                     (SUB1 X)))
         (EQUAL (ADD1 (TIMES (SUB1 X) 1)) X)).
```

Appealing to the lemma SUB1.ELIM, replace X by (ADD1 Z) to eliminate (SUB1 X). We use the type restriction lemma noted when SUB1 was introduced to constrain the new variable. We thus obtain:

```
(IMPLIES (AND (NUMBERP Z)
              (NUMBERP (ADD1 Z))
              (NOT (EQUAL (ADD1 Z) 0))
              (NOT (NUMBERP Y))
              (EQUAL (TIMES Z 1) Z))
         (EQUAL (ADD1 (TIMES Z 1)) (ADD1 Z))),
```

which we further simplify, using the lemma ADD1.-EQUAL, to:

(TRUE).

That finishes the proof of *1. Q.E.D.

CPU time (devoted to theorem-proving): 7.086 seconds


## H. ASSOCIATIVITY.OF.TIMES

**Theorem** ASSOCIATIVITY.OF.TIMES (rewrite):

```
(EQUAL (TIMES (TIMES X Y) Z)
       (TIMES X (TIMES Y Z))).
```

Name the conjecture *1.

Let us appeal to the induction principle. The recursive terms in the conjecture suggest three inductions. They merge into two likely candidate inductions. However, only one is unflawed. We will induct according to the following scheme:

```
(AND (IMPLIES (NOT (NUMBERP X)) (p X Y Z))
     (IMPLIES (EQUAL X 0) (p X Y Z))
```

```
(IMPLIES (AND (NUMBERP X)
              (NOT (EQUAL X 0))
              (p (SUB1 X) Y Z))
         (p X Y Z))).
```

The inequality SUB1.LESSP establishes that the measure (COUNT X) decreases according to the well-founded relation LESSP in the induction step of the scheme. The above induction scheme produces the following three new conjectures:

*Case 1.*
```
(IMPLIES (NOT (NUMBERP X))
         (EQUAL (TIMES (TIMES X Y) Z)
                (TIMES X (TIMES Y Z)))),
```
which simplifies, expanding the definition of TIMES, to:

```
(TRUE).
```

*Case 2.*
```
(IMPLIES (EQUAL X 0)
         (EQUAL (TIMES (TIMES X Y) Z)
                (TIMES X (TIMES Y Z)))),
```
which simplifies, opening up TIMES, to:

```
(TRUE).
```

*Case 3.*
```
(IMPLIES (AND (NUMBERP X)
              (NOT (EQUAL X 0))
              (EQUAL (TIMES (TIMES (SUB1 X) Y) Z)
                     (TIMES (SUB1 X) (TIMES Y Z))))
         (EQUAL (TIMES (TIMES X Y) Z)
                (TIMES X (TIMES Y Z)))).
```

This simplifies, expanding TIMES, to:

```
(IMPLIES (AND (NUMBERP X)
              (NOT (EQUAL X 0))
              (EQUAL (TIMES (TIMES (SUB1 X) Y) Z)
                     (TIMES (SUB1 X) (TIMES Y Z))))
         (EQUAL (TIMES (PLUS Y (TIMES (SUB1 X) Y)) Z)
                (PLUS (TIMES Y Z)
                      (TIMES (SUB1 X) (TIMES Y Z))))).
```

Applying the lemma SUB1.ELIM, replace X by (ADD1 V)
to eliminate (SUB1 X). We use the type restriction
lemma noted when SUB1 was introduced to constrain the
new variable. The result is:

```
(IMPLIES (AND (NUMBERP V)
              (NUMBERP (ADD1 V))
              (NOT (EQUAL (ADD1 V) 0))
              (EQUAL (TIMES (TIMES V Y) Z)
                     (TIMES V (TIMES Y Z))))
         (EQUAL (TIMES (PLUS Y (TIMES V Y)) Z)
                (PLUS (TIMES Y Z)
                      (TIMES V (TIMES Y Z))))).
```

Of course, this simplifies further, trivially, to:

```
(IMPLIES (AND (NUMBERP V)
              (EQUAL (TIMES (TIMES V Y) Z)
                     (TIMES V (TIMES Y Z))))
         (EQUAL (TIMES (PLUS Y (TIMES V Y)) Z)
                (PLUS (TIMES Y Z)
                      (TIMES V (TIMES Y Z))))).
```

We use the above equality hypothesis by
cross-fertilizing (TIMES (TIMES V Y) Z) for (TIMES V
(TIMES Y Z)) and throwing away the equality. We would
thus like to prove the conjecture:

```
(IMPLIES (NUMBERP V)
         (EQUAL (TIMES (PLUS Y (TIMES V Y)) Z)
                (PLUS (TIMES Y Z)
                      (TIMES (TIMES V Y) Z)))),
```

which we generalize by replacing (TIMES V Y) by A. We
restrict the new variable by recalling the type
restriction lemma noted when TIMES was introduced. We
would thus like to prove the new goal:

```
(IMPLIES (AND (NUMBERP A) (NUMBERP V))
         (EQUAL (TIMES (PLUS Y A) Z)
                (PLUS (TIMES Y Z) (TIMES A Z)))).
```

However, this simplifies further, using the lemma
COMMUTATIVITY.OF.PLUS, to the conjecture:

```
(IMPLIES (AND (NUMBERP A) (NUMBERP V))
         (EQUAL (TIMES (PLUS A Y) Z)
                (PLUS (TIMES A Z) (TIMES Y Z)))),
```

which has an irrelevant term in it. By eliminating
this term we get:

(IMPLIES (NUMBERP A)
         (EQUAL (TIMES (PLUS A Y) Z)
                (PLUS (TIMES A Z) (TIMES Y Z)))),

which we will finally name *1.1.

  Let us appeal to the induction principle. There are
three plausible inductions. They merge into two
likely candidate inductions. However, only one is
unflawed. We will induct according to the following
scheme:

(AND (IMPLIES (NOT (NUMBERP A)) (p A Y Z))
     (IMPLIES (EQUAL A 0) (p A Y Z))
     (IMPLIES (AND (NUMBERP A)
                   (NOT (EQUAL A 0))
                   (p (SUB1 A) Y Z))
              (p A Y Z))).

The inequality SUB1.LESSP establishes that the
measure (COUNT A) decreases according to the
well-founded relation LESSP in the induction step of
the scheme. The above induction scheme produces the
following three new conjectures:

*Case 1.*
(IMPLIES (AND (EQUAL A 0) (NUMBERP A))
         (EQUAL (TIMES (PLUS A Y) Z)
                (PLUS (TIMES A Z)
                      (TIMES Y Z)))).

This simplifies, expanding the functions PLUS and
TIMES, to:

        (TRUE).

*Case 2.*
(IMPLIES (AND (NOT (EQUAL A 0))
              (NOT (NUMBERP (SUB1 A)))
              (NUMBERP A))
         (EQUAL (TIMES (PLUS A Y) Z)
                (PLUS (TIMES A Z)
                      (TIMES Y Z)))),

which we simplify, obviously, to:

        (TRUE).

*Case 3.*
(IMPLIES (AND (NOT (EQUAL A 0))
              (EQUAL (TIMES (PLUS (SUB1 A) Y) Z)
                     (PLUS (TIMES (SUB1 A) Z)
                           (TIMES Y Z)))
              (NUMBERP A))
         (EQUAL (TIMES (PLUS A Y) Z)
                (PLUS (TIMES A Z) (TIMES Y Z)))),

which we simplify, rewriting with COMMUTATIVITY.OF.-
PLUS, SUB1.ADD1 and ASSOCIATIVITY.OF.PLUS, and
unfolding the functions PLUS and TIMES, to:

(IMPLIES (AND (NOT (EQUAL A 0))
              (EQUAL (TIMES (PLUS Y (SUB1 A)) Z)
                     (PLUS (TIMES Y Z)
                           (TIMES (SUB1 A) Z)))
              (NUMBERP A))
         (EQUAL (PLUS Z (TIMES (PLUS (SUB1 A) Y) Z))
                (PLUS Z
                      (PLUS (TIMES Y Z)
                            (TIMES (SUB1 A) Z))))),

which we again simplify, rewriting with
COMMUTATIVITY.OF.PLUS, to:

(IMPLIES (AND (NOT (EQUAL A 0))
              (EQUAL (TIMES (PLUS Y (SUB1 A)) Z)
                     (PLUS (TIMES Y Z)
                           (TIMES (SUB1 A) Z)))
              (NUMBERP A))
         (EQUAL (PLUS Z (TIMES (PLUS Y (SUB1 A)) Z))
                (PLUS Z
                      (PLUS (TIMES Y Z)
                            (TIMES (SUB1 A) Z))))).

Applying the lemma SUB1.ELIM, we now replace A by
(ADD1 X) to eliminate (SUB1 A). We use the type
restriction lemma noted when SUB1 was introduced to

constrain the new variable. We thus obtain:

```
(IMPLIES (AND (NUMBERP X)
              (NOT (EQUAL (ADD1 X) 0))
              (EQUAL (TIMES (PLUS Y X) Z)
                     (PLUS (TIMES Y Z) (TIMES X Z)))
              (NUMBERP (ADD1 X)))
         (EQUAL (PLUS Z (TIMES (PLUS Y X) Z))
                (PLUS Z
                      (PLUS (TIMES Y Z)
                            (TIMES X Z))))).
```

This further simplifies, applying COMMUTATIVITY.OF.-PLUS, to:

```
(IMPLIES (AND (NUMBERP X)
              (EQUAL (TIMES (PLUS X Y) Z)
                     (PLUS (TIMES X Z) (TIMES Y Z))))
         (EQUAL (PLUS Z (TIMES (PLUS X Y) Z))
                (PLUS Z
                      (PLUS (TIMES X Z)
                            (TIMES Y Z))))).
```

We use the above equality hypothesis by cross-fertilizing:

```
        (PLUS (TIMES X Z) (TIMES Y Z))
```

for (TIMES (PLUS X Y) Z) and throwing away the equality. We thus obtain:

```
(IMPLIES (NUMBERP X)
         (EQUAL (PLUS Z
                      (PLUS (TIMES X Z) (TIMES Y Z)))
                (PLUS Z
                      (PLUS (TIMES X Z)
                            (TIMES Y Z))))),
```

which we further simplify, clearly, to:

```
        (TRUE).
```

That finishes the proof of *1.1, which, consequently, finishes the proof of *1. Q.E.D.

CPU time (devoted to theorem-proving): 10.277 seconds

## I. DIFFERENCE

**Definition**

```
(DIFFERENCE I J)
    =
(IF (ZEROP I)
    0
    (IF (ZEROP J)
        I
        (DIFFERENCE (SUB1 I) (SUB1 J)))).
```

The lemma SUB1.LESSP can be used to show that (COUNT I) decreases according to the well-founded relation LESSP in each recursive call. Hence, DIFFERENCE is accepted under the definition principle. The definition of DIFFERENCE can be justified in another way. The lemma SUB1.LESSP informs us that (COUNT J) decreases according to the well-founded relation LESSP in each recursive call. Note that:

```
(NUMBERP (DIFFERENCE I J))
```

is a theorem.

CPU time (devoted to theorem-proving): 1.594 seconds


## J. RECURSION.BY.DIFFERENCE

**Theorem** RECURSION.BY.DIFFERENCE (induction):

```
(IMPLIES (AND (NUMBERP I)
              (NUMBERP N)
              (NOT (EQUAL I 0))
              (NOT (EQUAL N 0)))
         (LESSP (DIFFERENCE I N) I)).
```

Name the conjecture *1.

Perhaps we can prove it by induction. Three inductions are suggested by terms in the conjecture. However, they merge into one likely candidate

induction. We will induct according to the following
scheme:
```
(AND (IMPLIES (NOT (NUMBERP N)) (p I N))
     (IMPLIES (EQUAL N 0) (p I N))
     (IMPLIES (NOT (NUMBERP I)) (p I N))
     (IMPLIES (EQUAL I 0) (p I N))
     (IMPLIES (AND (NUMBERP N)
                   (NOT (EQUAL N 0))
                   (NUMBERP I)
                   (NOT (EQUAL I 0))
                   (p (SUB1 I) (SUB1 N)))
              (p I N))).
```
The inequality SUB1.LESSP establishes that the
measure (COUNT I) decreases according to the
well-founded relation LESSP in the induction step of
the scheme. Note, however, the inductive instance
chosen for N. The above induction scheme generates
five new conjectures:

*Case 1.* (IMPLIES (AND (NOT (NUMBERP (SUB1 I)))
                   (NUMBERP I)
                   (NUMBERP N)
                   (NOT (EQUAL I 0))
                   (NOT (EQUAL N 0)))
              (LESSP (DIFFERENCE I N) I)).

This simplifies, clearly, to:
      (TRUE).

*Case 2.* (IMPLIES (AND (NOT (NUMBERP (SUB1 N)))
                   (NUMBERP I)
                   (NUMBERP N)
                   (NOT (EQUAL I 0))
                   (NOT (EQUAL N 0)))
              (LESSP (DIFFERENCE I N) I)),

which simplifies, clearly, to:
      (TRUE).

*Case 3.* (IMPLIES (AND (EQUAL (SUB1 I) 0)
                   (NUMBERP I)
                   (NUMBERP N)
                   (NOT (EQUAL I 0))
                   (NOT (EQUAL N 0)))
              (LESSP (DIFFERENCE I N) I)).

Applying the lemma SUB1.ELIM, replace I by (ADD1 X)
to eliminate (SUB1 I). We use the type restriction
lemma noted when SUB1 was introduced to constrain the
new variable. The result is:

```
(IMPLIES (AND (NUMBERP X)
              (EQUAL X 0)
              (NUMBERP (ADD1 X))
              (NUMBERP N)
              (NOT (EQUAL (ADD1 X) 0))
              (NOT (EQUAL N 0)))
         (LESSP (DIFFERENCE (ADD1 X) N)
                (ADD1 X))),
```

which simplifies, appealing to the lemma SUB1.ADD1,
and expanding the definitions of DIFFERENCE and
LESSP, to:

```
(TRUE).
```

*Case 4.* (IMPLIES (AND (EQUAL (SUB1 N) 0)
              (NUMBERP I)
              (NUMBERP N)
              (NOT (EQUAL I 0))
              (NOT (EQUAL N 0)))
         (LESSP (DIFFERENCE I N) I)).

Applying the lemma SUB1.ELIM, we now replace N by
(ADD1 X) to eliminate (SUB1 N). We use the type
restriction lemma noted when SUB1 was introduced to
restrict the new variable. We would thus like to
prove:

```
(IMPLIES (AND (NUMBERP X)
              (EQUAL X 0)
              (NUMBERP I)
              (NUMBERP (ADD1 X))
              (NOT (EQUAL I 0))
              (NOT (EQUAL (ADD1 X) 0)))
         (LESSP (DIFFERENCE I (ADD1 X)) I)),
```

which simplifies, applying SUB1.ADD1, and opening up

the definitions of DIFFERENCE and LESSP, to the new goal:

```
(IMPLIES (AND (NUMBERP I)
              (NOT (EQUAL I 0))
              (NOT (EQUAL (SUB1 I) 0)))
         (LESSP (SUB1 I) I)).
```

Applying the lemma SUB1.ELIM, we now replace I by (ADD1 X) to eliminate (SUB1 I). We use the type restriction lemma noted when SUB1 was introduced to restrict the new variable. We thus obtain:

```
(IMPLIES (AND (NUMBERP X)
              (NUMBERP (ADD1 X))
              (NOT (EQUAL (ADD1 X) 0))
              (NOT (EQUAL X 0)))
         (LESSP X (ADD1 X))),
```

which further simplifies, applying SUB1.ADD1, and expanding LESSP, to:

```
(IMPLIES (AND (NUMBERP X) (NOT (EQUAL X 0)))
         (LESSP (SUB1 X) X)).
```

Give the above formula the name *1.1.

*Case 5.* (IMPLIES (AND (LESSP (DIFFERENCE (SUB1 I)
                                            (SUB1 N))
                               (SUB1 I))
                        (NUMBERP I)
                        (NUMBERP N)
                        (NOT (EQUAL I 0))
                        (NOT (EQUAL N 0)))
                   (LESSP (DIFFERENCE I N) I)).

This simplifies, expanding the definition of DIFFERENCE, to:

```
(IMPLIES (AND (LESSP (DIFFERENCE (SUB1 I)
                                  (SUB1 N))
                     (SUB1 I))
              (NUMBERP I)
              (NUMBERP N)
              (NOT (EQUAL I 0))
              (NOT (EQUAL N 0)))
         (LESSP (DIFFERENCE (SUB1 I) (SUB1 N))
                I)).
```

Appealing to the lemma SUB1.ELIM, we now replace I by
(ADD1 X) to eliminate (SUB1 I). We employ the type
restriction lemma noted when SUB1 was introduced to
restrict the new variable. This produces the new
formula:
```
(IMPLIES (AND (NUMBERP X)
              (LESSP (DIFFERENCE X (SUB1 N)) X)
              (NUMBERP (ADD1 X))
              (NUMBERP N)
              (NOT (EQUAL (ADD1 X) 0))
              (NOT (EQUAL N 0)))
         (LESSP (DIFFERENCE X (SUB1 N))
                (ADD1 X))),
```
which further simplifies, applying the lemma
SUB1.ADD1, and opening up the definition of LESSP, to
the conjecture:
```
(IMPLIES (AND (NUMBERP X)
              (LESSP (DIFFERENCE X (SUB1 N)) X)
              (NUMBERP N)
              (NOT (EQUAL N 0))
              (NOT (EQUAL (DIFFERENCE X
                                     (SUB1 N))
                          0)))
         (LESSP (SUB1 (DIFFERENCE X (SUB1 N)))
                X)).
```
Appealing to the lemma SUB1.ELIM, we now replace N by
(ADD1 Z) to eliminate (SUB1 N). We use the type
restriction lemma noted when SUB1 was introduced to
restrict the new variable. This generates:
```
(IMPLIES (AND (NUMBERP Z)
              (NUMBERP X)
              (LESSP (DIFFERENCE X Z) X)
              (NUMBERP (ADD1 Z))
              (NOT (EQUAL (ADD1 Z) 0))
              (NOT (EQUAL (DIFFERENCE X Z) 0)))
         (LESSP (SUB1 (DIFFERENCE X Z)) X)),
```
which further simplifies, trivially, to:
```
(IMPLIES (AND (NUMBERP Z)
              (NUMBERP X)
              (LESSP (DIFFERENCE X Z) X)
              (NOT (EQUAL (DIFFERENCE X Z) 0)))
         (LESSP (SUB1 (DIFFERENCE X Z)) X)).
```

We will try to prove the above conjecture by
generalizing it, replacing (DIFFERENCE X Z) by Y. We
restrict the new variable by appealing to the type
restriction lemma noted when DIFFERENCE was
introduced. The result is:

    (IMPLIES (AND (NUMBERP Y)
                (NUMBERP Z)
                (NUMBERP X)
                (LESSP Y X)
                (NOT (EQUAL Y 0)))
         (LESSP (SUB1 Y) X)).

Appealing to the lemma SUB1.ELIM, we now replace Y by
(ADD1 V) to eliminate (SUB1 Y). We employ the type
restriction lemma noted when SUB1 was introduced to
restrict the new variable. We would thus like to
prove the goal:

    (IMPLIES (AND (NUMBERP V)
                (NUMBERP (ADD1 V))
                (NUMBERP Z)
                (NUMBERP X)
                (LESSP (ADD1 V) X)
                (NOT (EQUAL (ADD1 V) 0)))
         (LESSP V X)),

which we further simplify, using the lemma SUB1.ADD1,
and opening up the definition of LESSP, to:

    (IMPLIES (AND (NUMBERP V)
                (NUMBERP Z)
                (NUMBERP X)
                (NOT (EQUAL X 0))
                (LESSP V (SUB1 X)))
         (LESSP V X)).

Eliminate the irrelevant term. This generates:

    (IMPLIES (AND (NUMBERP V)
                (NUMBERP X)
                (NOT (EQUAL X 0))
                (LESSP V (SUB1 X)))
         (LESSP V X)).

Finally give the above formula the name *1.2.

We will appeal to induction. Three inductions are suggested by terms in the conjecture. However, they merge into one likely candidate induction. We will induct according to the following scheme:

```
(AND (IMPLIES (NOT (NUMBERP X)) (p V X))
     (IMPLIES (EQUAL X 0) (p V X))
     (IMPLIES (NOT (NUMBERP V)) (p V X))
     (IMPLIES (EQUAL V 0) (p V X))
     (IMPLIES (AND (NUMBERP X)
                   (NOT (EQUAL X 0))
                   (NUMBERP V)
                   (NOT (EQUAL V 0))
                   (p (SUB1 V) (SUB1 X)))
              (p V X))).
```

The inequality SUB1.LESSP establishes that the measure (COUNT V) decreases according to the well-founded relation LESSP in the induction step of the scheme. Note, however, the inductive instance chosen for X. The above induction scheme generates six new conjectures:

*Case 1.* (IMPLIES (AND (EQUAL V 0)
                  (NUMBERP V)
                  (NUMBERP X)
                  (NOT (EQUAL X 0))
                  (LESSP V (SUB1 X)))
             (LESSP V X)),

which we simplify, expanding the definition of LESSP, to:
      (TRUE).

*Case 2.* (IMPLIES (AND (NOT (EQUAL V 0))
                  (NOT (NUMBERP (SUB1 V)))
                  (NUMBERP V)
                  (NUMBERP X)
                  (NOT (EQUAL X 0))
                  (LESSP V (SUB1 X)))
             (LESSP V X)).

This simplifies, trivially, to:
      (TRUE).

*Case 3.* (IMPLIES (AND (NOT (EQUAL V 0))
                    (NOT (NUMBERP (SUB1 X)))
                    (NUMBERP V)
                    (NUMBERP X)
                    (NOT (EQUAL X 0))
                    (LESSP V (SUB1 X)))
               (LESSP V X)).

Of course, this simplifies, trivially, to:

   (TRUE).

*Case 4.* (IMPLIES (AND (NOT (EQUAL V 0))
                    (EQUAL (SUB1 X) 0)
                    (NUMBERP V)
                    (NUMBERP X)
                    (NOT (EQUAL X 0))
                    (LESSP V (SUB1 X)))
               (LESSP V X)).

This simplifies, expanding LESSP, to:

   (TRUE).

*Case 5.* (IMPLIES (AND (NOT (EQUAL V 0))
                    (NOT (LESSP (SUB1 V)
                                (SUB1 (SUB1 X))))
                    (NUMBERP V)
                    (NUMBERP X)
                    (NOT (EQUAL X 0))
                    (LESSP V (SUB1 X)))
               (LESSP V X)),

which simplifies, opening up LESSP, to:

   (TRUE).

*Case 6.* (IMPLIES (AND (NOT (EQUAL V 0))
                    (LESSP (SUB1 V) (SUB1 X))
                    (NUMBERP V)
                    (NUMBERP X)
                    (NOT (EQUAL X 0))
                    (LESSP V (SUB1 X)))
               (LESSP V X)).

This simplifies, expanding LESSP, to:

   (TRUE).

That finishes the proof of *1.2.

So let us turn our attention to:

        (IMPLIES (AND (NUMBERP X) (NOT (EQUAL X 0)))
                 (LESSP (SUB1 X) X)),

which we named *1.1 above. Perhaps we can prove it by induction. There is only one plausible induction. We will induct according to the following scheme:

        (AND (IMPLIES (EQUAL X 0) (p X))
             (IMPLIES (NOT (NUMBERP X)) (p X))
             (IMPLIES (EQUAL (SUB1 X) 0) (p X))
             (IMPLIES (NOT (NUMBERP (SUB1 X)))
                      (p X))
             (IMPLIES (AND (NOT (EQUAL X 0))
                           (NUMBERP X)
                           (NOT (EQUAL (SUB1 X) 0))
                           (NUMBERP (SUB1 X))
                           (p (SUB1 X)))
                      (p X))).

The inequality SUB1.LESSP establishes that the measure (COUNT X) decreases according to the well-founded relation LESSP in the induction step of the scheme. The above induction scheme leads to three new goals:

*Case 1.* (IMPLIES (AND (NOT (NUMBERP (SUB1 X)))
                        (NUMBERP X)
                        (NOT (EQUAL X 0)))
                   (LESSP (SUB1 X) X)).

This simplifies, trivially, to:

        (TRUE).

*Case 2.* (IMPLIES (AND (EQUAL (SUB1 X) 0)
                        (NUMBERP X)
                        (NOT (EQUAL X 0)))
                   (LESSP (SUB1 X) X)).

This simplifies, expanding the definition of LESSP, to:

        (TRUE).

### 242 / XVI. ILLUSTRATIONS OF OUR TECHNIQUES VIA ELEMENTARY NUMBER THEORY

*Case 3.* (IMPLIES (AND (NOT (EQUAL (SUB1 X) 0))
      (NUMBERP (SUB1 X))
      (LESSP (SUB1 (SUB1 X)) (SUB1 X))
      (NUMBERP X)
      (NOT (EQUAL X 0)))
    (LESSP (SUB1 X) X)),

which simplifies, opening up the definition of LESSP, to:
  (TRUE).

That finishes the proof of *1.1, which, consequently, also finishes the proof of *1. Q.E.D.
CPU time (devoted to theorem-proving): 10.241 seconds

### K. REMAINDER
**Definition**

  (REMAINDER I J)
   =
  (IF (ZEROP J)
    (FIX I)
    (IF (LESSP I J)
      (FIX I)
      (REMAINDER (DIFFERENCE I J) J))).

The lemma RECURSION.BY.DIFFERENCE, together with the definition of LESSP, establish that (COUNT I) goes down according to the well-founded relation LESSP in each recursive call. Hence, REMAINDER is accepted under the definition principle. Observe that (NUMBERP (REMAINDER I J)) is a theorem.
CPU time (devoted to theorem-proving): 1.446 seconds

### L. QUOTIENT
**Definition**

  (QUOTIENT I J)
   =
  (IF (ZEROP J)
    0
    (IF (LESSP I J)
      0
      (ADD1 (QUOTIENT (DIFFERENCE I J) J)))).

The lemma RECURSION.BY.DIFFERENCE, together with the definition of LESSP, inform us that (COUNT I) goes down according to the well-founded relation LESSP in each recursive call. Hence, QUOTIENT is accepted under the principle of definition. Observe that (NUMBERP (QUOTIENT I J)) is a theorem.

CPU time (devoted to theorem-proving): 1.381 seconds

## M. REMAINDER.QUOTIENT.ELIM

**Theorem** REMAINDER.QUOTIENT.ELIM (rewrite):
```
            (IMPLIES (AND (NOT (ZEROP Y)) (NUMBERP X))
                     (EQUAL (PLUS (REMAINDER X Y)
                                  (TIMES Y (QUOTIENT X Y)))
                            X)).
```

This formula simplifies, expanding the definitions of ZEROP, NOT, AND and IMPLIES, to:
```
            (IMPLIES (AND (NOT (EQUAL Y 0))
                          (NUMBERP Y)
                          (NUMBERP X))
                     (EQUAL (PLUS (REMAINDER X Y)
                                  (TIMES Y (QUOTIENT X Y)))
                            X)).
```

Call the above conjecture *1.

Perhaps we can prove it by induction. There are three plausible inductions. They merge into two likely candidate inductions. However, only one is unflawed. We will induct according to the following scheme:
```
            (AND (IMPLIES (NOT (NUMBERP X)) (p X Y))
                 (IMPLIES (NOT (NUMBERP Y)) (p X Y))
                 (IMPLIES (EQUAL X 0) (p X Y))
                 (IMPLIES (EQUAL Y 0) (p X Y))
                 (IMPLIES (AND (NUMBERP X)
                               (NUMBERP Y)
                               (NOT (EQUAL X 0))
                               (NOT (EQUAL Y 0))
                               (p (DIFFERENCE X Y) Y))
                          (p X Y))).
```

The inequality RECURSION.BY.DIFFERENCE establishes that the measure (COUNT X) decreases according to the well-founded relation LESSP in the induction step of the scheme. The above induction scheme generates the following three new goals:

*Case 1.*

```
(IMPLIES (AND (EQUAL X 0)
              (NOT (EQUAL Y 0))
              (NUMBERP Y)
              (NUMBERP X))
         (EQUAL (PLUS (REMAINDER X Y)
                      (TIMES Y (QUOTIENT X Y)))
                X)),
```

which simplifies, applying the lemma TIMES.ZERO, and expanding the definitions of LESSP, REMAINDER, QUOTIENT and PLUS, to:

(TRUE).

*Case 2.*

```
(IMPLIES (AND (NOT (EQUAL X 0))
              (NOT (NUMBERP (DIFFERENCE X Y)))
              (NOT (EQUAL Y 0))
              (NUMBERP Y)
              (NUMBERP X))
         (EQUAL (PLUS (REMAINDER X Y)
                      (TIMES Y (QUOTIENT X Y)))
                X)),
```

which we simplify, trivially, to:

(TRUE).

*Case 3.*

```
(IMPLIES
     (AND (NOT (EQUAL X 0))
          (EQUAL (PLUS (REMAINDER (DIFFERENCE X Y) Y)
                       (TIMES Y
                              (QUOTIENT (DIFFERENCE X Y)
                                        Y)))
                 (DIFFERENCE X Y))
          (NOT (EQUAL Y 0))
          (NUMBERP Y)
          (NUMBERP X))
```

```
          (EQUAL (PLUS (REMAINDER X Y)
                      (TIMES Y (QUOTIENT X Y)))
                 X)).
```
This simplifies, applying the lemmas TIMES.ZERO,
PLUS.RIGHT.ID, TIMES.ADD1 and COMMUTATIVITY2.OF.PLUS,
and expanding the definitions of REMAINDER and
QUOTIENT, to:
```
(IMPLIES
    (AND (NOT (EQUAL X 0))
         (EQUAL (PLUS (REMAINDER (DIFFERENCE X Y) Y)
                      (TIMES Y
                             (QUOTIENT (DIFFERENCE X Y)
                                       Y)))
                (DIFFERENCE X Y))
         (NOT (EQUAL Y 0))
         (NUMBERP Y)
         (NUMBERP X)
         (NOT (LESSP X Y)))
    (EQUAL (PLUS Y
                 (PLUS (REMAINDER (DIFFERENCE X Y) Y)
                       (TIMES Y
                              (QUOTIENT (DIFFERENCE X Y)
                                        Y))))
           X)).
```
We use the above equality hypothesis by
cross-fertilizing (DIFFERENCE X Y) for:
```
          (PLUS (REMAINDER (DIFFERENCE X Y) Y)
                (TIMES Y
                       (QUOTIENT (DIFFERENCE X Y) Y)))
```
and throwing away the equality. This produces:
```
          (IMPLIES (AND (NOT (EQUAL X 0))
                        (NOT (EQUAL Y 0))
                        (NUMBERP Y)
                        (NUMBERP X)
                        (NOT (LESSP X Y)))
                   (EQUAL (PLUS Y (DIFFERENCE X Y)) X)),
```
which we will name *1.1.

Let us appeal to the induction principle. Five

inductions are suggested by terms in the conjecture.
However, they merge into one likely candidate
induction. We will induct according to the following
scheme:

```
(AND (IMPLIES (NOT (NUMBERP Y)) (p Y X))
     (IMPLIES (EQUAL Y 0) (p Y X))
     (IMPLIES (NOT (NUMBERP X)) (p Y X))
     (IMPLIES (EQUAL X 0) (p Y X))
     (IMPLIES (AND (NUMBERP Y)
                   (NOT (EQUAL Y 0))
                   (NUMBERP X)
                   (NOT (EQUAL X 0))
                   (p (SUB1 Y) (SUB1 X)))
              (p Y X))).
```

The inequality SUB1.LESSP establishes that the
measure (COUNT X) decreases according to the
well-founded relation LESSP in the induction step of
the scheme. Note, however, the inductive instance
chosen for Y. The above induction scheme generates
six new goals:

*Case 1.* (IMPLIES (AND (EQUAL (SUB1 X) 0)
                       (NOT (EQUAL X 0))
                       (NOT (EQUAL Y 0))
                       (NUMBERP Y)
                       (NUMBERP X)
                       (NOT (LESSP X Y)))
                  (EQUAL (PLUS Y (DIFFERENCE X Y)) X)).

Applying the lemma SUB1.ELIM, we now replace X by
(ADD1 Z) to eliminate (SUB1 X). We use the type
restriction lemma noted when SUB1 was introduced to
restrict the new variable. The result is:

```
(IMPLIES (AND (NUMBERP Z)
              (EQUAL Z 0)
              (NOT (EQUAL (ADD1 Z) 0))
              (NOT (EQUAL Y 0))
              (NUMBERP Y)
              (NUMBERP (ADD1 Z))
              (NOT (LESSP (ADD1 Z) Y)))
         (EQUAL (PLUS Y (DIFFERENCE (ADD1 Z)
                                    Y))
                (ADD1 Z))).
```

But this simplifies, appealing to the lemmas
SUB1.ADD1 and PLUS.RIGHT.ID, and unfolding the
definitions of LESSP and DIFFERENCE, to:

          (IMPLIES (AND (NOT (EQUAL Y 0))
                        (NUMBERP Y)
                        (EQUAL (SUB1 Y) 0))
                   (EQUAL Y 1)).

Applying the lemma SUB1.ELIM, we now replace Y by
(ADD1 Z) to eliminate (SUB1 Y). We employ the type
restriction lemma noted when SUB1 was introduced to
restrict the new variable. We would thus like to
prove:

          (IMPLIES (AND (NUMBERP Z)
                        (NOT (EQUAL (ADD1 Z) 0))
                        (NUMBERP (ADD1 Z))
                        (EQUAL Z 0))
                   (EQUAL (ADD1 Z) 1)).

This further simplifies, obviously, to:

          (TRUE).

*Case 2.* (IMPLIES (AND (EQUAL (SUB1 Y) 0)
                        (NOT (EQUAL X 0))
                        (NOT (EQUAL Y 0))
                        (NUMBERP Y)
                        (NUMBERP X)
                        (NOT (LESSP X Y)))
                   (EQUAL (PLUS Y (DIFFERENCE X Y)) X)).

This simplifies, unfolding the definition of PLUS, to
the new conjecture:

          (IMPLIES (AND (EQUAL (SUB1 Y) 0)
                        (NOT (EQUAL X 0))
                        (NOT (EQUAL Y 0))
                        (NUMBERP Y)
                        (NUMBERP X)
                        (NOT (LESSP X Y)))
                   (EQUAL (ADD1 (PLUS (SUB1 Y)
                                      (DIFFERENCE X Y)))
                          X)),

which we again simplify, opening up PLUS, to:
```
(IMPLIES (AND (EQUAL (SUB1 Y) 0)
              (NOT (EQUAL X 0))
              (NOT (EQUAL Y 0))
              (NUMBERP Y)
              (NUMBERP X)
              (NOT (LESSP X Y)))
         (EQUAL (ADD1 (DIFFERENCE X Y)) X)).
```
Appealing to the lemma SUB1.ELIM, replace Y by (ADD1 Z) to eliminate (SUB1 Y). We use the type restriction lemma noted when SUB1 was introduced to constrain the new variable. The result is:
```
(IMPLIES (AND (NUMBERP Z)
              (EQUAL Z 0)
              (NOT (EQUAL X 0))
              (NOT (EQUAL (ADD1 Z) 0))
              (NUMBERP (ADD1 Z))
              (NUMBERP X)
              (NOT (LESSP X (ADD1 Z))))
         (EQUAL (ADD1 (DIFFERENCE X (ADD1 Z)))
                X)).
```
This simplifies further, using the lemmas SUB1.ADD1 and ADD1.SUB1, and expanding the definitions of LESSP, DIFFERENCE, NOT and AND, to:
```
(IMPLIES (AND (NOT (EQUAL X 0))
              (NUMBERP X)
              (EQUAL (SUB1 X) 0))
         (EQUAL 1 X)).
```
Applying the lemma SUB1.ELIM, replace X by (ADD1 Z) to eliminate (SUB1 X). We employ the type restriction lemma noted when SUB1 was introduced to restrict the new variable. This produces:
```
(IMPLIES (AND (NUMBERP Z)
              (NOT (EQUAL (ADD1 Z) 0))
              (NUMBERP (ADD1 Z))
              (EQUAL Z 0))
         (EQUAL 1 (ADD1 Z))).
```
Of course, this further simplifies, obviously, to:
```
(TRUE).
```

*Case 3.* (IMPLIES (AND (NOT (NUMBERP (SUB1 Y)))
                        (NOT (EQUAL X 0))
                        (NOT (EQUAL Y 0))
                        (NUMBERP Y)
                        (NUMBERP X)
                        (NOT (LESSP X Y)))
                   (EQUAL (PLUS Y (DIFFERENCE X Y)) X)),

which simplifies, obviously, to:

    (TRUE).

*Case 4.* (IMPLIES (AND (NOT (NUMBERP (SUB1 X)))
                        (NOT (EQUAL X 0))
                        (NOT (EQUAL Y 0))
                        (NUMBERP Y)
                        (NUMBERP X)
                        (NOT (LESSP X Y)))
                   (EQUAL (PLUS Y (DIFFERENCE X Y)) X)).

Of course, this simplifies, obviously, to:

    (TRUE).

*Case 5.* (IMPLIES (AND (LESSP (SUB1 X) (SUB1 Y))
                        (NOT (EQUAL X 0))
                        (NOT (EQUAL Y 0))
                        (NUMBERP Y)
                        (NUMBERP X)
                        (NOT (LESSP X Y)))
                   (EQUAL (PLUS Y (DIFFERENCE X Y)) X)),

which we simplify, unfolding the function LESSP, to:

    (TRUE).

*Case 6.*
(IMPLIES (AND (EQUAL (PLUS (SUB1 Y)
                           (DIFFERENCE (SUB1 X)
                                       (SUB1 Y)))
                     (SUB1 X))
              (NOT (EQUAL X 0))
              (NOT (EQUAL Y 0))
              (NUMBERP Y)
              (NUMBERP X)
              (NOT (LESSP X Y)))
         (EQUAL (PLUS Y (DIFFERENCE X Y)) X)).

This simplifies, expanding the definitions of LESSP, DIFFERENCE and PLUS, to:

```
(IMPLIES (AND (EQUAL (PLUS (SUB1 Y)
                           (DIFFERENCE (SUB1 X)
                                       (SUB1 Y)))
                     (SUB1 X))
              (NOT (EQUAL X 0))
              (NOT (EQUAL Y 0))
              (NUMBERP Y)
              (NUMBERP X)
              (NOT (LESSP (SUB1 X) (SUB1 Y))))
         (EQUAL (ADD1 (PLUS (SUB1 Y)
                            (DIFFERENCE (SUB1 X)
                                        (SUB1 Y))))
                X)).
```

Applying the lemma SUB1.ELIM, we now replace Y by (ADD1 Z) to eliminate (SUB1 Y). We use the type restriction lemma noted when SUB1 was introduced to constrain the new variable. This generates:

```
(IMPLIES (AND (NUMBERP Z)
              (EQUAL (PLUS Z (DIFFERENCE (SUB1 X) Z))
                     (SUB1 X))
              (NOT (EQUAL X 0))
              (NOT (EQUAL (ADD1 Z) 0))
              (NUMBERP (ADD1 Z))
              (NUMBERP X)
              (NOT (LESSP (SUB1 X) Z)))
         (EQUAL (ADD1 (PLUS Z
                            (DIFFERENCE (SUB1 X) Z)))
                X)).
```

Of course, this simplifies further, trivially, to:

```
(IMPLIES (AND (NUMBERP Z)
              (EQUAL (PLUS Z (DIFFERENCE (SUB1 X) Z))
                     (SUB1 X))
              (NOT (EQUAL X 0))
              (NUMBERP X)
              (NOT (LESSP (SUB1 X) Z)))
         (EQUAL (ADD1 (PLUS Z
                            (DIFFERENCE (SUB1 X) Z)))
                X)).
```

Appealing to the lemma SUB1.ELIM, we now replace X by
(ADD1 V) to eliminate (SUB1 X). We use the type
restriction lemma noted when SUB1 was introduced to
constrain the new variable. This produces:

(IMPLIES (AND (NUMBERP V)
              (NUMBERP Z)
              (EQUAL (PLUS Z (DIFFERENCE V Z)) V)
              (NOT (EQUAL (ADD1 V) 0))
              (NUMBERP (ADD1 V))
              (NOT (LESSP V Z)))
         (EQUAL (ADD1 (PLUS Z (DIFFERENCE V Z)))
                (ADD1 V))).

However this simplifies further, applying the lemma
ADD1.EQUAL, to:

(TRUE).

That finishes the proof of *1.1, which, in turn,
also finishes the proof of *1. Q.E.D.

CPU time (devoted to theorem-proving): 12.334 seconds

# XVII

## The Correctness of a Simple Optimizing Expression Compiler

In this and the next two chapters, we construct in our theory the proofs of three interesting theorems: the correctness of an expression compiler, the correctness of a fast string searching algorithm, and the fundamental theorem of arithmetic. The proofs we discuss are those discovered by our theorem-proving program.

The examples are presented in the same format as the TAUTOL-OGY.CHECKER example in Chapter IV. We first present an informal explanation of the problem at hand. Then we present the formal development of the problem and describe the mechanical proofs.

Recall the discussion in Chapter IV, where we presented the roles of the buyer, implementor, mathematician user, and theorem-prover. In particular, recall that the mathematician user presents to the theorem-prover a series of function definitions and conjectures. The theorem-prover is responsible for verifying that each definition meets the requirements of our definition principle and for constructing a proof of each conjecture (possibly relying on previously proved conjectures). While the mathematician user is of inestimable value in proposing definitions and conjectures that structure the problem, the burden of proof rests entirely on the theorem-prover.

In this chapter we apply the theory and proof techniques to establish the correctness of an algorithm for converting expressions such as

(PLUS (QUOTIENT X 2) (TIMES X (TIMES Y Z)))

into a sequence of instructions, such as might be executed by a digital computer to compute the value of the expression in an environment in which X, Y, and Z have numeric values. The algorithm is a very simple example of an important class of algorithms known as "compilers." A compiler is just a translator from a "high-level" programming language to a "machine" language. In comparison to most compilers, our example is a toy.[31] In particular, we are dealing with only a part of the compiler problem: the compilation of expressions. We do not address the compilation of conditionals, assignments, jumps, and other control primitives.

We will here exploit the observation by McCarthy [32] that any program can be translated into a general recursive function with the same input/output behavior. Because programs can obtain information other than that "passed" explicitly to them in the form of arguments (e.g., they can refer to "global variables" or arbitrary memory locations), and because they can pass on information in ways other than by simply returning an answer (e.g., by modifying global variables or memory locations), it is necessary to understand "input/output" in a general sense. In general, the recursive function in question must map from the machine state at the beginning of the program's execution to the machine state at the end.

We will ultimately exhibit a recursive function that "is" the compiler and another one that "is" the machine upon which we expect to "run" the compiled code, and we shall prove that the state of this machine is correctly transformed by running the compiled code.

## A. INFORMAL DEVELOPMENT

In this section we illustrate the basic idea of an "expression compiler" by describing how one might use a certain pocket calculator to evaluate arithmetic expressions.

Suppose we have an expression such as

$(X \div 2) + (X * (Y * Z))$

composed entirely of variables, natural numbers, and numerically valued function symbols of two arguments.

Further, suppose that we want to evaluate the expression for given values of the variables, using a rudimentary hand-held calculator of the following description:

Your new Model 0.0 is illustrated in Fig. 1.

---

[31] It is useful to keep in mind other toys: dime-store gyroscopes, kites, building blocks, crayons, frictionless planes, ideal gases.

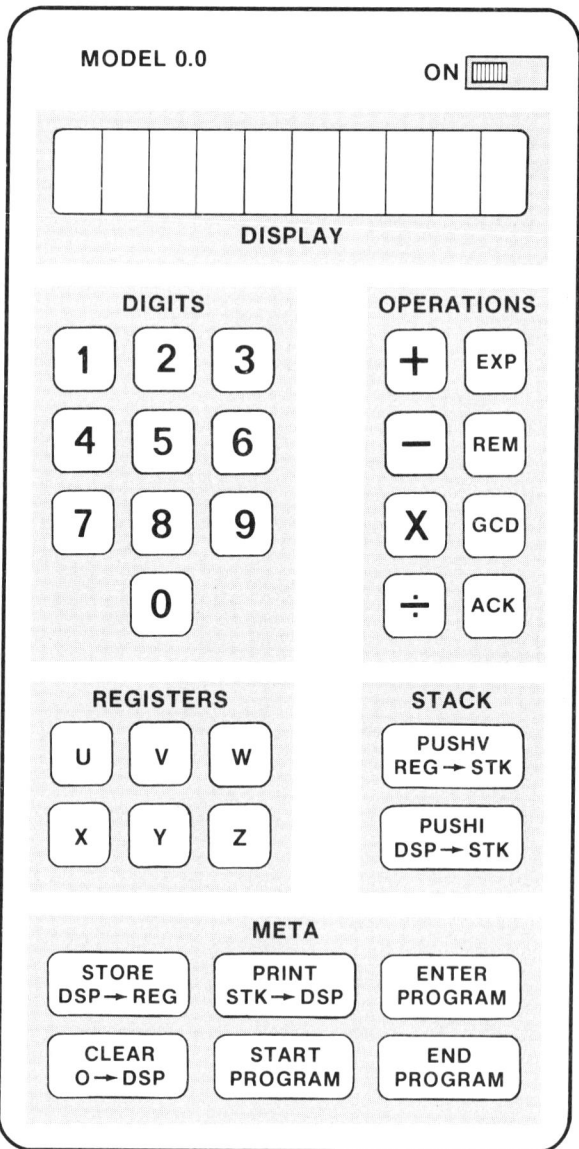

**Fig. 1.** The Model 0.0.

Integers may be entered into the display by pressing the digit buttons. The display may be set to 0 by depressing the CLEAR button. The contents of the display may be stored in any of the registers U, V, W, X, Y, and Z by first pressing STORE and then pressing the register key. The Model 0.0 has a stack. If any of the operation buttons (+, −, *, ÷, EXP, REM, GCD, ACK) is pressed, the top two occupants of the stack are removed from the stack, the indicated operation is performed on them, and the result is pushed on the stack. If a register button is pressed and then the PUSHV button is pressed, the contents of the register is pushed on the stack. If the PUSHI button is pressed, the contents of the display is pushed.

The Model 0.0 is also programmable. If the ENTER button is pressed, then until the END button is pressed, the intervening digit, operation, register, and stack key strokes are executed as described above and furthermore the key strokes are remembered in a program memory. Pressing the START button causes the most recently remembered sequence of key strokes to be reexecuted.

How do we use the Model 0.0 to compute the value of our expression,

$$(X \div 2) + (X * (Y * Z)),$$

assuming that we have already loaded the Model 0.0's registers with the appropriate values of the variables?

We first compute the value of X÷2 and push it on the stack. We then compute X*(Y*X) and push that on the stack on top of our previous result. Then we press the + key, which causes the Model 0.0 to pop two things off the stack, add them, and put the result back on the stack. Thus, we use a sequence of Model 0.0 instructions like this:

.
.
.  } (compute and push the value of X÷2)
.
.

.
.  } (compute and push the value of X*(Y*Z))
.
.

+ (press the + key)

## XVII. THE CORRECTNESS OF A SIMPLE OPTIMIZING EXPRESSION COMPILER

But to compute the value of $X \div 2$ we proceed as follows. We first press the X and PUSHV buttons to push the value of X on the stack. Next, we put 2 into the display and press the PUSHI button to push 2 on the stack. Finally, we press the $\div$ button to remove the top two numbers from the stack, compute their quotient, and push the result.

By performing the analogous decomposition of $X*(Y*Z)$, we arrive at the following sequence of key strokes for causing the Model 0.0 to compute the value of the given expression:

```
X
PUSHV
2
PUSHI
÷
X
PUSHV
Y
PUSHV
Z
PUSHV
*
*
+
```

The above sequence of instructions is a Model 0.0 "program" for computing the value of the expression. The process of translating the expression into such a sequence is called "compiling" the expression.

We are interested in defining an algorithm that will correctly compile any such expression. That is, we want an algorithm that when applied to an expression in our "high-level" language of expressions produces a sequence of instructions whose execution causes our calculator to compute and push the value of the expression (as defined in the usual mathematical sense with respect to the assignment reflecting the state of the calculator's registers). We do not care how much of the stack the calculator uses during its computation. However, we do not want the computation to disturb the entries present on the stack when the computation begins (we might have several important subtotals saved there).

If the calculator is very slow or if we wish to evaluate the expression repeatedly under different assignments to the variables, it might be useful to optimize the expression before compiling it. Optimization is allowed by the specification sketched, so long as the calculator ends up pushing the correct final value.

The compiler we will describe implements a very simple optimization scheme called "constant folding." The basic idea is that if asked to compile a subexpression with constant arguments, we compute and "compile in" the value of that subexpression. Thus, given

```
(2*3)+X,
```

we generate the following Model 0.0 code,

```
6
PUSHI
X
PUSHV
+
```

rather than force the Model 0.0 to do the arithmetic with

```
2
PUSHI
3
PUSHI
*
X
PUSHV
+
```

## B. FORMAL SPECIFICATION OF THE PROBLEM

To specify an expression compiler formally, we must define the "high-level" language of expressions and we must define the behavior of our calculator. The calculator that we formalize is actually a generalization of the Model 0.0. In our formalization, we do not limit the calculator to have a finite stack, finite precision, a finite number of operations, or a finite number of registers.

### 1. The High-Level Language

We need to specify how we shall represent expressions and how we define the value of an expression. These are fairly straightforward problems and we will dispense with them quickly since we discussed similar problems in the TAUTOLOGY.CHECKER example, Chapter IV.

## a. Representing Expressions

We are interested in tree structured expressions composed of integers, variables, and applications of numerically valued dyadic function symbols. An example expression is (PLUS (QUOTIENT X 2) (TIMES X (TIMES Y Z))). In the tautology-checker example, we employed a special shell to represent terms. For variety, we here employ the list representation mentioned in Chapter III.

We define the recursive function FORMP (see Appendix A) to return T if its argument is a form suitable for compiling, and F otherwise. We will consider all non-LISTP (i.e., "atomic") objects to be forms. Those that satisfy NUMBERP will stand for numbers and the remaining atomic objects will represent variables. We will consider a LISTP object a form if and only if it has the structure (CONS fn (CONS x (CONS y tail))), where fn is atomic, and x and y are (recursively) forms. We think of such a list as representing the term (fn x y). We do not care what tail is because it will be ignored by the value assignment function. Thus, if x is a LISTP object and also a form, then (CAR x) is the function symbol, (CADR x) is the form representing the first argument, and (CADDR x) is the form representing the second.

By assuming that a form is tree structured, we are glossing over the important problem of parsing, i.e., converting a sequence of symbols into a tree structure. In [20], Gloess describes a proof by our theorem-prover of the correctness of a simple expression parser for a language with infix operations.

## b. The Value of Forms

The value of a form is defined by the function EVAL (of two arguments, the form and an "environment," envrn, specifying the values of the variables). The value of a number is itself. The value of a variable, x, is defined to be (GETVALUE x envrn). The value of a form representing (fn u v) is defined to be (APPLY fn u' v'), where u' and v' are the recursively obtained values of u and v in envrn. The formal definition of EVAL is exhibited in Appendix A.

The two functions GETVALUE and APPLY are undefined.

GETVALUE is undefined because we do not care what the structure of an "environment" is, so long as every variable has some value "in" it. If the reader would care to think of GETVALUE as the function AS-SIGNMENT is Chapter IV and "environments" as association lists, that is acceptable.[32]

---

[32] In the tautology example, we specified the structure of assignments because we wanted to be able to write functions for constructing assignments with specific properties.

## B. FORMAL SPECIFICATION OF THE PROBLEM / 259

As for `APPLY`, it supposedly takes an arbitrary function symbol and two arguments and returns the result of applying the function "denoted" by that symbol to those arguments. But we do not really care what the function "denoted" by any particular symbol is (so long as it is numerically valued).

To ensure that the function symbols denote numeric functions, we add the axiom

**Axiom** NUMBERP.APPLY:

(NUMBERP (APPLY FN X Y)),

which is consistent since `APPLY` is undefined (and until now unmentioned in the theory).

If we wanted to ensure that the function symbol "PLUS" appearing in a `FORMP` was to stand for the recursive function we call PLUS, we could add the axiom

(EQUAL (APPLY "PLUS" X Y) (PLUS X Y)),

or define `APPLY` to recognize some fixed set of function symbols (including "PLUS") and call the numeric function we had in mind. However, for the present purposes we do not need to specify the functions denoted by our symbols.

### 2. The Low-Level Language

We now specify our calculator. The calculator is an idealized version of the Model 0.0. The development of the specification is broken up into two parts, the representation of data within the calculator and the specification of the machine's "fetch and execute" cycle (i.e., how it interprets sequences of instructions).

*a. Representing Data*

i. The Named Registers

We represent the state of the registers as an "environment." The meaning of the "contents of the register named x in the environment envrn" is (GETVALUE x envrn).

ii. Push-Down Stacks

To represent the state of the push-down stack, we use the objects in a new shell class:

## Shell Definition

Add the shell PUSH of two arguments
with recognizer STACKP,
accessors TOP and POP,
default values 0 and 0, and
well-founded relation TOP.POPP.

The stack resulting from pushing x onto pds will be (PUSH x pds). The topmost element of pds is (TOP pds), which is 0 if pds is not the result of a PUSH. The stack resulting from "popping" the topmost element from pds is (POP pds). If called on a nonstack, POP returns 0. We could have introduced an "empty stack" and required that the second argument to PUSH always be a stack; however we did not (simply because there is no need for any additional constraints on our stacks).[33]

iii. The Instruction Set

Since the PUSHV button on the Model 0.0 is always used in combination with a register button, we treat the two key strokes as one instruction. We use (CONS "PUSHV" (CONS reg "NIL")) as the instruction to push the value of register reg.

Similarly, we use (CONS "PUSHI" (CONS n "NIL")) as the instruction to push n.

We use atomic (i.e., NLISTP) objects (e.g., "PLUS" and "TIMES") as the instructions corresponding to the Model 0.0's operation buttons. An atom fn instructs the calculator to pop x and y off the stack and push (APPLY fn x y).

iv. Programs

A program is a list of instructions.

### b. The Fetch-and-Execute Cycle

We will describe the behavior of our calculator with a recursive function that maps from an initial state of the machine to a final state. Formally, the state of the calculator at any moment is a triple consisting of the location in the program from which the next instruction will be fetched (usually called the "program counter" or "pc"), the current state of the push-down stack, and the current settings of all the named registers.

---

[33] We could have used lists to represent stacks. CONS would be PUSH, LISTP would be STACKP, and CAR and CDR would be TOP and POP. We used the shell mechanism for variety.

## B. FORMAL SPECIFICATION OF THE PROBLEM / 261

The recursive function EXEC, describing the calculator, thus takes three arguments:

| | |
|---|---|
| PC | the current list of instructions, the first of which is the "next" to be executed, |
| PDS | the state of the push-down stack, |
| ENVRN | the state of the named registers. |

The function returns the state of the push-down stack at the conclusion of the sequence of instructions. In the final state, the PC will always be the end of the initial list of instructions, and the environment will be the original one since no programmable instruction affects the registers.[34] Thus, returning only the final push-down stack is sufficient.

We want EXEC to iterate down the list of instructions, interpreting each instruction on the way. Each instruction causes some modification to PDS. When the instruction is a (PUSHI x), EXEC should reset PDS to (PUSH x PDS). When it is a (PUSHV x) instruction, EXEC should reset PDS to (PUSH (GETVALUE x ENVRN) PDS). When the instruction is atomic (i.e., an operation button), fn, EXEC should (a) pop the stack once obtaining some value x (i.e., (TOP PDS)), (b) pop the stack again, obtaining some value y (i.e., (TOP (POP PDS))), (c) apply the indicated function, obtaining some value v (i.e., (APPLY fn y x) —note that the first argument to the function is the one that was deepest on the stack), and (d) push v on the stack (i.e., set PDS to (PUSH v (POP (POP PDS)))).

The way EXEC "iterates" down the PC is to recurse, each time replacing PC by (CDR PC), PDS by the stack produced by interpreting (CAR PC) as above and leaving ENVRN unchanged. When PC is no longer a list, EXEC returns PDS.

The formal definition of EXEC is thus

**Definition**

```
(EXEC PC PDS ENVRN)
    =
(IF (NLISTP PC)
    PDS
    (IF (LISTP (CAR PC))
        (IF (EQUAL (CAR (CAR PC)) "PUSHI")
            (EXEC (CDR PC)
                  (PUSH (CAR (CDR (CAR PC))) PDS)
                  ENVRN)
```

---

[34] The registers of our calculator may be set arbitrarily by the user, but they must be set before a program is activated.

```
       (EXEC (CDR PC)
             (PUSH (GETVALUE (CAR (CDR (CAR PC)))
                             ENVRN)
                   PDS)
             ENVRN))
       (EXEC (CDR PC)
             (PUSH (APPLY (CAR PC)
                          (TOP (POP PDS))
                          (TOP PDS))
                   (POP (POP PDS)))
             ENVRN))).
```

Let us step back from the problem of evaluating expressions to consider the problem of designing a calculator to meet these specifications. The designer is free to represent data in the machine any way he sees fit. For example, if the hardware factory is having trouble producing blue 1-tuples, the designer may choose to implement our NUMBERPs as sequences of binary digits. The designer would then wire the "PLUS" key to a binary adddition algorithm rather than Peano's recursive function. If he desired to *prove* that his design met these specifications, he would have to establish the correctness of his algorithms with respect to his representation. For example, he would have to prove that when given the binary representation of two Peano numbers his addition algorithm yields the binary representation of their Peano sum. The reader is referred to the theorem CORRECTNESS.OF.BIG.PLUS in Appendix A, which states the above result formally for an algorithm that adds numbers represented as sequences of digits in an arbitrary base (binary addition being a special case of the more general digit-by-digit algorithm with carry).

Readers interested in a method for designing machines and implementing them on other machines are referred to Robinson and Levitt [51].

### 3. The Formal Statement of Correctness

Suppose that COMPILE is a function of one argument, namely, a form to be compiled. Then, to be correct, COMPILE must have the following property. If X is a FORMP, then the push-down stack resulting from executing the output of (COMPILE X) on our calculator, with some initial push-down stack PDS and some environment ENVRN, is the push-down stack obtained by pushing (EVAL X ENVRN) on PDS. That is, when the compilation and execution have concluded, the calculator and COMPILE are completely out of the picture: the mathematical value of the expression has been PUSHed on the original stack.

Stated formally this is

**Theorem** CORRECTNESS.OF.OPTIMIZING.COMPILER:

```
(IMPLIES (FORMP X)
         (EQUAL (EXEC (COMPILE X) PDS ENVRN)
                (PUSH (EVAL X ENVRN) PDS))).
```

Let us return for a moment to the fact that APPLY and GETVALUE are undefined (but used by EVAL and EXEC). The job of the compiler is to cause EXEC to compute the same thing EVAL does, regardless of the semantics of GETVALUE and the numeric APPLY.

## C. FORMAL DEFINITION OF THE COMPILER

Now that we know exactly what the semantics of forms are, and we know exactly how the calculator behaves on a sequence of instructions, we consider the problem of compiling forms for the calculator. As noted in our informal development of the problem, we wish to do "constant folding" optimization of the expression. Thus, we break our compiler into two "passes." The first pass, performed by the function OPTIMIZE, takes the expression to be compiled and returns a possibly simpler form with the same value under all assignments. OPTIMIZE replaces constant subexpressions with their values. The second pass of the compiler, CODEGEN, generates the compiled code for the optimized form. We compose the two to obtain our compiler.

### 1. The Optimization Pass

The function OPTIMIZE, of one argument, optimizes a form by replacing any subexpression, (fn u v), where u and v are specific numeric constants, by the result of applying fn to u and v. The precise (but informal) definition of (OPTIMIZE X) is as follows:

If X does not represent a function call, return X.

Otherwise, suppose X represents (fn u v).
Let u' be (OPTIMIZE u) and let v' be (OPTIMIZE v).
If (NUMBERP u') and (NUMBERP v'),
   then return (APPLY fn u' v').

Otherwise,
> return (CONS fn (CONS u' (CONS v' "NIL")))
> (i.e., return a representation of the
> term (fn u' v')).

The formal definition of OPTIMIZE is in Appendix A.

## 2. The Code Generation Pass

Given the ability to optimize an expression as above, we now turn our attention to generating correct code for forms in general (i.e., we ignore the fact that we know the code generator will be called on optimized forms).

Our objective is to generate code for a form x that will cause the calculator to push the value of x on the stack. Let us suppose we are compiling x by hand, writing down the instructions we generate on a note pad. Clearly, if x is a variable, we write down (PUSHV x). If x is a number, we write down (PUSHI x). Finally, if x represents (fn u v), we recursively write down the instructions for pushing the value of u on the stack, then write down the instructions for pushing the value of v, and finally write down the single atomic instruction fn. When the calculator executes the code for u, it may push and pop many times, but it never pops the initial stack. When the code for u has been executed, the net effect will be to push the value of u on the initial stack. Then the calculator will begin to execute the code for v with the value of u safely pushed. The v computation may push and pop many times but eventually it will push the value of v immediately on top of the value for u. Thus, by the time the calculator sees the atomic instruction fn, the two argument values will be the topmost entries on the stack.

To formalize the code generation pass, we must decide how to represent our "note pad" above. We use a global collection site, maintained as a list of instructions, and initialized to "NIL". To "write down" an instruction on the "note pad," we CONS it onto the front of the site. The final value of the site will thus be in the reverse of the order in which the instructions should be executed (e.g., the topmost function symbol in the expression will be the first element of the final list, but should be the last instruction executed). Thus, the list should be reversed before being used as a program.

Since this algorithm side-effects a global variable (the collection site), its functional representation is as a function of two arguments:

## C. FORMAL DEFINITION OF THE COMPILER / 265

the form being compiled, FORM, and the global collection site, INS. The function returns the value of the collection site.[35] The formal definition is

**Definition**

```
(CODEGEN FORM INS)
    =
(IF (NUMBERP FORM)
    (CONS (CONS "PUSHI" (CONS FORM "NIL"))
          INS)
    (IF (LISTP (CDDR FORM))
        (CONS (CAR FORM)
              (CODEGEN (CADDR FORM)
                       (CODEGEN (CADR FORM) INS)))
        (CONS (CONS "PUSHV" (CONS FORM "NIL"))
              INS))).
```

If FORM is a FORMP, then the test (LISTP (CDDR FORM)) is a cheap way to ask whether FORM has the structure (CONS fn (CONS u (CONS v tail))). Note that we compile the first argument of function calls first (in the innermost recursive call) using the input value of the collection site. Then we compile the second argument, using the collection site resulting from the first recursive call. Finally, we CONS the function symbol onto the front and return the resulting collection site.

### 3. Compile

The compiler is the composition of OPTIMIZE and CODEGEN. We must remember to initialize the collection site to "NIL" before starting the code generation pass, and we must remember to reverse the final list of instructions.

The formal definition of the compiler is

**Definition**

```
        (COMPILE FORM)
            =
        (REVERSE (CODEGEN (OPTIMIZE FORM) "NIL")).
```

---

[35] Note the similarity to the function MC.FLATTEN of Chapter II.

## D. THE MECHANICAL PROOF OF CORRECTNESS

In this section, we describe the proof of the correctness of the optimizing compiler. The presentation is divided into three parts. We first decompose the main theorem into three lemmas to be proved. Then we prove the main theorem from those lemmas. Finally we present the proofs of the lemmas.

### 1. Decomposition of the Main Goal

The proof of the correctness of our optimizing compiler can be naturally decomposed into two parts: show that the optimizer is correct and show that the code generator is correct.

The correctness of the optimizer requires two clauses to state: if given a form, OPTIMIZE returns a form, and if given a form, OPTIMIZE returns something with the same value.

These two lemmas are stated formally as

**Theorem** FORMP.OPTIMIZE:

(IMPLIES (FORMP X)
        (FORMP (OPTIMIZE X))),

**Theorem** CORRECTNESS.OF.OPTIMIZE:

(IMPLIES (FORMP X)
        (EQUAL (EVAL (OPTIMIZE X) ENVRN)
              (EVAL X ENVRN))).

To state the correctness of CODEGEN, we must explain the use of the global "collection site" and recall that the output of CODEGEN will be reversed. Exactly what is the state into which our calculator should be driven if given the program produced by (REVERSE (CODEGEN X INS)) with some initial PDS and ENVRN? Since CODEGEN is supposed to concatenate the reverse of the code for X onto INS, and since that concatenation is reversed before running it, the code in (REVERSE INS) will be executed before the code for X is encountered. When the code for X is encountered, the push-down stack will be that produced by executing the instructions in (REVERSE INS) on PDS and ENVRN. The code for X is supposed to push the value of X on that push-down stack. Thus, the statement of the correctness of CODEGEN is

## D. THE MECHANICAL PROOF OF CORRECTNESS / 267

**Theorem** CORRECTNESS.OF.CODEGEN:

```
(IMPLIES (FORMP X)
         (EQUAL (EXEC (REVERSE (CODEGEN X
                                       INS))
                      PDS ENVRN)
                (PUSH (EVAL X ENVRN)
                      (EXEC (REVERSE INS)
                            PDS ENVRN)))).
```

### 2. Proof of the Main Goal

For the moment, let us suppose we have proved these three lemmas. Now let us consider the theorem-prover's proof of the main result:

**Theorem** CORRECTNESS.OF.OPTIMIZING.COMPILER:

```
(IMPLIES (FORMP X)
         (EQUAL (EXEC (COMPILE X) PDS ENVRN)
                (PUSH (EVAL X ENVRN) PDS))).
```

The proof from our three lemmas is immediate (i.e., involves only simplification):

Consider the left-hand side of the conclusion of the main goal:

```
(EXEC (COMPILE X) PDS ENVRN).
```

Since COMPILE is nonrecursive, we expand it to get

```
(EXEC (REVERSE (CODEGEN (OPTIMIZE X) "NIL"))
      PDS ENVRN).
```

We then rewrite this using CORRECTNESS.OF.CODEGEN after establishing the hypothesis (FORMP (OPTIMIZE X)) by backwards chaining through FORMP.OPTIMIZE and appealing to the (FORMP X) hypothesis in our main goal. The result is

```
(PUSH (EVAL (OPTIMIZE X) ENVRN)
      (EXEC (REVERSE "NIL") PDS ENVRN)).
```

However, this can be further simplified. We may rewrite the term (EVAL (OPTIMIZE X) ENVRN) to (EVAL X ENVRN) by CORRECT-NESS.OF.OPTIMIZE, appealing again to the assumption (FORMP X) to relieve the hypothesis of the rewrite rule. In addition, (REVERSE "NIL") computes to "NIL", and the resulting (EXEC "NIL" PDS ENVRN) then computes to PDS. The result of simplifying the left-hand side of the conclusion of our main goal is thus

    (PUSH (EVAL X ENVRN) PDS),

which is the right-hand side of the conclusion. Thus, the main goal has been proved.

## 3. Proofs of the Lemmas

Let us now discuss the proofs of the three lemmas used.

The first two lemmas, FORMP.OPTIMIZE and CORRECTNESS.-OF.OPTIMIZE, are proved by straightforward induction on the structure of forms, as unanimously suggested by all applicable induction templates. The various cases produced by the inductions are proved by simplification (using list axioms and function definitions), elimination of CARs and CDRs, and equality substitution.

The proof of CORRECTNESS.OF.CODEGEN is more interesting. In developing the statement of the correctness of CODEGEN, we used an important fact about our calculator: if called upon to execute the concatenation of two programs, X and Y, the state of the push-down stack when Y is encountered is that produced by executing X. This fact is important to the proof of the correctness of CODEGEN. Stated formally, the lemma is

**Theorem** SEQUENTIAL.EXECUTION:

    (EQUAL (EXEC (APPEND X Y) PDS ENVRN)
           (EXEC Y
                 (EXEC X PDS ENVRN)
                 ENVRN)).

SEQUENTIAL.EXECUTION can be proved by induction on X, as suggested by the result of merging the inductions from (APPEND X Y) and (EXEC X PDS ENVRN). Note that there is a suggested induction on Y; however, it is flawed by the (APPEND X Y) term. The proof of SEQUENTIAL.EXECUTION appeals only to axioms and definitions.

Once SEQUENTIAL.EXECUTION is proved, the CORRECTNESS.-OF.CODEGEN can be proved by induction on the structure of the form X. Note that an induction on INS is suggested by the (REVERSE INS) term, but that induction is flawed.

Below is the theorem-prover's proof of the correctness of CODEGEN. We have included only the induction analysis, the proof of the first base case, and the proof of the induction step. We have extensively annotated the theorem-prover's own output with remarks set in this typeface.

**Theorem** CORRECTNESS.OF.CODEGEN:
```
(IMPLIES
      (FORMP X)
      (EQUAL (EXEC (REVERSE (CODEGEN X INS))
                   PDS ENVRN)
             (PUSH (EVAL X ENVRN)
                   (EXEC (REVERSE INS) PDS ENVRN)))).
```
Name the conjecture *1.

We will appeal to induction. There are four plausible inductions. They merge into two likely candidate inductions. However, only one is unflawed. We will induct according to the following scheme:
```
(AND (IMPLIES (NOT (LISTP X))
              (p X INS PDS ENVRN))
     (IMPLIES (NOT (LISTP (CDR X)))
              (p X INS PDS ENVRN))
     (IMPLIES (NOT (LISTP (CDR (CDR X))))
              (p X INS PDS ENVRN))
     (IMPLIES (AND (LISTP X)
                   (LISTP (CDR X))
                   (LISTP (CDR (CDR X)))
                   (p (CAR (CDR (CDR X)))
                      (CODEGEN (CAR (CDR X)) INS)
                      PDS ENVRN)
                   (p (CAR (CDR X)) INS PDS ENVRN))
              (p X INS PDS ENVRN))).
```
The inequalities CAR.LESSP and CDR.LESSP establish that the measure (COUNT X) decreases according to the well-founded relation LESSP in the induction step of the scheme. Note, however, the inductive instances chosen for INS. The above induction scheme produces the following seven new goals:

*Case 1.*
```
(IMPLIES
     (AND (NOT (LISTP X)) (FORMP X))
     (EQUAL (EXEC (REVERSE (CODEGEN X INS))
                  PDS ENVRN)
            (PUSH (EVAL X ENVRN)
                  (EXEC (REVERSE INS) PDS ENVRN)))),
```

## 270 / XVII. THE CORRECTNESS OF A SIMPLE OPTIMIZING EXPRESSION COMPILER

which simplifies, applying CDR.NLISTP, CAR.CONS, CDR.CONS and SEQUENTIAL.EXECUTION, and opening up the definitions of FORMP, CODEGEN, EVAL, REVERSE and EXEC, to:

(TRUE).

Case 1 is the most interesting base case. We now sketch the series of simplifications that reduce it to (TRUE).

Consider the left-hand side of the conclusion:

(EXEC (REVERSE (CODEGEN X INS)) PDS ENVRN).

Since X is not a list, it represents a variable or a number. The CODE-GEN term in the left-hand side of the conclusion thus opens up to

```
(IF (NUMBERP X)
    (CONS (CONS "PUSHI" (CONS X "NIL"))
          INS)
    (CONS (CONS "PUSHV" (CONS X "NIL"))
          INS)).
```

The introduction of the above IF-expression splits the theorem into two cases. In one case, the REVERSE expression in the left-hand side of the conclusion has (CONS push INS) as its argument, where push is a (PUSHI X) instruction and X is known to be numeric. In the other case, the REVERSE expression has (CONS push INS) as its argument, where push is a (PUSHV X) instruction and X is known to be not a number.

In both cases, the REVERSE expression (which occupies the first argument of the EXEC-expression in the left-hand side of the conclusion) opens up to

(APPEND (REVERSE INS) (CONS push "NIL")),

allowing the SEQUENTIAL.EXECUTION lemma to rewrite the resulting EXEC-expression to

```
(EXEC (CONS push "NIL")
      (EXEC (REVERSE INS) PDS ENVRN)
      ENVRN).
```

The outermost EXEC-expression above then opens up, and indeed computes, to either:

(PUSH X (EXEC (REVERSE INS) PDS ENVRN)),

or to

    (PUSH (GETVALUE X ENVRN)
        (EXEC (REVERSE INS) PDS ENVRN))

depending on whether we are in the numeric or nonnumeric case. That completes the simplification of the left-hand side of the conclusion.

Now consider the right-hand side of the conclusion. The term (EVAL X ENVRN) opens up to

    (IF (NUMBERP X)
        X
        (GETVALUE X ENVRN)).

Thus, the right-hand side of the conclusion becomes

    (PUSH (IF (NUMBERP X)
           X
           (GETVALUE X ENVRN))
        (EXEC (REVERSE INS) PDS ENVRN)).

When the IF is distributed out, two cases are produced, depending on whether X is numeric:

    (PUSH X (EXEC (REVERSE INS) PDS ENVRN))

and

    (PUSH (GETVALUE X ENVRN)
        (EXEC (REVERSE INS) PDS ENVRN)).

That completes the simplification of the right-hand side of the conclusion.

Now consider the concluding equality itself. In the case where X is a number, the left- and right-hand sides simplified to the same term, namely, (PUSH X (EXEC (REVERSE INS) PDS ENVRN)). In the other case, when X is not a number, both sides simplified to the same term, namely, (PUSH (GETVALUE X ENVRN) (EXEC (REVERSE INS) PDS ENVRN)). So Case 1 above has been proved.

The proofs of Cases 2 through 6 are similar and are thus not exhibited here. We resume the system's proof on Case 7, the induction step.

.
.
.

*Case 7.*
```
(IMPLIES
 (AND
  (LISTP X)
  (LISTP (CDR X))
  (LISTP (CDR (CDR X)))
  (EQUAL
   (EXEC
     (REVERSE (CODEGEN (CAR (CDR (CDR X)))
                       (CODEGEN (CAR (CDR X)) INS)))
     PDS ENVRN)
   (PUSH (EVAL (CAR (CDR (CDR X))) ENVRN)
         (EXEC (REVERSE (CODEGEN (CAR (CDR X)) INS))
               PDS ENVRN)))
  (EQUAL (EXEC (REVERSE (CODEGEN (CAR (CDR X)) INS))
               PDS ENVRN)
         (PUSH (EVAL (CAR (CDR X)) ENVRN)
               (EXEC (REVERSE INS) PDS ENVRN)))
  (FORMP X))
 (EQUAL (EXEC (REVERSE (CODEGEN X INS))
              PDS ENVRN)
        (PUSH (EVAL X ENVRN)
              (EXEC (REVERSE INS) PDS ENVRN)))),
```
which we simplify, applying CAR.CONS, CDR.CONS, SE-QUENTIAL.EXECUTION and PUSH.EQUAL, and expanding the definitions of FORMP, CODEGEN, REVERSE, EXEC and EVAL, to two new formulas:

The conditions governing Case 7 imply that X represents a function call. Let us suppose it represents the function call (v d z), where v is an atomic symbol and d and z are forms (we have used the letters that the system will introduce in a moment when it eliminates the CARs and CDRs).

We have two induction hypotheses. The first is about the code produced for z after d has been compiled. The second is about the code produced for d. In particular, the first hypothesis tells us that the execution of the code for z pushes the value of z on the stack that is produced by the execution of the code for d. The second hypothesis tells us that the execution of the code for d pushes the value of d on the stack produced by the execution of whatever instructions were on INS to start with.

Now let us look at the left-hand side of the conclusion:

```
(EXEC (REVERSE (CODEGEN X INS))
      PDS ENVRN).
```

The CODEGEN term opens up to

```
(CONS v (CODEGEN z (CODEGEN d INS))).
```

Thus the REVERSE term opens up to

```
(APPEND (REVERSE (CODEGEN z (CODEGEN d INS)))
        (CONS v "NIL")).
```

This allows the SEQUENTIAL.EXECUTION lemma to rewrite the resulting EXEC term to

```
(EXEC (CONS v "NIL") pds ENVRN),
```

where we have used pds to denote the term

```
(EXEC (REVERSE (CODEGEN z (CODEGEN d INS)))
      PDS ENVRN).
```

(EXEC (CONS v "NIL") pds ENVRN) computes to

```
(PUSH (APPLY v (TOP (POP pds)) (TOP pds))
      (POP (POP pds))).
```

There is nothing more we can do to the left-hand side of the conclusion.

However, on the right-hand side of the conclusion, the (EVAL X ENVRN) term opens up to (APPLY v (EVAL d ENVRN) (EVAL z ENVRN)), reducing the right-hand side to

```
(PUSH (APPLY v (EVAL d ENVRN) (EVAL z ENVRN))
      (EXEC (REVERSE INS) PDS ENVRN)).
```

Since now the left- and right-hand sides of the conclusion are PUSH-expressions, we can apply one of the theorems added by the shell principle when PUSH was defined,

**Theorem** PUSH.EQUAL:

```
(EQUAL (EQUAL (PUSH X1 X2) (PUSH Y1 Y2))
       (AND (EQUAL X1 Y1)
            (EQUAL X2 Y2))),
```

to split the conjecture into two parts. In the first (called Case 1 below), we must prove that the two APPLY expressions produced by the left- and right-hand sides are equal, and in the second (called Case 2), we must prove that the stacks upon which they are pushed are equal.

## 274 / XVII. THE CORRECTNESS OF A SIMPLE OPTIMIZING EXPRESSION COMPILER

The proofs are similar, both involving appeals to the induction hypotheses. We follow the proof of the equivalence of the APPLY expressions.

*Case 1.*

```
(IMPLIES
 (AND
  (LISTP X)
  (LISTP (CDR X))
  (LISTP (CDR (CDR X)))
  (EQUAL
   (EXEC
    (REVERSE
           (CODEGEN (CAR (CDR (CDR X)))
                    (CODEGEN (CAR (CDR X)) INS)))
    PDS ENVRN)
   (PUSH
       (EVAL (CAR (CDR (CDR X))) ENVRN)
       (EXEC (REVERSE (CODEGEN (CAR (CDR X)) INS))
             PDS ENVRN)))
  (EQUAL
       (EXEC (REVERSE (CODEGEN (CAR (CDR X)) INS))
             PDS ENVRN)
       (PUSH (EVAL (CAR (CDR X)) ENVRN)
             (EXEC (REVERSE INS) PDS ENVRN)))
  (NOT (LISTP (CAR X)))
  (FORMP (CAR (CDR X)))
  (FORMP (CAR (CDR (CDR X)))))
 (EQUAL
  (APPLY
   (CAR X)
   (TOP
    (POP
     (EXEC
      (REVERSE
             (CODEGEN (CAR (CDR (CDR X)))
                      (CODEGEN (CAR (CDR X)) INS)))
      PDS ENVRN)))
   (TOP
    (EXEC
     (REVERSE
```

```
                    (CODEGEN (CAR (CDR (CDR X)))
                    (CODEGEN (CAR (CDR X)) INS)))
     PDS ENVRN)))
  (APPLY (CAR X)
         (EVAL (CAR (CDR X)) ENVRN)
         (EVAL (CAR (CDR (CDR X))) ENVRN)))).
```
Applying the lemma CAR/CDR.ELIM, we now replace X by
(CONS V Z) to eliminate (CDR X) and (CAR X), Z by
(CONS D W) to eliminate (CDR Z) and (CAR Z) and W by
(CONS Z C) to eliminate (CAR W) and (CDR W). This
produces the new conjecture:
```
(IMPLIES
 (AND
  (LISTP (CONS V (CONS D (CONS Z C))))
  (LISTP (CONS D (CONS Z C)))
  (LISTP (CONS Z C))
  (EQUAL
       (EXEC (REVERSE (CODEGEN Z (CODEGEN D INS)))
             PDS ENVRN)
       (PUSH (EVAL Z ENVRN)
             (EXEC (REVERSE (CODEGEN D INS))
                   PDS ENVRN)))
  (EQUAL (EXEC (REVERSE (CODEGEN D INS))
               PDS ENVRN)
         (PUSH (EVAL D ENVRN)
               (EXEC (REVERSE INS) PDS ENVRN)))
  (NOT (LISTP V))
  (FORMP D)
  (FORMP Z))
 (EQUAL
  (APPLY V
   (TOP
    (POP
      (EXEC (REVERSE (CODEGEN Z (CODEGEN D INS)))
            PDS ENVRN)))
   (TOP
      (EXEC (REVERSE (CODEGEN Z (CODEGEN D INS)))
            PDS ENVRN)))
  (APPLY V
         (EVAL D ENVRN)
         (EVAL Z ENVRN)))),
```

which further simplifies, clearly, to:
```
(IMPLIES
 (AND
  (EQUAL
       (EXEC (REVERSE (CODEGEN Z (CODEGEN D INS)))
             PDS ENVRN)
       (PUSH (EVAL Z ENVRN)
             (EXEC (REVERSE (CODEGEN D INS))
                   PDS ENVRN)))
  (EQUAL (EXEC (REVERSE (CODEGEN D INS))
               PDS ENVRN)
         (PUSH (EVAL D ENVRN)
               (EXEC (REVERSE INS) PDS ENVRN)))
  (NOT (LISTP V))
  (FORMP D)
  (FORMP Z))
 (EQUAL
  (APPLY V
   (TOP
    (POP
       (EXEC (REVERSE (CODEGEN Z (CODEGEN D INS)))
             PDS ENVRN)))
   (TOP
      (EXEC (REVERSE (CODEGEN Z (CODEGEN D INS)))
            PDS ENVRN)))
  (APPLY V
         (EVAL D ENVRN)
         (EVAL Z ENVRN)))).
```

Note that the first equality hypothesis tells us that the stack we called pds above (i.e., the result of executing INS followed by the code for D followed by the code for Z) is ( PUSH ( EVAL Z ENVRN) pds' ) , where pds' is the stack resulting from executing INS followed by the code for D. The hypothesis can be used by cross-fertilization because pds occurs in the conclusion.

We use the first equality hypothesis by cross-fertilizing:
```
        (PUSH (EVAL Z ENVRN)
              (EXEC (REVERSE (CODEGEN D INS))
                    PDS ENVRN))
for:
```

## D. THE MECHANICAL PROOF OF CORRECTNESS / 277

```
         (EXEC (REVERSE (CODEGEN Z (CODEGEN D INS)))
               PDS ENVRN)
```

and throwing away the equality. This generates:

```
(IMPLIES
 (AND
     (EQUAL (EXEC (REVERSE (CODEGEN D INS))
                  PDS ENVRN)
            (PUSH (EVAL D ENVRN)
                  (EXEC (REVERSE INS) PDS ENVRN)))
     (NOT (LISTP V))
     (FORMP D)
     (FORMP Z))
 (EQUAL
   (APPLY V
    (TOP (POP (PUSH (EVAL Z ENVRN)
                    (EXEC (REVERSE (CODEGEN D INS))
                          PDS ENVRN))))
    (TOP (PUSH (EVAL Z ENVRN)
               (EXEC (REVERSE (CODEGEN D INS))
                     PDS ENVRN))))
   (APPLY V
          (EVAL D ENVRN)
          (EVAL Z ENVRN)))).
```

Having substituted an explicit PUSH-expression for pds we can now use the axioms about (TOP (PUSH x y)) and (POP (PUSH x y)) to clean up the conjecture.

This further simplifies, applying POP.PUSH and TOP.PUSH, to:

```
(IMPLIES
 (AND
     (EQUAL (EXEC (REVERSE (CODEGEN D INS))
                  PDS ENVRN)
            (PUSH (EVAL D ENVRN)
                  (EXEC (REVERSE INS) PDS ENVRN)))
     (NOT (LISTP V))
     (FORMP D)
     (FORMP Z))
 (EQUAL
        (APPLY V
```

```
                (TOP (EXEC (REVERSE (CODEGEN D INS))
                           PDS ENVRN))
                (EVAL Z ENVRN))
      (APPLY V
             (EVAL D ENVRN)
             (EVAL Z ENVRN)))).
```

We now use the second induction hypothesis. It tells us that pds' (i.e., the result of executing INS followed by the code for D) is (PUSH (EVAL D ENVRN) pds''), where pds'' is the result of executing INS. We use it by cross-fertilizing the PUSH-expression for pds'.

```
We use the above equality hypothesis by
cross-fertilizing:

        (PUSH (EVAL D ENVRN)
              (EXEC (REVERSE INS) PDS ENVRN))
```

for (EXEC (REVERSE (CODEGEN D INS)) PDS ENVRN) and throwing away the equality. We would thus like to prove:

```
(IMPLIES
 (AND (NOT (LISTP V))
      (FORMP D)
      (FORMP Z))
 (EQUAL
  (APPLY V
         (TOP (PUSH (EVAL D ENVRN)
                    (EXEC (REVERSE INS) PDS ENVRN)))
         (EVAL Z ENVRN))
  (APPLY V
         (EVAL D ENVRN)
         (EVAL Z ENVRN)))),
```

Having substituted an explicit PUSH-expression for pds', we can use the TOP.PUSH axiom to reduce (TOP pds') to (EVAL D ENVRN). Note that after performing the rewrite, both sides of the concluding equality are identical.

```
which finally simplifies, using the lemma TOP.PUSH,
to:

        (TRUE).
```

*Case 2.*
.
.
.


Recall that the above is one of two cases split off the induction step. The second case is the analogous theorem about the stacks upon which the two APPLY-expressions are pushed. Its proof is exactly analogous to the one above and is thus not exhibited here.

.
.
.

```
  That finishes the proof of *1.  Q.E.D.
CPU time (devoted to theorem-proving): 40.03 seconds
```
As described, the above proof may seem complicated. But it should be noted that it was carried out entirely automatically.

Having proved all the lemmas in the decomposition of the main goal, and having proved the main goal from our lemmas, we have completed the proof of the correctness of our optimizing expression compiler.

## E. NOTES

In this section, we describe three bugs our theorem-prover uncovered in earlier versions of our optimizing expression compiler, and we present a brief history of the expression compiler problem.

### 1. Bugs Uncovered by the Theorem-Prover

In our original efforts to formalize this problem, we made three mistakes that did not come to light until the theorem-prover failed to prove the relevant lemmas and essentially exhibited counterexamples.

The reader might wonder where in the proofs the axiom

```
       (NUMBERP (APPLY FN X Y))
```

was used. The answer is that it is crucial in the proof of the correctness of OPTIMIZE. Recall that if the optimizer encounters a form representing (fn x y), where x and y are specific constants, then it replaces it by (APPLY fn x y). Consider what would happen if (APPLY "FN" 3 4) were "Z". Then the representation of the expression (FN 3 4) would be "optimized" to the representation of the expression Z. But these two expressions do not have the same value in all environments. In particular, the first expression has the value "Z", and the second expression has whatever value the variable Z is assigned in the environment. This problem does not arise if APPLY is known to be numeric since the value of a number is itself. We did not realize we had confined ourselves to numerically valued functions until the system, in trying to prove CORRECTNESS.OF.OPTIMIZE, without the numeric APPLY axiom, produced the goal

```
(IMPLIES (NOT (LISTP Z))
         (EQUAL Z (GETVALUE Z ENVRN))).
```

The second bug concerned the instruction set of the calculator and the legal function symbols in expressions. When we first formalized the problem, we did not require that the function symbols of expressions be atomic. We thought it did not matter what they were. However, consider the behavior of CODEGEN when the form being compiled represents ((PUSHI 7) X Y). The compiled code is

```
(PUSHV X)
(PUSHV Y)
(PUSHI 7)
```

This code pushes three things on the stack and halts. But the value of the representation of the expression ((PUSHI 7) X Y), defined by EVAL, is the result of applying the representation of (PUSHI 7) (whatever that is) to the values of the two arguments. Thus, in FORMP, we had to specify that function symbols were atomic; that fact was actually used several times in the foregoing proof of the correctness of CODEGEN, although we brushed over it in our description of the proof.

The third bug was in our definition of the calculator. Recall that when it encounters an atomic function symbol it pops x off the stack, then it pops y off the stack, and then it pushes (APPLY fn y x). In our original definition of the machine, we forgot that the first argument to fn would naturally have been pushed first (and thus be the second thing popped off). Thus, we specified that the machine push (APPLY fn x y). Our error showed up when we tried to prove the correctness of CODEGEN. If we reconstruct the proof of CODEGEN (described above) with the faulty version of EXEC (or CODEGEN, depending on your point of view), the goal that finally reduced to (TRUE) above would be

```
(IMPLIES (AND (NOT (LISTP V))
              (FORMP D)
              (FORMP Z))
         (EQUAL (APPLY V (EVAL Z ENVRN)
                         (EVAL D ENVRN))
                (APPLY V (EVAL D ENVRN)
                         (EVAL Z ENVRN)))).
```

We could then generalize the two EVAL-expressions and then eliminate the two FORMP hypotheses as irrelevant. The resulting goal would be

```
(IMPLIES (NOT (LISTP V))
         (EQUAL (APPLY V X Y)
                (APPLY V Y X))).
```

Since this contains no recursive functions, we would have eliminated it as irrelevant (i.e., falsifiable) and quit. Note that we thus reduced the correctness of CODEGEN to proving that all functions are commutative!

One final note about these bugs. Our interactive system allows functions to be evaluated so that after they are defined the user can test them. We tested our three different (and faulty) versions of the compiler and calculator on a variety of forms and failed to expose any bugs because we limited our tests to well-formed expressions involving only PLUS and TIMES (both of which are numeric and commutative).

## 2. The History of the Problem

To the best of our knowledge, our mechanical theorem-prover was the first to undertake the proof of the correctness of an *optimizing* expression compiler. However, the problem of the correctness of a function similar to our CODEGEN was first raised by McCarthy and Painter [36] in 1967. They proved, by hand, the correctness of an expression compiler for an idealized machine that contained addressable registers rather than a push-down stack like our calculator. Milner and Weyhrauch [37], in 1972, mechanically proof-checked a version of the McCarthy–Painter correctness proof with a considerable amount of user interaction. Cartwright [14], in 1977, produced a mechanical proof requiring somewhat less user assistance than the Milner–Weyhrauch proof. Aubin [2], in 1976, produced a mechanical proof exactly analogous to our proof of the correctness of CODEGEN, requiring no user help other than a single lemma similar to our SE-QUENTIAL.EXECUTION lemma.

# XVIII

## The Correctness of a Fast String Searching Algorithm

Both the tautology-checker and the compiler examples dealt with recursive algorithms for handling tree structured data. We here apply the theory and proof techniques to establish the correctness of an iterative program processing linear, indexed arrays. In particular, we prove the correctness of a program for finding the first occurrence of one character string in another. String searching algorithms can be easily written; the problem becomes more difficult and more interesting, however, if one considers implementing an efficient algorithm. We prove the correctness of one of the fastest known ways to solve the string searching problem.

In further contrast to the compiler example, where we used the functional approach to program semantics, we here attach meaning to our program in another way. The method we use is called the "inductive assertion" method (see Floyd [18] and also Naur [44], Hoare [24], and Manna and Pnueli [30]). We will explain the method when we use it. In fact, the primary intent of this chapter is to demonstrate, with a realistic example, that a theory based on recursive functions (rather than quantification) may be profitably used to specify programs by the inductive assertion method.

The structure of this chapter is as follows. We first describe the string searching problem informally and derive and explain a very efficient string searching algorithm. Then we formally specify the string searching problem and, using the inductive assertion method, derive the formulas (called "verification conditions") that we must prove to establish the algorithm correct. Then we sketch the proofs of these verification conditions. We conclude with some remarks addressing commonly held misconceptions about the inductive assertion method.

## A. INFORMAL DEVELOPMENT

Throughout this chapter we are concerned with sequences of characters, usually called "character strings." We are interested only in strings of finite length, containing characters from some finite alphabet. We enumerate the characters in a string from the left, starting at 0. Thus, the leftmost character has "position" (or "index") 0, the second from the left has position 1, etc.

### 1. The Naive String Searching Algorithm

Suppose we have two strings, PAT and STR, and we want to find out whether PAT is a substring of STR. That is, we want to know whether there is a position i in STR such that if the 0th character of PAT is placed on top of the ith character of STR, each of the characters of PAT matches the corresponding character of STR. If PAT does so occur in STR, we would like to determine the smallest i at which such a match occurs, and otherwise we would like to indicate that no match occurs.

For example, given the two strings

```
PAT:     EXAMPLE
STR:     LET_US_CONSIDER_A_SIMPLE_EXAMPLE.
```

the appropriate answer is 25:

```
PAT:                                     EXAMPLE
STR:     LET_US_CONSIDER_A_SIMPLE_EXAMPLE.
                                  ↑
                           (position 25)
```

There is an obvious way to compute the smallest such i. Consider each of the successive values of i, starting at 0, and ask whether each of the successive characters of PAT (starting at 0) is equal to the corresponding character of STR (starting at the current i). If a match is found, then we stop and i is the position of the leftmost match. But if the end of STR is encountered, then since we tried all possible positions, no match occurs.

We will eventually define a recursive function, STRPOS, of two arguments, that embodies this naive algorithm. We will take STRPOS as the definition of what it means to find the position of the leftmost occurrence of PAT in STR. It is convenient to define STRPOS to return the length of STR (an "illegal" position in STR since we begin indexing at 0) to indicate that PAT does not occur in STR. (This is useful in our specification of the problem because we can pretend that PAT occurs just beyond the end of STR.)

## 2. An Example of a Faster Method

"String searching" is fairly common in everyday computing. For example, string searching is crucial to the on-line preparation of text (e.g., editing programs stored in text files, preparing letters and memos on "word processors," and inputting and correcting text for large scale computer typesetting applications). Thus, it is advantageous to use an efficient algorithm.

A standard way to measure the efficiency of an algorithm is to count the number of machine instructions required to execute it on the average. For a string searching algorithm, this number is usually proportional to the number of times a character from STR is fetched before a match is found or STR is exhausted at position i. The naive algorithm, the one embodied by STRPOS, suffers from the fact that it looks at each of the first i characters of STR at least once.[36]

At first sight, it is not obvious that one can do better than to look at each of the first i characters.[37] To see how it can be done, let us look again at the example:

```
PAT:    EXAMPLE
STR:    LET_US_CONSIDER_A_SIMPLE_EXAMPLE.
```

Rather than focus our attention on the beginning of STR, consider the character of STR aligned with the *rightmost* character of PAT, namely, the "_" indicated by the arrow below:

---

[36] In the worse case, the naive algorithm is "quadratic" in that it looks at k*i characters, where k is the length of the pattern and i is the location of the winning match. Consider the case where PAT is "AAAB" and STR is "AAAAA...AAAAAB". Knuth, Morris, and Pratt [27] developed a "linear" algorithm that always makes order i+k comparisons. However, the worst-case behavior of the simple algorithm rarely occurs in real string searching applications. On the average, the simple algorithm is practically linear.

[37] In fact, Rivest [48] has shown that for each string searching algorithm there exist PAT and STR for which that algorithm inspects at least i characters.

```
PAT:     EXAMPLE
STR:     LET_US_CONSIDER_A_SIMPLE_EXAMPLE.
            ↑
```

Since "_" is not equal to the last character of PAT, we know we do not have a match at the current juxtaposition of PAT and STR. More than that, since "_" does not even occur in PAT, we know that we can slide PAT to the right by its length and not miss a match. If sliding PAT by less than its length would produce a match, then the "_" at which we are pointing would be involved in the match. In particular, "_" would have to be equal to (i.e., *be*) some character in PAT. So we can slide PAT down by its length, to put it just past the position of the arrow above. Then we move the arrow down STR so that it is once again under the rightmost character of PAT:

```
PAT:            EXAMPLE
STR:     LET_US_CONSIDER_A_SIMPLE_EXAMPLE.
                   ↑
```

Now let us repeat the process. Consider the character indicated by the arrow. It matches the rightmost character of PAT. Thus we may have a match. To check this, we move the arrow to the *left* by 1:

```
PAT:            EXAMPLE
STR:     LET_US_CONSIDER_A_SIMPLE_EXAMPLE.
                  ↑
```

The character "D" does not occur in PAT. Thus we can slide PAT down by its length again (from the current position of the arrow):

```
PAT:                   EXAMPLE
STR:     LET_US_CONSIDER_A_SIMPLE_EXAMPLE.
                          ↑
```

Once again we shift our attention to the end of PAT and find that the corresponding character of STR, "I", does not occur in PAT. So we slide PAT down again by its length:

```
PAT:                          EXAMPLE
STR:     LET_US_CONSIDER_A_SIMPLE_EXAMPLE.
                                 ↑
```

This time we fetch an "X". It does not match the last character of PAT, so we are not at a match. But "X" does occur in PAT, so we may not slide PAT down by its length. However, the *rightmost* occurrence of "X" in PAT is five characters from the right-hand end of PAT. Thus, if we were to slide PAT down by just one, or two, or any number less than five, we would not find a match: if a match were to occur at such a juxtaposition, then the "X" at which we were pointing would be involved in the match and would have to be equal to a character in PAT (namely, an "X") to the right of the rightmost "X" in PAT! So we may slide PAT down by five (so as to align the current position of the arrow with the right-most "X" in PAT) without risking a miss. Then we move the arrow so that it once again points to the end of PAT:

```
PAT:                          EXAMPLE
STR:    LET_US_CONSIDER_A_SIMPLE_EXAMPLE.
                                      ↑
```

Finally, we see that the indicated character of STR is equal to the corresponding one of PAT. By looking for a mismatch backward, as before, we find that all the characters of PAT are similarly matched. Thus, we have found the leftmost occurrence of PAT in STR.

Here are the characters we fetched from STR, up to the time we had made the final alignment and were about to confirm the match:

```
STR:    LET_US_CONSIDER_A_SIMPLE_EXAMPLE.
            ↑    ↑↑        ↑       ↑
```

Of course, we had to spend seven more comparisons confirming the final match. But in skimming past the first 25 characters of STR, we only had to look at five of them.[38]

## 3. Preprocessing the Pattern

One might wonder why the measure of the number of characters fetched from STR is relevant here, given that every time we fetched

---

[38] The algorithm just illustrated has quadratic worst-case behavior. However, on the average it looks at fewer than i characters of STR before finding a match at i, and its behavior improves as patterns get longer (because it may slide the pattern further each move), and deteriorates as the alphabet gets smaller (because the chances are increased that the character just fetched from STR occurs close to the end of PAT). The algorithm can be implemented so that if searching for English patterns of length five or more, through English text, fewer than i machine instructions are executed on the average before the pattern is found at position i. The algorithm is actually a simplification of the Boyer–Moore fast string searching algorithm [9], which treats the finite set of terminal substrings of PAT in a way analogous to the way we just treated the characters of the alphabet. The Boyer–Moore fast string searching algorithm is, on the average, much faster than the simplified version on small alphabets (e.g., binary ones) but only marginally faster on large alphabets (e.g., English text). The worst-case behavior of the Boyer–Moore algorithm is linear, as proved by Knuth in [27]. Guibas and Odlyzko in [23] also prove (and improve upon) the linearity result.

one we had to ask whether and where it occurred in PAT. However, because the alphabet is finite, we can, for a given PAT, precompute the answers to all those questions.

The algorithm requires that whenever we fetch a character C from STR that does not match the corresponding character of PAT, we be able to determine how far down we can slide the pattern without missing a match.

If we are standing under the right end of PAT when we fetch C, then we want to know how many characters there are between the right end of PAT and the rightmost occurrence of C in PAT. We know we may slide PAT that far forward, so as to align the C we just discovered in STR with the rightmost C in PAT. If C does not occur in PAT at all, then we may slide PAT forward by its length (i.e., we can pretend C occurs at "position" $-1$). This number, called the "delta1" for C and PAT in [9], can be obtained by scanning PAT from right to left, counting the number of characters seen before encountering C (or running off the beginning of PAT).

For example, if PAT were the string "EXAMPLE", then the following table contains all the information we need (over the alphabet "A" through "Z" and "_"):

| C | delta1 for C and "EXAMPLE" |
|---|---|
| A | 4 |
| B | 7 |
| C | 7 |
| D | 7 |
| E | 0 |
| F | 7 |
| . | |
| . | |
| . | |
| K | 7 |
| L | 1 |
| M | 3 |
| N | 7 |
| O | 7 |
| P | 2 |
| Q | 7 |
| . | |
| . | |
| . | |

*(continued)*

| | |
|---|---|
| W | 7 |
| X | 5 |
| Y | 7 |
| Z | 7 |
| – | 7 |

It is possible to set up this table in order (k + the alphabet size) instructions where k is the length of PAT.[39]

We will not discuss the preprocessing further. Instead, we assume we have a function DELTA1, that takes C and PAT and returns the table entry as defined above.

### 4. How To Do a Fast String Search

We now tell the reader how to carry out a fast string search, assuming the reader knows how to compute DELTA1 as above.

The directions involve several variable names. We imagine that when the reader is following these directions to carry out an actual search, he has in mind an environment that associates some specific mathematical value (such as an integer or string) to each of the variables mentioned. Every time he encounters an expression during the course of following these instructions, he is to evaluate the expression with respect to the current environment. We sometimes direct the reader to "Let var be expr." By this we mean for the reader to construct the new environment in which all the variables except var have the same values they had in the old environment. The value of var under the new environment is to be the value of expr under the old environment. We expect the reader to use this new environment in the evaluation of all subsequently encountered expressions (until we instruct him to change the environment again).

We assume that (LENGTH STR) is the number of characters in STR, and that (NTHCHAR I STR) is the Ith character of STR.

We occasionally make "claims" regarding what we believe to be true every time the reader arrives at an indicated step in the process. These claims can be ignored while using the description as a means of finding a pattern in a string. Suppose we wish to find the leftmost occurrence of PAT* in STR* if one exists.

1. The initial environment should associate with the variable PAT

---

[39] The table can be set up by filling it with the number k (as though no character occurs in PAT), and then sweeping through PAT once from left to right filling in the correct value for each occurrence of each character. Thus, in the example above, the "E" entry first has a 7 in it (i.e., the length of PAT), then 6 (as a result of seeing the first "E"), and then finally 0 (as a result of seeing the last "E").

the string PAT* and with the variable STR the string STR*. Our object is to produce a "final answer" equal to the one produced by the naive algorithm: (STRPOS PAT* STR*).

2. We are interested only in the values of the variables PAT and STR, and the following additional variables: I, J, PATLEN, STRLEN, NEXTI, and C. We are indifferent to the initial values of the additional variables.
3. Let PATLEN be (LENGTH PAT).
4. If (EQUAL PATLEN 0) is true, then do the following (and otherwise continue at line 5):

> Claim 1. You will never arrive at this point while following these directions unless 0 is equal to (STRPOS PAT* STR*).
>
> Therefore you should stop and consider 0 the final answer.

5. Let STRLEN be (LENGTH STR).
6. Let I be (SUB1 PATLEN). The value of I will always be the position in STR of the "↑" we have drawn while illustrating the algorithm at work. It now points to the character of STR directly under the rightmost character of PAT (with PAT left aligned with the beginning of STR).
7. Claim 2. Every time you come past this point there are at least as many characters to the left of I in STR as there are characters to the left of the rightmost in PAT (so we may compare them pairwise). Furthermore, either I marks the right-hand end of the winning match of PAT in STR, or else the right-hand end of the winning match is somewhere to the right of I (or else there is no winning match). In any case, we have not yet passed the winning match but might be standing at its right-hand end.

    However, it is in general possible that I is already beyond the end of STR. Therefore we must check that I is indeed a legal index into STR before we start investigating whether I marks the end of a match.

8. If (GREATEREQP I STRLEN) is true (i.e., I≥STRLEN), then do the following (and otherwise continue at line 9):

> Claim 3. You will never get here while following these directions unless STRLEN is (STRPOS PAT* STR*).
>
> Therefore you should stop and consider STRLEN the final answer.

9. Let J be (SUB1 PATLEN).
10. Let NEXTI be (ADD1 I). We are about to start backing up, comparing characters from PAT with characters from STR. I (the

"↑") will mark the position in STR from which we will fetch characters, and J will mark the corresponding position in PAT. We may later need to know the position currently to the right of I, so we have saved it in NEXTI.

11. Claim 4. Every time you come past this point, we claim that everything we say in Claim 2 is true (except that (SUB1 NEXTI) should be used in place of I); furthermore J is the same distance to the left of PATLEN as I is to the left of NEXTI, NEXTI is no bigger than STRLEN, J is less than or equal to I, and it is the case that the terminal substring of PAT starting at position (ADD1 J) matches the terminal substring of STR starting at position (ADD1 I).

The last part of his claim is vacuously true initially, but as we back up establishing that the characters from PAT are equal to their counterparts in STR it is more interesting.

12. Let C be (NTHCHAR I STR).
13. If (EQUAL C (NTHCHAR J PAT)) is true, then do the following (and otherwise continue at line 18):
14. If (EQUAL J 0) is true, then do the following (and otherwise continue at line 15):

Claim 5. You will never get here while following these directions unless I is (STRPOS PAT* STR*).

Therefore, you should stop and consider I the final answer.

15. Let I be (SUB1 I). Note that this backs up the " ↑ " by 1.
16. Let J be (SUB1 J).
17. Go to line 11 and continue.
18. Let I be the maximum of the two integers (PLUS I (DELTA1 C PAT)) and NEXTI. This step slides the pattern down. At first sight one is tempted to slide the pattern down by incrementing I with (DELTA1 C PAT), for that quantity can be regarded as the sum of (a) the distance we must slide the pattern to align the current " ↑ " with the right-most C in PAT, plus (b) the distance we must move the " ↑ " to put it at the end of the new location of PAT. This reasoning is entirely accurate but it ignores the fact that the rightmost occurrence of C might have already been passed:

```
PAT:              EXAMPLE
STR:              ...IT_IS_ELEMENTARY
I:                         ↑ **
NEXTI:                      ↑
```

In the example above, we have matched the two characters marked with *'s and have backed up to the I indicated. But since the rightmost "E" in PAT has already been passed, incrementing I by (DELTA1 "E" PAT) = 0 would slide PAT backwards. Instead, we choose to slide PAT forward by one, namely, to the position marked by NEXTI.
19. Go to line 7 and continue.

This concludes the informal presentation of the algorithm. Most programmers could now go away and implement it. However, we are interested in proving it correct.

## B. FORMAL SPECIFICATION OF THE PROBLEM

In this section, we specify the string searching problem formally by defining STRPOS. In the next section, we use the inductive assertion method to specify the fast string searching algorithm and derive its "verification conditions."

### 1. Strings

From the mathematical point of view, what is a string? It is a sequence of objects (that we will think of as characters in this application). Consequently, from the mathematical point of view we can regard a string as a list. The only two operations on strings required by our string searching algorithm are LENGTH and NTHCHAR. These are recursively defined in Appendix A. For example, the 0th character of X is ( CAR X) and the i + 1st character of X is the ith character of ( CDR X).

A program semanticist may object that our definition of the mathematical object "string" with CONS, CAR, and CDR is naive because it dictates an implementation of strings less efficient than the usual one using indexed byte operations. This objection is as unfounded as the analogous objection to defining the mathematical object "integer" with ADD1 and SUB1. An engineer or systems programmer is as free to implement strings efficiently as he is to implement integers using twos-complement arithmetic.

## 2. The String Matching Problem

We wish to define the notion of whether PAT occurs as a substring of STR, and if so, what is the position in STR of the leftmost such occurrence.

We first define the function MATCH that determines whether PAT is equal to an initial piece of STR. This is the case precisely if PAT is empty or if both PAT and STR are nonempty and (a) their first characters are identical and (b) (CDR PAT) is (recursively) an initial piece of (CDR STR).

**Definition**

```
(MATCH PAT STR)
    =
(IF (LISTP PAT)
    (IF (LISTP STR)
        (IF (EQUAL (CAR PAT) (CAR STR))
            (MATCH (CDR PAT) (CDR STR))
            F)
        F)
    T).
```

Now we define STRPOS. It is to tell us how many characters in STR must be "stepped over" before finding a terminal substring of STR that has PAT as an initial piece (or STR is exhausted):

**Definition**

```
(STRPOS PAT STR)
    =
(IF (MATCH PAT STR)
    0
    (IF (LISTP STR)
        (ADD1 (STRPOS PAT (CDR STR)))
        0)).
```

## C. DEVELOPING THE VERIFICATION CONDITIONS FOR THE ALGORITHM

We want to prove that the fast string searching algorithm, exhibited above, always computes (STRPOS PAT* STR*). This requires that we formally define DELTA1 and that we somehow formalize what it means for the algorithm to compute (STRPOS PAT* STR*).

## C. DEVELOPING THE VERIFICATION CONDITIONS FOR THE ALGORITHM / 293

The first task is trivial. As for the second, note that if our claims are correct, specifically Claims 1, 3, and 5, the algorithm is correct: it never returns an answer except when the answer is (claimed) equal to (STRPOS PAT* STR*). Therefore, we could prove the algorithm correct by proving our claims. This raises two problems: (1) our claims have not been written down formally, and (2) they involve the values of the variables in the environment current at the time the claims are encountered.

### 1. The Formal Definition of DELTA1

Recall that (DELTA1 C PAT) is to return the number of characters to the right of the rightmost C in PAT. The number can be obtained by scanning PAT from right to left, counting the number of characters stepped over before the first C is encountered (or the beginning of the pattern is reached). This is just a STRPOS string search for the singleton string containing C, over the reverse of the pattern.

We thus use the following definition of DELTA1:

**Definition**

```
(DELTA1 C PAT)
=
(STRPOS (CONS C "NIL") (REVERSE PAT)).
```

Since (STRPOS PAT STR) is the length of STR if PAT does not occur, this definition of DELTA1 returns the length of the pattern if C does not occur in it.

### 2. Formalizing the Claims

We here formalize each of the five claims. We reproduce (the relevant part of) each claim before expressing it formally. All the functions mentioned are defined in Appendix A.

*a. Claim 1*

Claim 1 is that "0 is equal to (STRPOS PAT* STR*)." Thus, the formal statement of Claim 1 is

```
*Claim 1
    (EQUAL 0 (STRPOS PAT* STR*)).
```

## b. Claim 2

Claim 2 is irrelevant to the correctness of the algorithm; we care only about the truth of the "exit" Claims 1, 3, and 5. However, we formalize Claim 2 because it will be involved in the proofs of the other claims. For example, we will prove Claim 3 by proving that if Claim 2 is true, then, when we reach Claim 3, it is true.

Claim 2 is that "there are at least as many characters to the left of I in STR as there are characters to the left of the rightmost in PAT" and that "I marks the right-hand end of the winning match of PAT in STR, or else the right-hand end of the winning match is somewhere to the right of I (or else there is no winning match)."

The first part of this can be formally phrased (LESSEQP (SUB1 PATLEN) I). The second can be expressed as (LESSP I (PLUS PATLEN (STRPOS PAT STR))). The PLUS expression is the position, in STR, of the character just to the right of the right-hand end of the first occurrence of PAT in STR. The claim is that I is strictly less than that position. If PAT does not occur in STR, then (STRPOS PAT STR) is the length of STR and the statement still handles our claim.

Claim 2, as currently stated, is inadequate to prove Claim 3. Among other things, we have not said that PAT is (still) PAT*, that PATLEN is (LENGTH PAT), and that PATLEN is nonzero. Thus, we actually strengthen Claim 2 to the following:

```
*Claim 2
     (AND (EQUAL PAT PAT*)
          (EQUAL STR STR*)
          (EQUAL PATLEN (LENGTH PAT))
          (LISTP PAT)
          (EQUAL STRLEN (LENGTH STR))
          (NUMBERP I)
          (LESSEQP (SUB1 PATLEN) I)
          (LESSP I
                 (PLUS PATLEN (STRPOS PAT STR)))).
```

We make *Claim 2 the body of the definition of the function TOP.ASSERT, to make future discussion more succinct.

As noted, *Claim 2 is irrelevant to the correctness of the algorithm, so the reader should not be bothered by its complicated nature (except insofar as it affects the difficulty of proof).

## c. Claim 3

Our third claim is that "STRLEN is (STRPOS PAT* STR*)." The formal statement of this is

# C. DEVELOPING THE VERIFICATION CONDITIONS FOR THE ALGORITHM

```
*Claim 3
     (EQUAL STRLEN (STRPOS PAT* STR*)).
```

## d. Claim 4

Like Claim 2, Claim 4 is irrelevant to the correctness of the algorithm, but it is important in establishing the other claims.

Claim 4 is that "Claim 2 is true (except that (SUB1 NEXTI) should be used in place of I); furthermore, J is the same distance to the left of PATLEN as I is to the left of NEXTI, NEXTI is no bigger than STRLEN, J is less than or equal to I, and it is the case that the terminal substring of PAT starting at position (ADD1 J) matches the terminal substring of STR starting at position (ADD1 I)."

To formalize the relationship between J, PATLEN, I, and NEXTI, we claim that NEXTI is equal to PATLEN plus the difference between I and J. To formalize the relationship between the terminal substrings of PAT and STR, we use the function NTH (which underlies the definition of NTHCHAR) to define the terminal substring starting at a given position, and our previously defined MATCH to characterize when one string matches the initial part of another.

For the same reasons that we had to strengthen Claim 2, we have to strengthen Claim 4. Its formal statement is

```
*Claim 4
     (AND (TOP.ASSERT PAT STR (SUB1 NEXTI)
                      PATLEN STRLEN PAT* STR*)
          (NUMBERP I)
          (NUMBERP J)
          (NUMBERP NEXTI)
          (LESSP J PATLEN)
          (LESSP I STRLEN)
          (EQUAL NEXTI
                 (PLUS PATLEN (DIFFERENCE I J)))
          (LESSEQP NEXTI STRLEN)
          (LESSEQP J I)
          (MATCH (NTH PAT (ADD1 J))
                 (NTH STR (ADD1 I)))).
```

We make *Claim 4 the body of the definition of the function LOOP.ASSERT.

## e. Claim 5

Our final claim is that "I is (STRPOS PAT* STR*)."

*Claim 5
    (EQUAL I (STRPOS PAT* STR*)).

## 3. Applying the Inductive Assertion Method

Now that the claims have been formalized, we must eliminate their implicit reliance upon the flow of control through the procedure. It is at this point that we employ Floyd's inductive assertion method.

### a. A Sketch of the Inductive Assertion Method

To explain how we eliminate the "dynamic" nature of the claims, it is useful to have a copy of the algorithm with the formal claims in place of the informal ones. Since the description of the algorithm given above is so long, we abbreviate it here. We have numbered the steps the same way we did earlier.

```
1.   Procedure FSTRPOS(PAT,STR);
2.        Variables I,J,PATLEN,STRLEN,NEXTI,C;
3.        PATLEN ← LENGTH(PAT);
4.        If PATLEN=0
              then
              [*Claim 1: (EQUAL 0 (STRPOS PAT* STR*))]
              return 0;
              close;
5.        STRLEN ← LENGTH(STR);
6.        I ← PATLEN-1;
7.  top:  [*Claim 2: (TOP.ASSERT PAT STR I PATLEN
                              STRLEN PAT* STR*)]
8.        If I ≥ STRLEN
              then
              [*Claim 3: (EQUAL STRLEN
                              (STRPOS PAT* STR*))]
              return STRLEN;
              close;
9.        J ← PATLEN-1;
10.       NEXTI ← I+1;
11. loop: [*Claim 4: (LOOP.ASSERT PAT STR I J PATLEN
                              STRLEN NEXTI PAT* STR*)]
12.       C ← STR(I);
13.       If C=PAT(J)
              then
```

## C. DEVELOPING THE VERIFICATION CONDITIONS FOR THE ALGORITHM / 297

```
14.             If J=0
                    then
                    [*Claim 5: (EQUAL I
                                    (STRPOS PAT* STR*))]
                    return I;
                    close;
15.             I ← I-1;
16.             J ← J-1;
17.             goto loop;
                close;
18.         I ← MAX(I+DELTA1(C,PAT),NEXTI);
19.         goto top;
            end;
```

We desire to prove that each claim is true every time it is encountered while using the above procedure to search for PAT* in STR*.

The inductive assertion method may be applied to a program, provided the program has been annotated with a sufficient number of claims so that every entrance and exit has a claim and so that in traversing any loop at least one claim is encountered. If a sufficient number of claims has been supplied, then we can prove that the program is correct by considering the finite number of paths that begin and end at a claim (and have no interior claims). If for each such path the claim at the beginning of the path (together with the tests along the path) implies the claim at the end of the path (under the environment produced by the "Let" statements along the path), then, by induction on the number of steps in the computation, every claim is true every time it is encountered. Thus the exit claims, in particular, are true whenever the program exits.[40]

We have provided a sufficient number of claims to apply the method to FSTRPOS.

### b. The Paths through FSTRPOS

The relevant paths through FSTRPOS are:

*Path 1.* From the entrance to *Claim 1, performing the assignment on line 3 and assuming the test on line 4 to be true. (Since we did not

---

[40] Note, however, that we have only proved partial correctness: *if* the program exits, it exits with the correct answer. The inductive assertion method we have described can be easily adapted to include proofs of termination: one needs to check that some well-founded relation is decreasing on each of the finite paths.

explicitly annotate the entrance to FSTRPOS with a claim, we assume the entrance claim in T.)

*Path 2.* From the entrance to *Claim 2, performing the assignment on line 3, assuming the test on line 4 to be false, and performing the assignments on lines 5 and 6.

*Path 3.* From *Claim 2 to *Claim 3, assuming the test on line 8 to be true.

*Path 4.* From *Claim 2 to *Claim 4, assuming the test on line 8 to be false and performing the assignments on lines 9 and 10.

*Path 5.* From *Claim 4 to *Claim 5, performing the assignment on line 12 and assuming the tests on lines 13 and 14 to be true.

*Path 6.* From *Claim 4 to *Claim 4, performing the assignment on line 12, assuming the test on line 13 to be true, assuming the test on line 14 to be false, and performing the assignments on lines 15 and 16.

*Path 7.* From *Claim 4 to *Claim 2, performing the assignment on line 12, assuming the test on line 13 to be false, and performing the assignment on line 18.

*c. Generating the Verification Conditions*

For each of these seven paths, we must prove its "verification condition"; that is, we must prove that if the starting claim is true and the tests along the path are true, then the final claim is true (in the environment produced by the assignment statements along the path).

Below we present the generation of two of the seven verification conditions.[41]

i. The Generation of FSTRPOS.VC1

Consider the first path. Starting with an environment in which PAT is PAT* and STR is STR*, we are to perform the assignment on line 3, assume that the test at line 4 is true, and then prove *Claim 1 under the environment thus produced.

The statement at line 3 is

3. PATLEN ← LENGTH( PAT );

---

[41] To produce formal verification conditions, one must have a formal semantics for his programming language. We do not present such a semantics here, but we have precisely formalized such a semantics in [10]. We generated the verification conditions here using an implementation of that semantics.

## C. DEVELOPING THE VERIFICATION CONDITIONS FOR THE ALGORITHM / 299

This means: change the environment so that PAT and STR still have their old values, but PATLEN has the value that (LENGTH PAT) has in the current environment. Thus, in our new environment, PAT is PAT*, STR is STR*, and PATLEN is (LENGTH PAT*).

Next we hit the test at line 4 and are to assume it true:

4. If PATLEN=0
      then ...

This means that we should assume the current value of (EQUAL PATLEN 0) to be true. That is, we should assume (EQUAL (LENGTH PAT*) 0).

Finally, we hit *Claim 1:

[*Claim 1: (EQUAL 0 (STRPOS PAT* STR*))]

which we must prove given the assumption above.

The resulting verification condition for Path 1 is

**Theorem** FSTRPOS.VC1:

(IMPLIES (EQUAL (LENGTH PAT*) 0)
         (EQUAL 0 (STRPOS PAT* STR*))).

ii. The Generation of FSTRPOS.VC7

Let us look at a more interesting path, namely Path 7. Starting at *Claim 4, we are to perform the assignment on line 12, fail the test at line 13, perform the assignment on line 18, and prove *Claim 2 under the resulting environment.

Assume that we have an initial environment in which the program variable I has the value I, the program variable J has the value J, etc. Assume *Claim 4 is true in that environment:

(LOOP.ASSERT PAT STR I J PATLEN
             STRLEN NEXTI PAT* STR*).

At line 12

12. C ← STR(I);

C receives the value (NTHCHAR I STR).

Next we encounter the test at line 13 and are to assume it false:

13. If C=PAT(J)
      then...

This means we assume that the current value of (EQUAL C (NTHCHAR J PAT)) is false:

## 300 / XVIII. THE CORRECTNESS OF A FAST STRING SEARCHING ALGORITHM

```
*TEST
        (NOT (EQUAL (NTHCHAR I STR) (NTHCHAR J PAT))).
```

Finally we hit line 18

```
18.  I ← MAX(I+DELTA1(C,PAT),NEXTI);
```

and change the environment to

```
*ENVRN
          variable      value
                        after line 18

          PAT           PAT
          STR           STR
          I             (IF (LESSP (PLUS I (DELTA1
                                    (NTHCHAR I STR) PAT)) NEXTI)
                            NEXTI
                            (PLUS I (DELTA1 (NTHCHAR I STR)
                                            PAT)))  ⁴²
          J             J
          PATLEN        PATLEN
          STRLEN        STRLEN
          NEXTI         NEXTI
          C             (NTHCHAR I STR)
```

After line 18 we encounter the "goto top" and return to *Claim 2, which we must prove:

```
7. top:[*Claim 2: (TOP.ASSERT PAT STR I PATLEN
                              STRLEN PAT* STR*)]
```

after instantiating it with the environment *ENVRN and assuming *Claim 4 and *TEST.

That is, the verification condition for Path 7 is

**Theorem** FSTRPOS.VC7:

```
(IMPLIES (AND (LOOP.ASSERT PAT STR I J PATLEN
                            STRLEN NEXTI PAT* STR*)
              (NOT (EQUAL (NTHCHAR I STR)
                          (NTHCHAR J PAT)))))
```

---

[42] Our mechanical verification condition generator is driven off the compiled code for our high-level language. Our compiler compiles (MAX x y) "open" in the sense that it is treated as though it were (IF (LESSP y x) x y). Hence the IF in this value where a MAX was expected.

```
(TOP.ASSERT
 PAT
 STR
 (IF (LESSP
        (PLUS I (DELTA1 (NTHCHAR I STR)
                        PAT))
        NEXTI)
     NEXTI
     (PLUS I (DELTA1 (NTHCHAR I STR)
                     PAT)))
 PATLEN
 STRLEN
 PAT*
 STR*)).
```

This formula requires us to prove that if the assertion in the inner loop is true and we find a mismatch on some character at position I in STR, then the assertion at the outer loop holds for the value of I obtained by skipping ahead by DELTA1 (or to NEXTI, as appropriate).

iii. The Remaining Verification Conditions

The five remaining verification conditions are similarly generated and are listed in Appendix A under the names FSTRPOS.VCi, for i from 2 to 6.

It requires induction to prove that FSTRPOS.VC1 through FSTRPOS.VC7 are sufficient to establish that all our claims are true every time they are encountered. In particular, one must induct on the number of steps in the computation [18]. This induction is crucial in order to unwind the iteration inherent in a procedure described the way FSTRPOS is described. It is from this use of induction that the "inductive assertion" method gets its name.

But now we must prove FSTRPOS.VC1 through FSTRPOS.VC7. They involve the natural numbers (which are inductively defined), sequences (which are inductively defined), and functions such as NTH, MATCH, and STRPOS (which are recursively defined). To prove them, we also need induction. This second use of induction is crucial because of the nature of the mathematical objects with which programs deal.

## D. THE MECHANICAL PROOFS OF THE VERIFICATION CONDITIONS

The theorem-prover has proved the seven verification conditions for FSTRPOS. We will not go into the proofs in detail since this chapter was intended to demonstrate, by realistic example, that our theory is applicable to the specification of programs by the inductive assertion method.

We sketch briefly the proof of each of the verification conditions. Our sketches will concern themselves mainly with the string processing lemmas that have to be proved in order to set up the proofs of the verification conditions. A fair amount of arithmetic is generally involved. Since the proof of the correctness of FSTRPOS is actually conducted after the theorem-prover has proved the unique prime factorization theorem (and remembered all the theorems along the way), it knows a good deal of arithmetic by the time it starts proving FSTRPOS.VC1. However, we had it prove several additional theorems about arithmetic, almost all of them involving LESSP, because of its rather poor handling of transitivity. All of the lemmas mentioned below are in Appendix A.

### 1. Proofs of FSTRPOS.VC1 and FSTRPOS.VC2

FSTRPOS.VC1 and FSTRPOS.VC2 can be reduced to true by simplifications alone (in the presence of all the previously proved arithmetic theorems).

### 2. Proof of FSTRPOS.VC3

FSTRPOS.VC3 is the verification condition that establishes that if the algorithm exits because I is eventually pushed beyond the end of STR, then (STRPOS PAT* STR*) is STRLEN. This has to be proved assuming *Claim2. But *Claim2 provides the hypothesis that (LESSP I (PLUS PATLEN (STRPOS PAT STR))). If I is greater than or equal to STRLEN, then by transitivity of LESSP, we can conclude that (PLUS PATLEN (STRPOS PAT STR)) is greater than STRLEN. The proof can then be completed if we have previously proved that when (STRPOS PAT STR) is not equal to (LENGTH STR), it is at least (LENGTH PAT) shy of (LENGTH STR). This lemma, called STRPOS.BOUNDARY.CONDITION, must be proved by induction on the length of STR, and requires the inductively proved theorem that if PAT MATCHes STR, then the length of STR is at least that of PAT.

### 3. Proof of FSTRPOS.VC4

FSTRPOS.VC4, the verification condition for the path from *Claim 2 to *Claim 4, requires nothing more than arithmetic.

## 4. Proof of FSTRPOS.VC5

FSTRPOS.VC5 is more interesting. This verification condition corresponds to the winning exit from the procedure. In particular, assuming that *Claim 4 holds and that the Ith character of STR is equal to the Jth character of PAT, and that J is 0, we must establish that I is equal to (STRPOS PAT* STR*). *Claim 4 informs us that we have a MATCH established between the terminal substrings of PAT and STR, and the conditions on the path establish that the first character of PAT is equal to the Ith character of STR. Thus, we certainly have a MATCH at position I in STR. However, this does not immediately imply that (STRPOS PAT* STR*) is I because there might be an earlier match. But *Claim 4 tells us that *Claim 2 holds, and *Claim 2 tells us that there is no match to our left. Thus, we can complete the proof if we have proved

**Theorem** STRPOS.EQUAL:

```
(IMPLIES (AND (LESSP I (LENGTH STR))
              (NOT (LESSP (STRPOS PAT STR) I))
              (NUMBERP I)
              (MATCH PAT (NTH STR I)))
         (EQUAL (STRPOS PAT STR) I).
```

This lemma says that if I is a legal index into STR and (STRPOS PAT STR) is greater than or equal to I, and PAT matches the Ith terminal substring of STR, then (STRPOS PAT STR) *is* I. This is one of the obvious properties of the intuitive definition of the "position of the leftmost match" and must be proved by induction (on I and STR) from the definition of STRPOS.

## 5. Proof of FSTRPOS.VC6

FSTRPOS.VC6 is the verification condition for the path from *Claim 4 back to *Claim 4. On that path, we find that the Ith character of STR is the Jth character of PAT but that J is not 0. It is straightforward to confirm that *Claim 4 still holds after the match has been extended by one character in the backward direction.

## 6. Proof of FSTRPOS.VC7

FSTRPOS.VC7 is the most interesting verification condition; it is the only one involving DELTA1. It requires us to prove that if *Claim 4 holds and we find a mismatch at position I, then we can increment I by (DELTA1 C PAT) (or set it to NEXTI) and still prove *Claim 2. Of course, the interesting part of *Claim 2 is that we have not missed a match. The proof rests mainly on two lemmas, both of which require induction to prove, and both of which themselves rest on inductively proved lemmas.

The first is called EQ.CHARS.AT.STRPOS. It assures us that if the Ith character of STR is not equal to the Jth character of PAT, under certain obvious restrictions on I and J, then (STRPOS PAT STR) is not the difference between I and J. In particular, this lemma tells us that if the inner loop finds a mismatch of two corresponding characters anywhere in the region of interest, then we are not currently in a match. Therefore, we can move I down by at least one.

The second lemma is called DELTA1.LEMMA. It states that if I is a legal index into STR, and the right-hand end of the winning match of PAT in STR is at I or to the right, then one does not miss a match by incrementing I by (DELTA1 (NTHCHAR I STR) PAT). Proving this lemma requires substantial reasoning about STRPOS and MATCH, in addition to many lemmas about list processing and arithmetic. For example, one key fact is DELTA1.LESSP.IFF.MEMBER, which states that (DELTA1 CHAR PAT) is strictly less than the length of PAT if and only if CHAR is a MEMBER of PAT (a crucial fact if incrementing I by DELTA1 is not going to skip over a possible alignment).[43]

The induction argument to prove DELTA1.LEMMA was exhibited at the conclusion of the discussion on induction, Chapter XV. It was the example in which the field of ten candidate inductions was narrowed to one by merging and the consideration of flaws.

## 7. What Have We Proved?

We have thus completed sketching the proofs of the verification conditions. What exactly have we proved about the program FSTRPOS?

Appealing to the induction argument behind the Floyd method, we have proved that each of our claims is true every time it is encountered during the execution of the procedure. In particular, the "exit"

---

[43] We were pleased to see that many "toy" theorems the theorem-prover has proved for years actually get used in this nontoy problem. For example, proving DELTA1.LESSP.IFF.MEMBER requires using the lemmas that the length of (REVERSE X) is the length of X, and that CHAR is a MEMBER of (REVERSE X) if and only if it is a MEMBER of X.

claims establish that whenever the procedure returns an answer, it is equal to that computed by STRPOS.

But one should always ask of a proof, Upon what axioms or assumptions does the proof rest? That is, if someone is going to "buy" the correctness of FSTRPOS, what must he accept? The answer is that he must accept:

> our axioms of Peano arithmetic, lists, and literal atoms as provided by the shell addition scheme (literal atoms are involved only because of the use of "NIL");
>
> the definition of STRPOS (and thus of MATCH) as being the meaning of "leftmost match";
>
> the correctness of the verification condition generator with respect to the programming language used in FSTRPOS;
>
> the soundness of our theorem-prover.[44]

One might ask, But what of all the definitions? How do I know that PLUS is right? That LESSP is right? That NTH and NTHCHAR and MEMBER and REVERSE and DELTA1 are all right? The answer is that they are all *defined*, and because they are defined and are not involved in the statement of the theorem (namely that FSTRPOS(PAT*,STR*) returns (STRPOS PAT* STR*)), they are eliminable. It is our opinion that this is the most important reason why people should *define* concepts whenever possible (with a mechanically checked definition principle when the logic is mechanized) rather than merely add arbitrary axioms. For example, had DELTA1 been constrained to have certain properties by the addition of nondefinitional axioms, then the "buyer" of the correctness of FSTRPOS would have had to understand and believe in those axioms.

## E. NOTES

We here comment on the difficulty of the proofs and on the use of recursive functions with the inductive assertion specification method.

---

[44] This is not an assumption to be taken lightly, given the complexity of the system. However, it is an assumption that could be relieved for all time and for all future proofs by careful scrutiny of the theorem-prover by the mathematics community. While such an endeavor is clearly not cost effective for our current theorem-prover (because of its rudimentary skills), it would be a cheap price to pay for the reliable service of a truly powerful mechanical theorem-prover.

## 1. Intuitive Simplicity versus Formal Difficulty

The string searching algorithm is easier to explain than it is to prove formally correct. The algorithm is based on some obvious visual intuitions about strings, but the formalization of those intuitions involves ugly arithmetic expressions.

However, this is not a condemnation of formal reasoning but a recommendation of it. The program implementing those visual intuitions about strings does not rely upon those intuitions at all, but rather upon their translation into arithmetic operations. In particular, it is very easy to make "±1" and boundary errors in defining or using `DELTA1`. Furthermore, the string searching algorithm, even when mistakes of this sort are present, often produces the correct answer in many test cases because, except in very close proximity to the winning match, one can afford to skip ahead by too much.

In order to run our verification condition generator on the `FSTRPOS` procedure, we had to code the procedure in our "high-level" programming language. We did this by working from the published version of the algorithm [9]. We nevertheless made several translation mistakes that did not show up until we tried to prove the resulting verification conditions mechanically.

We learned one interesting fact about our algorithm from the theorem-prover. Until we tried proving `FSTRPOS` correct, we held the mistaken belief that the `MAX` expression involved in the incrementing of `I` was necessary for the termination of the algorithm but not for its correctness. That is, we believed that if one accidentally moved the pattern backward it would not hurt anything except that the algorithm might not terminate. Upon reflection (or careful scrutiny by a theorem-prover), this can be seen to be wrong. Consider what happens if in moving the pattern backward its left-hand end is moved off the left-hand end of the string. Then it is possible to find a match out there by fetching "illegal" characters from the string. Many readers may object that a well-designed, high-level language would expose this bug as soon as it arose (because of the attempt to index illegally into the string). However, well-designed, high-level languages are almost never used to code algorithms of this sort because the overhead involved in ensuring legal access to strings offsets the advantage of using the algorithm in the first place.

## 2. Common Misconceptions about the Inductive Assertion Method

We comment here on three misconceptions about the inductive assertion method. We have found these misconceptions rather widespread in the program verification community and feel that they should be addressed, even though the subject is technically beyond the scope of this book.

One common misconception is that the use of the inductive assertion method requires a "program specification language" providing the usual unbounded universal and existential quantifiers. Our intent has been to show that this is unnecessary and that the inductive assertion method is a way of attaching meaning to programs that is independent of what mathematical language one uses to express the specifications. We feel that the bounded quantification provided by recursive functions is not only sufficient for almost all specification needs, but often results in cleaner, better structured specifications, and aids mechanical generation of proofs.

Another common misconception is that the inductive assertion method eliminates the need for induction. The foregoing proof sketches illustrate that induction may be useful in the proofs of the verification conditions themselves. It happens, in the FSTRPOS example, that the correctness can be stated so that the verification conditions involve little more than linear arithmetic and quantification. Thus, it is possible, using, say, Presburger's decision procedure [47] and a little reasoning about function symbols, to prove the FSTRPOS verification conditions without induction. Of course, given the right lemmas, it is possible to prove anything without induction. The point is that whether induction is involved in the proofs of the verification conditions depends upon how much the theorem-prover knows or assumes about the particular mathematical operations involved, and not upon whether induction was used to attach meaning to the program.

A third misconception is that the inductive assertion method somehow magically "modernizes" mathematics so that it can deal with side effects such as destructive assignment to arrays. In fact, the method provides a very pretty, easy way of eliminating the dynamic nature of a program by reducing its correctness to the correctness of finite paths, each of which can be viewed statically. For example, many programming languages permit one to write

$$STR(I) \leftarrow 0;$$

When executed, this destructively deposits a 0 in the memory location containing the Ith element of STR. Let us assume that 0 was not already in that location. From the computational point of view, STR is the same object it was before, but now has a different Ith element. However, the inductive assertion method tells us that the correct way to view the situation, mathematically, is that before the assignment, STR denotes one (mathematical) sequence; after the assignment, it denotes another (mathematical) sequence. For a path containing such an assignment, the inductive assertion method would produce a verification condition (i.e., a mathematical conjecture) concerning two mathematical sequences, not a dynamically changing object.

These remarks are not meant to detract from the inductive assertion method. The method is a very natural way to assign meaning to certain kinds of programs. However, it should be recognized for what it is: a straightforward and well-defined way of mapping assertions about a program into logical formulas amenable to mathematical proof.

# XIX

## The Unique Prime Factorization Theorem

In this chapter, we discuss our system's proof of the unique prime factorization theorem, also known as the fundamental theorem of arithmetic. This theorem is certainly the deepest and hardest theorem yet proved by our theorem-prover. The principal difficulty behind the proof is that Euclid's greatest common divisor function (GCD) plays an important role, even though it is not involved in the statement of the theorem. A beautiful but surprising fact (that multiplication distributes over GCD) is used; the more obvious fact that the GCD of two numbers divides both of them is also used. No other theorem yet proved by the theorem-prover employs as a lemma a surprising fact about a function not involved in the statement of the theorem.

### A. THE CONTEXT

To prove the unique prime factorization theorem, we start at the very beginning with the axioms of Peano arithmetic (i.e., the ADD1 shell). We also need the axioms of lists and literal atoms (because we will eventually need to deal with lists of numbers). We then introduce, with the definition principle, the usual elementary functions on the natural numbers:

PLUS—the sum of its two arguments.
TIMES—the product of its two arguments.

DIFFERENCE—the first argument minus the second argument unless the first is less than the second, in which case the answer is 0.

QUOTIENT—the integer quotient of the first argument when divided by the second argument. If the divisor is 0, then the answer is 0.

REMAINDER—the remainder after dividing the first argument by the second. If the divisor is 0, then the answer is the first argument.

DIVIDES—the predicate whether the remainder of the second argument divided by the first argument is 0.

GCD—the greatest common divisor of the two arguments (0 if both are 0).

For all these functions, the convention is observed always to treat any argument that is not a number as if it were 0.

We also have the theorem-prover prove the usual elementary facts about these functions, such as that PLUS is associative and commutative and that TIMES distributes over PLUS.

We eventually reach less elementary theorems such as

**Theorem** REMAINDER.QUOTIENT.ELIM:
```
(IMPLIES (AND (NOT (ZEROP Y))
              (NUMBERP X))
         (EQUAL (PLUS (REMAINDER X Y)
                      (TIMES Y
                             (QUOTIENT X Y)))
                X)),
```

**Theorem** RECURSION.BY.QUOTIENT:
```
(IMPLIES (AND (NUMBERP I)
              (NOT (EQUAL I 0))
              (NUMBERP J)
              (NOT (EQUAL J 0))
              (NOT (EQUAL J 1)))
         (LESSP (QUOTIENT I J) I)),
```

**Theorem** DISTRIBUTIVITY.OF.TIMES.OVER.GCD:
```
(EQUAL (GCD (TIMES X Z) (TIMES Y Z))
       (TIMES Z (GCD X Y))),
```

**Theorem** GCD.DIVIDES.BOTH:

```
(AND (EQUAL (REMAINDER X (GCD X Y))
            0)
     (EQUAL (REMAINDER Y (GCD X Y))
            0)).
```

In Appendix A, we present the complete sequence of definitions and theorems leading to the unique prime factorization theorem. In this chapter, we do not further discuss those definitions and theorems preceding the definition of PRIME. However, we recommend those elementary theorems to any student of the foundations of mathematics and to anybody interested in mechanical theorem-proving. All the heuristics necessary to prove the unique prime factorization theorem were discovered and mechanized while proving those simpler theorems. In Chapter XVI, we presented proofs of some of those theorems.

Throughout the rest of this chapter, we will assume the reader is familiar with the theorems preceding the definition of PRIME in Appendix A. It is worth noting that the theorem-prover knows those theorems because it has proved them all from Peano's axioms (which it knows because it has been told to assume them).

## B. FORMAL DEVELOPMENT OF THE UNIQUE PRIME FACTORIZATION THEOREM

One may well ask how the uniqueness of prime factorizations may even be stated within our theory. McCarthy [34], for example, thought that the statement of the theorem was beyond the scope of his theory, which is very similar to ours. In our statement of the theorem, we use lists of numbers to represent factorizations of numbers, and we use explicit functions to overcome our lack of existential quantification. Our statement of the theorem has two parts. The first part states that an explicitly given function produces a prime factorization of any given nonzero integer. The second part states that any two lists of prime numbers whose "products" are the same are in fact permutations of one another. We now precisely define the necessary concepts and formally state these two parts of our version of the unique factorization theorem.

## 1. Definition of the Concepts

*a. The Definition of PRIME*

First, let us consider the definition of the concept of prime number. The usual definition is: X is prime provided it is an integer greater than 1 and its only divisors are itself and 1. This definition, strictly speaking, involves the consideration of all integers as possible divisors of X. We cannot use our principle of definition to introduce a concept that requires considering an infinite number of questions. Fortunately, we need consider only the finite number of integers less than X, since every divisor of a positive number X is less than or equal to X.[45] Therefore, we can write a definition of prime within our theory.

We need the auxiliary function PRIME1 that returns T if and only if X has no divisors less than or equal to Y (and greater than 1):

**Definition**

```
(PRIME1 X Y)
=
(IF (ZEROP Y)
    F
    (IF (EQUAL Y 1)
        T
        (AND (NOT (DIVIDES Y X))
             (PRIME1 X (SUB1 Y))))).
```

We then define PRIME as

**Definition**

```
(PRIME X)
=
(AND (NOT (ZEROP X))
     (NOT (EQUAL X 1))
     (PRIME1 X (SUB1 X))).
```

That is, PRIME checks that its argument is not 0, not 1, and has no divisor less than itself and greater than 1.

Readers unhappy about this constructive definition of prime number may be comforted to learn that in order to prove the unique prime

---

[45] In particular, the theorem
```
(IMPLIES (AND (NOT (ZEROP X)) (DIVIDES Y X)) (LESSEQP Y X))
```
can be proved by simplification alone, given the definitions of the functions and the rudimentary facts about LESSP.

## B. FORMAL DEVELOPMENT OF THE UNIQUE PRIME FACTORIZATION THEOREM / 313

factorization theorem, we will have to prove as lemmas the essential content of the ordinary definition of prime. These lemmas and their proofs are discussed in Section C.

### b. PRIME.LIST, TIMES.LIST, and PERM

We now introduce the other concepts needed in the statement of the theorem. The statement of the unique factorization theorem requires three new concepts: PRIME.LIST, TIMES.LIST, and PERM.

PRIME.LIST checks that every member of its argument is a prime. TIMES.LIST computes the product of the members of its argument.

**Definition**

```
(PRIME.LIST L)
    =
(IF (NLISTP L)
    T
    (AND (PRIME (CAR L))
         (PRIME.LIST (CDR L)))).
```

**Definition**

```
(TIMES.LIST L)
    =
(IF (NLISTP L)
    1
    (TIMES (CAR L)
           (TIMES.LIST (CDR L)))).
```

The function PERM computes whether its first argument, A, is a permutation of its second argument, B. If A is not a list (that is, A is empty), PERM requires that B not be a list; but if A is a list, then PERM requires that (CAR A) be a member of B and that (CDR A) be a permutation of the result of deleting (CAR A) from B. The definitions of DELETE and PERM are

**Definition**

```
(DELETE X L)
    =
(IF (NLISTP L)
    L
    (IF (EQUAL X (CAR L))
        (CDR L)
        (CONS (CAR L)
              (DELETE X (CDR L))))).
```

**Definition**

```
(PERM A B)
  =
(IF (NLISTP A)
    (NLISTP B)
    (AND (MEMBER (CAR A) B)
         (PERM (CDR A)
               (DELETE (CAR A) B)))).
```

## 2. The Statement of the Theorem

We are now able to state the two parts of our formulation of the unique factorization theorem. The first part states that there exists a prime factorization of any positive integer. We handle the existential quantification by exhibiting a recursive function, PRIME.FACTORS, that computes such a factorization. In particular, we prove that (PRIME.FACTORS X), when X is a nonzero integer, is a list of primes whose product is X:

**Theorem**  PRIME.FACTORIZATION.EXISTENCE:

```
(IMPLIES (NOT (ZEROP X))
         (AND (EQUAL (TIMES.LIST (PRIME.FACTORS X))
                     X)
              (PRIME.LIST (PRIME.FACTORS X)))).
```

The definition of PRIME.FACTORS has not yet been exhibited because its definition is irrelevant: if the above theorem is true, then regardless of the definition of PRIME.FACTORS, every positive integer has at least one prime decomposition. We have to define PRIME.FACTORS to prove the theorem, but we can define it any way we please.

The second part of the unique prime factorization theorem states the uniqueness of the decomposition. We will prove that if two lists of primes have the same product, then the lists are permutations of one another:

**Theorem**  PRIME.FACTORIZATION.UNIQUENESS:

```
(IMPLIES (AND (PRIME.LIST L1)
              (PRIME.LIST L2)
              (EQUAL (TIMES.LIST L1)
                     (TIMES.LIST L2)))
         (PERM L1 L2)).
```

## C. THE MECHANICAL PROOFS

Now that we have formally expressed the unique prime factorization theorem in our theory, we proceed to discuss proofs. In Appendix A, we present in complete detail the sequence of lemmas that the system is asked to prove on the way to the proof of PRIME.FACTORIZATION.UNIQUENESS and PRIME.FACTORIZATION.EXISTENCE. In the following discussion of the proofs, we do not discuss certain of these lemmas. Most of those we omit are proved only because they state obvious facts not "obvious" to the theorem-prover until they have been explicitly proved.

We describe the proofs much as they are described in number theory textbooks. We leave as exercises for the reader (or the theorem-prover) the details.

### 1. Elementary Facts about Primes

We first turn to the problem of proving the simplest facts about prime numbers, facts virtually equivalent to the ordinary definition of prime. The reason we have our theorem-prover prove these lemmas is that unless they are proved explicitly, the theorem-prover will never consider trying to prove that some number is prime by asking whether it has a factor. That is, the quantification implicit in the definition of prime is not apparent to the theorem-prover.

*a. Primes Have No Divisors*

There are two main parts to the ordinary definition of prime. The first tells us that if a number X has a divisor that is not 1 and not X, then X is not a prime. This part of the definition of prime is expressed in the lemma

**Theorem** PRIME.BASIC:

```
(IMPLIES (AND (NOT (EQUAL Z 1))
              (NOT (EQUAL Z X))
              (DIVIDES Z X))
         (NOT (PRIME1 X (SUB1 X)))).
```

The proof of this lemma involves a subsidiary lemma,

**Theorem** PRIME1.BASIC:
```
(IMPLIES (AND (NOT (EQUAL Z 1))
              (NOT (EQUAL Z (ADD1 X)))
              (EQUAL (REMAINDER (ADD1 X) Z)
                     0))
         (NOT (PRIME1 (ADD1 X) (PLUS Z L)))).
```

PRIME1.BASIC is actually more general than PRIME.BASIC because the conclusion of PRIME.BASIC, the formula (PRIME1 X (SUB1 X)), suffers from having X occur in both arguments. This double occurrence of X effectively prohibits PRIME.BASIC's being proved by induction. PRIME1.BASIC is more general because the two arguments to PRIME1 in the conclusion have no variables in common. It can be proved in one induction.

*b. Nonprimes Do Have Divisors*

The second part of the ordinary definition of PRIME tells us that if a number X is not a prime and not 0 or 1, then it has a divisor that is neither 1 nor X. To express the existence of such a factor we define the function GREATEST.FACTOR and prove that it has the appropriate properties.

i. GREATEST.FACTOR

We want a function that returns a factor of X, less than X and greater than 1, if X is a nonprime greater than 1. We can define this function any way we wish (provided it is acceptable to our definition principle). We have chosen to compute the greatest factor of X in exactly the way that PRIME1 discovers whether X is prime. Namely, starting at Y (which, in our use of the function will usually be (SUB1 X)), we count down, looking for a number that divides X:

**Definition**
```
(GREATEST.FACTOR X Y)
    =
(IF (OR (ZEROP Y) (EQUAL Y 1))
    X
    (IF (DIVIDES Y X)
        Y
        (GREATEST.FACTOR X (SUB1 Y)))).
```

Intuitively, if PRIME1 returns F, it is because it ran down to and discovered the GREATEST.FACTOR.

ii. The Properties of GREATEST.FACTOR

We prove that if X is a nonprime number (and not 0 or 1), then the GREATEST.FACTOR of X divides X, is less than X, and is not 0, 1, or a nonnumber.

These last three properties are captured by the theorems GREAT-EST.FACTOR.0, GREATEST.FACTOR.1, and NUMBERP.GREAT-EST.FACTOR. These theorems state the (trivial) conditions under which GREATEST.FACTOR is 0, 1, or not a number. Each is proved easily in one induction.

We are left with the two main properties of GREATEST.FACTOR. The lemma GREATEST.FACTOR.DIVIDES establishes that nonprimes greater than 1 have at least one divisor, namely, that computed by GREATEST.FACTOR:

**Theorem** GREATEST.FACTOR.DIVIDES:
```
(IMPLIES (AND (LESSP Y (ID X))
              (NOT (PRIME1 X Y))
              (NOT (ZEROP X))
              (NOT (EQUAL X 1))
              (NOT (ZEROP Y)))
         (EQUAL (REMAINDER X
                           (GREATEST.FACTOR X Y))
                0)).
```

Note that this lemma is stated more generally than one might at first have thought necessary. Instead of proving a theorem about (PRIME X) and (GREATEST.FACTOR X (SUB1 X)), we prove a theorem about (PRIME1 X Y) and (GREATEST.FACTOR X Y). The only reason for this generality is that one cannot prove the less general theorem by induction directly. However, the more general theorem is easily proved by the theorem-prover in one induction.[46]

While the above theorem establishes that the GREATEST.FACTOR of X is indeed a factor of X (under the right hypotheses), it does not assure us that it is not X itself. The theorem GREATEST.FACTOR.LESSP establishes that in fact GREATEST.FACTOR returns something less than X (and thus not equal to X):

---

[46] The use of the function ID (which is just the identity function on numbers) was necessary to the mechanical proofs because otherwise the induction heuristic merges the suggested inductions on Y and X, flawing the "right" induction on Y. This problem was mentioned in the discussion of induction, Chapter XV. The use of ID illustrates a subtle way in which the user can influence the theorem-prover's behavior.

**Theorem** GREATEST.FACTOR.LESSP:

```
(IMPLIES (AND (LESSP Y (ID X))
              (NOT (PRIME1 X Y))
              (NOT (ZEROP X))
              (NOT (EQUAL X 1))
              (NOT (ZEROP Y)))
         (LESSP (GREATEST.FACTOR X Y) X)).
```

The generality of GREATEST.FACTOR.LESSP was desired for precisely the same reasons that we stated GREATEST.FACTOR.DIVIDES as generally as we did.

This concludes our description of the rather elaborate groundwork that must be laid in the form of explicit lemmas, so that the theorem-prover is aware of the facts implicit in the ordinary definition of prime. We now proceed to the more interesting theorems concerning the existence and uniqueness of the prime factorization.

## 2. The Existence of a Prime Factorization

The proof of the existence of a prime factorization of every number greater than 0 is simple in comparison to the (coming) proof of the uniqueness of such a factorization. Recall that our objective is to prove

**Theorem** PRIME.FACTORIZATION.EXISTENCE:

```
(IMPLIES (NOT (ZEROP X))
         (AND (EQUAL (TIMES.LIST (PRIME.FACTORS X))
                     X)
              (PRIME.LIST (PRIME.FACTORS X)))).
```

The key to making a proof of this theorem difficult or easy lies in the definition of the function PRIME.FACTORS. Recall that we are happy, for the purposes of the unique prime factorization theorem, with any function whatsoever for computing an appropriate factorization.

Our description of the proof of PRIME.FACTORIZATION.EXISTENCE is divided into three parts. The first presents the definition of PRIME.FACTORS that we choose to use. The second describes the proof that PRIME.FACTORS returns a list of primes. The third describes the proof that the product over that list is the desired integer.

*a. The Definition of PRIME.FACTORS*

How can we compute a prime decomposition of a nonzero integer X? We first find the greatest factor of X less than X. Then we recursively compute a prime decomposition of that factor and a prime decomposition of the result of dividing X by that factor. Finally, we return the concatenation of the two decompositions.

## Definition

```
(PRIME.FACTORS X)
   =
(IF
 (OR (ZEROP X)
     (EQUAL (SUB1 X) 0))
 "NIL"
 (IF
  (PRIME1 X (SUB1 X))
  (CONS X "NIL")
  (APPEND
      (PRIME.FACTORS (GREATEST.FACTOR X (SUB1 X)))
      (PRIME.FACTORS
          (QUOTIENT X
                    (GREATEST.FACTOR X
                                     (SUB1 X))))))).
```

We chose to define PRIME.FACTORS the way we have because it makes one half of the proof of PRIME.FACTORIZATION.EXISTENCE trivial and the other half easy.

The trivial half of the proof is that PRIME.FACTORS always returns a list of PRIMES. The easy half is showing that the product of (PRIME.FACTORS X) is X. Both these halves are proved by induction. The induction used is the one analogous to the recursion involved in the definition of PRIME.FACTORS. Hence, it is doubly important that we first look at the justification of the definition of PRIME.FACTORS.

The theorem-prover decides that the definition PRIME.FACTORS is acceptable by determining that the COUNT of the argument decreases on each recursive call. There are two calls. In the first, X is replaced by a GREATEST.FACTOR expression, and in the second, X is replaced by the QUOTIENT of X by that greatest factor.

The proof that X is getting smaller in the first call relies on GREATEST.FACTOR.LESSP, mentioned earlier.[47] The proof that X is getting smaller in the second call relies on RECURSION.BY.QUOTIENT, noted in Section A.

---

[47] Actually, the definitional facility must be provided with a trivial variant of it called GREATEST.FACTOR.LESSP.IND, proved as an induction lemma, that makes explicit use of COUNT since GREATEST.FACTOR is not always numeric.

### b. PRIME.FACTORS *Returns a List of Primes*

The proof that PRIME.FACTORS returns a list of primes requires the easy lemma

**Theorem** PRIME.LIST.APPEND:

```
(EQUAL (PRIME.LIST (APPEND X Y))
       (AND (PRIME.LIST X)
            (PRIME.LIST Y))).
```

Once this theorem is known, a simple induction proves that the output of PRIME.FACTORS is always a list of primes. "NIL" is a list of primes, (CONS X "NIL") is a list of primes if X is a prime, and (by induction) the recursive calls of PRIME.FACTORS return lists of primes, so their concatenation is a list of primes, by PRIME.LIST.APPEND.

### c. PRIME.FACTORS *Factors Its Argument*

All that remains to establish the existence of a prime factorization of any nonzero number X is the proof that (TIMES.LIST (PRIME.FACTORS X)) is X. This theorem is proved by one induction (according to the way that PRIME.FACTORS recurses) after proving the two lemmas

**Theorem** QUOTIENT.TIMES1:

```
(IMPLIES (AND (NUMBERP Y)
              (NUMBERP X)
              (NOT (EQUAL X 0))
              (DIVIDES X Y))
         (EQUAL (TIMES X (QUOTIENT Y X))
                Y))
```

and

**Theorem** TIMES.LIST.APPEND:

```
(EQUAL (TIMES.LIST (APPEND X Y))
       (TIMES (TIMES.LIST X)
              (TIMES.LIST Y)))
```

and observing that the lemma GREATEST.FACTOR.DIVIDES helps to relieve the hypotheses of QUOTIENT.TIMES1.

### 3. The Uniqueness of a Prime Factorization

We now move on to the more difficult proof that the prime decomposition of a number is unique, up to permutation. We first prove a key theorem about primes, after which we carry off the final proof.

### a. A Key Theorem about Primes

The key fact about primes necessary in our proof of the unique prime factorization theorem is the theorem that states that if p is a prime and p divides the product of a and b, then p divides a or p divides b. All the lemmas mentioned in the other parts of this chapter are, comparatively speaking, routine. However, this lemma is decidedly not routine because its proof employs the GCD function in a surprising way.

We formulate this key lemma thus:

**Theorem** PRIME.KEY:

```
(IMPLIES (AND (NUMBERP Z)
              (PRIME X)
              (NOT (DIVIDES X Z))
              (NOT (DIVIDES X B)))
         (NOT (EQUAL (TIMES X K)
                     (TIMES B Z)))).
```

To prove PRIME.KEY, we first need to prove the two lemmas

**Theorem** HACK1:

```
(IMPLIES (AND (NOT (DIVIDES X A))
              (EQUAL A
                     (GCD (TIMES X A)
                          (TIMES B A))))
         (NOT (EQUAL (TIMES K X)
                     (TIMES B A)))),
```

**Theorem** PRIME.GCD:

```
(IMPLIES (AND (NOT (DIVIDES X B))
              (PRIME1 X (SUB1 X)))
         (EQUAL (EQUAL (GCD B X) 1) T)).
```

Once these two lemmas have been proved, the proof of PRIME.KEY is as follows. We first rewrite the conclusion of PRIME.KEY to (EQUAL (TIMES K X) (TIMES B Z)) and try to use HACK1 to establish the conclusion of PRIME.KEY. To use HACK1, we must establish two hypotheses: (NOT (DIVIDES X Z)) and (EQUAL Z (GCD (TIMES Z X) (TIMES Z B))). We have the first of these hypotheses among the hypotheses of PRIME.KEY. To establish the equality, we first rewrite it using the DISTRIBUTIVITY.OF.TIMES.OVER.GCD to (EQUAL Z (TIMES Z (GCD X B))). A simple arithmetic lemma (called TIMES.IDENTITY) tells us that to establish such an equality we need to check that (EQUAL (GCD X B) 1). But this is a direct consequence of PRIME.GCD and the hypotheses in PRIME.KEY. Q.E.D.

Now let us prove the two lemmas we used, HACK1 and PRIME.GCD. First, we prove HACK1. The proof is most easily seen by reformulating HACK1 in the contrapositive:

```
(IMPLIES (AND (EQUAL (TIMES K X) (TIMES B A))
              (EQUAL A (GCD (TIMES X A)
                            (TIMES B A))))
         (DIVIDES X A)).
```

Since (TIMES K X) is (TIMES B A) by hypothesis, we obtain, by substituting into the other hypothesis, that A is equal to (GCD (TIMES X A) (TIMES K X)). By the commutativity of TIMES and the DISTRIBUTIVITY.OF.TIMES.OVER.GCD, we obtain that A is equal to (TIMES X (GCD A K)), which implies that X divides A.

Finally, we turn to the proof of the second lemma we used, PRIME.GCD; since (GCD X B) divides X (because of the lemma GCD.DIVIDES.BOTH) and since X is a PRIME by hypothesis, then (GCD X B) must be X or 1. If (GCD X B) were X, then X would divide B (since (GCD X B) divides both), but this contradicts the hypotheses that (NOT (DIVIDES X B)). So (GCD X B) must be 1.

## b. The Unique Factorization

Now that we have established the lemma PRIME.KEY, it is routine work to obtain the uniqueness of the prime factorization.

First we establish the lemma that tells us that if the product of a list of primes is divisible by a prime, then the prime is a member of the list:

**Theorem** PRIME.LIST.TIMES.LIST:

```
(IMPLIES (AND (PRIME C)
              (PRIME.LIST L2)
              (NOT (MEMBER C L2)))
         (NOT (EQUAL (REMAINDER (TIMES.LIST L2) C)
                     0))).
```

It is here that we make use of our key theorem about primes,

PRIME.KEY. The proof of PRIME.LIST.TIMES.LIST is by induction on the length of the list L2. If C divides (TIMES.LIST L2), then by PRIME.KEY, C divides (CAR L2) or C divides (TIMES.LIST (CDR L2)). If C divides (CAR L2), then (since L2 is a list of primes), C is (CAR L2). If C divides (TIMES.LIST (CDR L2)), then by the inductive hypothesis, C is a member of (CDR L2).

Given PRIME.LIST.TIMES.LIST, the unique factorization theorem,

**Theorem** PRIME.FACTORIZATION.UNIQUENESS:

        (IMPLIES (AND (PRIME.LIST L1)
                      (PRIME.LIST L2)
                      (EQUAL (TIMES.LIST L1)
                             (TIMES.LIST L2)))
                 (PERM L1 L2)),

can be proved by inducting according to the way that PERM recurses. That is, in the inductive step, we assume an instance of the theorem in which L1 is replaced by (CDR L1) and L2 is replaced by (DELETE (CAR L1) L2). If L1 and L2 are both nonempty lists of primes and (TIMES.LIST L1) is equal to (TIMES.LIST L2), then we wish to establish that L1 is a permutation of L2. But (CAR L1) divides (TIMES.LIST L2). By PRIME.LIST.TIMES.LIST, (CAR L1) must be a member of L2. Thus we need establish only that (CDR L1) is a permutation of (DELETE (CAR L1) L2). To do so, we try to use our induction hypothesis. Clearly, (CDR L1) is a list of primes. Further, (DELETE (CAR L1) L2) is a list of primes, by the easily proved lemma

**Theorem** PRIME.LIST.DELETE:

        (IMPLIES (PRIME.LIST L2)
                 (PRIME.LIST (DELETE X L2))).

Finally, (TIMES.LIST (CDR L1)) is equal to (TIMES.LIST (DELETE (CAR L1) L2)) by the lemma

**Theorem** TIMES.LIST.DELETE:

        (IMPLIES (AND (NOT (ZEROP C))
                      (MEMBER C L))
                 (EQUAL (TIMES.LIST (DELETE C L))
                        (QUOTIENT (TIMES.LIST L) C))).

Therefore, we can use the inductive hypothesis to conclude that (CDR L2) is a permutation of (DELETE (CAR L1) L2).   Q.E.D.

We conclude by presenting the output of the theorem-prover for
PRIME.FACTORIZATION.UNIQUENESS. In the output, the lemma
PRIME.LIST.TIMES.LIST is not actually mentioned, but a mild reformulation of it called PRIME.MEMBER is. Note that the proof generates an inductively proved subgoal, namely, that if the product over
a list of primes is 1, then the list is empty. Prime factorizations would
not be unique if 1 were a prime.

**Theorem** PRIME.FACTORIZATION.UNIQUENESS:
```
(IMPLIES (AND (PRIME.LIST L1)
              (PRIME.LIST L2)
              (EQUAL (TIMES.LIST L1)
                     (TIMES.LIST L2)))
         (PERM L1 L2)).
```
Name the above subgoal *1.

Let us appeal to the induction principle. The
recursive terms in the conjecture suggest five
inductions. They merge into two likely candidate
inductions. However, only one is unflawed. We will
induct according to the following scheme:

```
(AND (IMPLIES (NOT (LISTP L1)) (p L1 L2))
     (IMPLIES (AND (LISTP L1)
                   (p (CDR L1) (DELETE (CAR L1) L2)))
              (p L1 L2))).
```

The inequality CDR.LESSP establishes that the measure
(COUNT L1) decreases according to the well-founded
relation LESSP in the induction step of the scheme.
Note, however, the inductive instance chosen for L2.
The above induction scheme leads to five new goals:

*Case 1.* (IMPLIES (AND (NOT (LISTP L1))
                      (PRIME.LIST L1)
                      (PRIME.LIST L2)
                      (EQUAL (TIMES.LIST L1)
                             (TIMES.LIST L2)))
                 (PERM L1 L2)),

which simplifies, opening up the functions
PRIME.LIST, TIMES.LIST and PERM, to:

```
        (IMPLIES (AND (NOT (LISTP L1))
                      (PRIME.LIST L2)
                      (EQUAL 1 (TIMES.LIST L2)))
                 (NOT (LISTP L2))),
```
which has an irrelevant term in it. By eliminating this term we get:
```
        (IMPLIES (AND (PRIME.LIST L2)
                      (EQUAL 1 (TIMES.LIST L2)))
                 (NOT (LISTP L2))).
```
Name the above subgoal *1.1.

*Case 2.*
```
        (IMPLIES (AND (LISTP L1)
                      (NOT (PRIME.LIST (CDR L1)))
                      (PRIME.LIST L1)
                      (PRIME.LIST L2)
                      (EQUAL (TIMES.LIST L1)
                             (TIMES.LIST L2)))
                 (PERM L1 L2)),
```
which we simplify, opening up PRIME.LIST, to:
```
        (TRUE).
```

*Case 3.*
```
        (IMPLIES
                (AND (LISTP L1)
                     (NOT (PRIME.LIST (DELETE (CAR L1)
                                              L2)))
                     (PRIME.LIST L1)
                     (PRIME.LIST L2)
                     (EQUAL (TIMES.LIST L1)
                            (TIMES.LIST L2)))
                (PERM L1 L2)),
```
which we simplify, using the lemma PRIME.LIST.DELETE, to:
```
        (TRUE).
```

*Case 4.*
```
        (IMPLIES
           (AND
              (LISTP L1)
              (NOT (EQUAL (TIMES.LIST (CDR L1))
                          (TIMES.LIST (DELETE (CAR L1)
                                              L2))))
```

```
        (PRIME.LIST L1)
        (PRIME.LIST L2)
        (EQUAL (TIMES.LIST L1)
               (TIMES.LIST L2)))
     (PERM L1 L2)),
```

which we simplify, appealing to the lemma
PRIME.MEMBER, and expanding the definitions of
PRIME.LIST, TIMES.LIST and PERM, to the formula:

```
     (IMPLIES
      (AND
        (LISTP L1)
        (NOT (EQUAL (TIMES.LIST (CDR L1))
                    (TIMES.LIST (DELETE (CAR L1)
                                        L2))))
        (NOT (EQUAL (CAR L1) 0))
        (NUMBERP (CAR L1))
        (NOT (EQUAL (CAR L1) 1))
        (PRIME1 (CAR L1) (SUB1 (CAR L1)))
        (PRIME.LIST (CDR L1))
        (PRIME.LIST L2)
        (EQUAL (TIMES (CAR L1)
                      (TIMES.LIST (CDR L1)))
               (TIMES.LIST L2)))
      (PERM (CDR L1) (DELETE (CAR L1) L2))),
```

which again simplifies, applying the lemmas
TIMES.LIST.DELETE, PRIME.MEMBER and
DIVIDES.IMPLIES.TIMES, to:

    (TRUE).

*Case 5.* (IMPLIES (AND (LISTP L1)
                        (PERM (CDR L1) (DELETE (CAR L1)
                                               L2))
                        (PRIME.LIST L1)
                        (PRIME.LIST L2)
                        (EQUAL (TIMES.LIST L1)
                               (TIMES.LIST L2)))
                   (PERM L1 L2)).

This simplifies, appealing to the lemma PRIME.MEMBER,
and expanding PRIME.LIST, TIMES.LIST and PERM, to:

# C. THE MECHANICAL PROOFS / 327

(TRUE).

So we now return to:

```
(IMPLIES (AND (PRIME.LIST L2)
              (EQUAL 1 (TIMES.LIST L2)))
         (NOT (LISTP L2))),
```

which we named *1.1 above. Let us appeal to the induction principle. Two inductions are suggested by terms in the conjecture. However, they merge into one likely candidate induction. We will induct according to the following scheme:

```
(AND (IMPLIES (NOT (LISTP L2)) (p L2))
     (IMPLIES (AND (LISTP L2) (p (CDR L2)))
              (p L2))).
```

The inequality CDR.LESSP establishes that the measure (COUNT L2) decreases according to the well-founded relation LESSP in the induction step of the scheme. The above induction scheme generates the following three new formulas:

*Case 1.* `(IMPLIES (AND (NOT (PRIME.LIST (CDR L2)))`
```
              (PRIME.LIST L2)
              (EQUAL 1 (TIMES.LIST L2)))
         (NOT (LISTP L2))),
```

which simplifies, opening up the function PRIME.LIST, to:

(TRUE).

*Case 2.*
```
(IMPLIES (AND (NOT (EQUAL 1 (TIMES.LIST (CDR L2))))
              (PRIME.LIST L2)
              (EQUAL 1 (TIMES.LIST L2)))
         (NOT (LISTP L2))).
```

This simplifies, applying TIMES.EQUAL.1, and opening up the definitions of PRIME.LIST and TIMES.LIST, to:

```
(IMPLIES (AND (NOT (EQUAL 1 (TIMES.LIST (CDR L2))))
              (NOT (EQUAL (CAR L2) 0))
              (NUMBERP (CAR L2))
              (NOT (EQUAL (CAR L2) 1))
              (PRIME1 (CAR L2) (SUB1 (CAR L2))))
```

```
                    (PRIME.LIST (CDR L2))
                    (NOT (EQUAL 0 (TIMES.LIST (CDR L2))))
                    (EQUAL 0 (SUB1 (CAR L2)))
                    (EQUAL 0
                           (SUB1 (TIMES.LIST (CDR L2)))))
              (NOT (LISTP L2))),
```

which again simplifies, opening up the function PRIME1, to:

```
       (TRUE).
```

*Case 3.* (IMPLIES (AND (NOT (LISTP (CDR L2)))
                    (PRIME.LIST L2)
                    (EQUAL 1 (TIMES.LIST L2)))
              (NOT (LISTP L2))),

which we simplify, applying the lemmas TIMES.EQUAL.1 and SUB1.ADD1, and opening up the definitions of PRIME.LIST and TIMES.LIST, to:

```
       (IMPLIES (AND (NOT (LISTP (CDR L2)))
                    (NOT (EQUAL (CAR L2) 0))
                    (NUMBERP (CAR L2))
                    (NOT (EQUAL (CAR L2) 1))
                    (PRIME1 (CAR L2)
                            (SUB1 (CAR L2)))
                    (EQUAL 0 (SUB1 (CAR L2))))
              (NOT (LISTP L2))),
```

which again simplifies, expanding the function PRIME1, to:

```
       (TRUE).
```

That finishes the proof of *1.1, which, consequently, finishes the proof of *1. Q.E.D.

CPU time (devoted to theorem-proving): 94.717 seconds

# Appendix A

## Definitions Accepted and Theorems Proved by Our System

Here is a complete list (as of May, 1978), in the order processed by the system, of all our standard definitions and theorems. Whenever we add a new proof technique or change an old one, we make sure that the "improved" theorem-prover can still prove all these theorems. The current list contains 286 theorems and 110 definitions. After each theorem or axiom we list the user-supplied "types" (i.e., rewrite, elimination, generalization, and induction). A theorem with no types is proved but not made available for future use.

When the system begins processing the sequence, it knows only the axioms, definitions, lemmas, and principles described in Chapter III.

**1. Definition**

```
(NLISTP X)
   =
(NOT (LISTP X))
```

**2. Definition**

```
(APPEND X Y)
   =
(IF (LISTP X)
    (CONS (CAR X) (APPEND (CDR X) Y))
    Y)
```

**3. Definition**

```
(REVERSE X)
   =
(IF (LISTP X)
    (APPEND (REVERSE (CDR X))
            (CONS (CAR X) "NIL"))
    "NIL")
```

**4. Definition**
```
(TIMES I J)
    =
(IF (ZEROP I)
    0
    (PLUS J (TIMES (SUB1 I) J)))
```

**5. Theorem** ASSOCIATIVITY.OF.APPEND (rewrite):
```
(EQUAL (APPEND (APPEND X Y) Z)
       (APPEND X (APPEND Y Z)))
```

**6. Definition**
```
(PLISTP X)
    =
(IF (LISTP X)
    (PLISTP (CDR X))
    (EQUAL X "NIL"))
```

**7. Theorem** APPEND.RIGHT.ID (rewrite):
```
(IMPLIES (PLISTP X)
         (EQUAL (APPEND X "NIL") X))
```

**8. Theorem** PLISTP.REVERSE (generalize and rewrite):
```
(PLISTP (REVERSE X))
```

**9. Theorem** APPEND.REVERSE (rewrite):
```
(EQUAL (REVERSE (APPEND A B))
       (APPEND (REVERSE B) (REVERSE A)))
```

**10. Theorem** PLUS.RIGHT.ID (rewrite):
```
(EQUAL (PLUS X 0) (FIX X))
```

**11. Theorem** PLUS.ADD1 (rewrite):
```
(EQUAL (PLUS X (ADD1 Y))
       (IF (NUMBERP Y)
           (ADD1 (PLUS X Y))
           (ADD1 X)))
```

**12. Theorem** COMMUTATIVITY2.OF.PLUS (rewrite):
```
(EQUAL (PLUS X (PLUS Y Z))
       (PLUS Y (PLUS X Z)))
```

**13. Theorem** COMMUTATIVITY.OF.PLUS (rewrite):
```
(EQUAL (PLUS X Y) (PLUS Y X))
```

**14. Theorem** ASSOCIATIVITY.OF.PLUS (rewrite):
```
(EQUAL (PLUS (PLUS X Y) Z)
       (PLUS X (PLUS Y Z)))
```

**15. Theorem** TIMES.ZERO (rewrite):
```
(EQUAL (TIMES X 0) 0)
```

## DEFINITIONS ACCEPTED AND THEOREMS PROVED BY OUR SYSTEM / 331

**16. Theorem** DISTRIBUTIVITY.OF.TIMES.OVER.PLUS (rewrite):

```
(EQUAL (TIMES X (PLUS Y Z))
       (PLUS (TIMES X Y) (TIMES X Z)))
```

**17. Theorem** TIMES.ADD1 (rewrite):

```
(EQUAL (TIMES X (ADD1 Y))
       (IF (NUMBERP Y)
           (PLUS X (TIMES X Y))
           (FIX X)))
```

**18. Theorem** COMMUTATIVITY.OF.TIMES (rewrite):

```
(EQUAL (TIMES X Y) (TIMES Y X))
```

**19. Theorem** COMMUTATIVITY2.OF.TIMES (rewrite):

```
(EQUAL (TIMES X (TIMES Y Z))
       (TIMES Y (TIMES X Z)))
```

**20. Theorem** ASSOCIATIVITY.OF.TIMES (rewrite):

```
(EQUAL (TIMES (TIMES X Y) Z)
       (TIMES X (TIMES Y Z)))
```

**21. Theorem** EQUAL.TIMES.0 (rewrite):

```
(EQUAL (EQUAL (TIMES X Y) 0)
       (OR (ZEROP X) (ZEROP Y)))
```

**22. Shell Definition**

Add the shell PUSH of two arguments
with recognizer STACKP,
accessors TOP and POP,
default values 0 and 0, and
well-founded relation TOP.POPP.

**23. Undefined Function**

```
(APPLY FN X Y)
```

**24. Undefined Function**

```
(GETVALUE VAR ENVRN)
```

**25. Axiom** NUMBERP.APPLY (rewrite):

```
(NUMBERP (APPLY FN X Y))
```

**26. Definition**

```
(FORMP X)
   =
(IF (LISTP X)
    (IF (LISTP (CAR X))
        F
        (IF (LISTP (CDR X))
            (IF (LISTP (CDR (CDR X)))
```

```
                        (IF (FORMP (CAR (CDR X)))
                            (FORMP (CAR (CDR (CDR X)))))
                            F)
                    F)
                F))
        T)
```

## 27. Definition

```
(EVAL FORM ENVRN)
    =
(IF (NUMBERP FORM)
    FORM
    (IF (LISTP (CDDR FORM))
        (APPLY (CAR FORM)
               (EVAL (CADR FORM) ENVRN)
               (EVAL (CADDR FORM) ENVRN))
        (GETVALUE FORM ENVRN)))
```

## 28. Definition

```
(OPTIMIZE FORM)
    =
(IF (LISTP (CDDR FORM))
    (IF (NUMBERP (OPTIMIZE (CADR FORM)))
        (IF (NUMBERP (OPTIMIZE (CADDR FORM)))
            (APPLY (CAR FORM)
                   (OPTIMIZE (CADR FORM))
                   (OPTIMIZE (CADDR FORM)))
            (CONS (CAR FORM)
                  (CONS (OPTIMIZE (CADR FORM))
                        (CONS (OPTIMIZE (CADDR FORM))
                              "NIL"))))
        (CONS (CAR FORM)
              (CONS (OPTIMIZE (CADR FORM))
                    (CONS (OPTIMIZE (CADDR FORM)
                          "NIL"))))
    FORM)
```

## 29. Definition

```
(CODEGEN FORM INS)
    =
(IF (NUMBERP FORM)
    (CONS (CONS "PUSHI" (CONS FORM "NIL")) INS)
    (IF (LISTP (CDDR FORM))
        (CONS (CAR FORM)
              (CODEGEN (CADDR FORM)
                       (CODEGEN (CADR FORM) INS)))
        (CONS (CONS "PUSHV" (CONS FORM "NIL")) INS)))
```

## 30. Definition

```
(COMPILE FORM)
    =
(REVERSE (CODEGEN (OPTIMIZE FORM) "NIL"))
```

### DEFINITIONS ACCEPTED AND THEOREMS PROVED BY OUR SYSTEM / 333

**31. Theorem** FORMP.OPTIMIZE (rewrite):

```
(IMPLIES (FORMP X)
         (FORMP (OPTIMIZE X)))
```

**32. Theorem** CORRECTNESS.OF.OPTIMIZE (rewrite):

```
(IMPLIES (FORMP X)
         (EQUAL (EVAL (OPTIMIZE X) ENVRN)
                (EVAL X ENVRN)))
```

**33. Definition**

```
(EXEC PC PDS ENVRN)
  =
(IF
 (NLISTP PC)
 PDS
 (IF
  (LISTP (CAR PC))
  (IF
     (EQUAL (CAR (CAR PC)) "PUSHI")
     (EXEC (CDR PC)
           (PUSH (CAR (CDR (CAR PC))) PDS)
           ENVRN)
     (EXEC (CDR PC)
           (PUSH (GETVALUE (CAR (CDR (CAR PC))) ENVRN)
                 PDS)
           ENVRN))
    (EXEC (CDR PC)
          (PUSH (APPLY (CAR PC)
                       (TOP (POP PDS))
                       (TOP PDS))
                (POP (POP PDS)))
          ENVRN)))
```

**34. Theorem** SEQUENTIAL.EXECUTION (rewrite):

```
(EQUAL (EXEC (APPEND X Y) PDS ENVRN)
       (EXEC Y (EXEC X PDS ENVRN) ENVRN))
```

**35. Theorem** CORRECTNESS.OF.CODEGEN (rewrite):

```
(IMPLIES
         (FORMP X)
         (EQUAL (EXEC (REVERSE (CODEGEN X INS))
                      PDS ENVRN)
                (PUSH (EVAL X ENVRN)
                      (EXEC (REVERSE INS) PDS ENVRN))))
```

**36. Theorem** CORRECTNESS.OF.OPTIMIZING.COMPILER:

```
(IMPLIES (FORMP X)
         (EQUAL (EXEC (COMPILE X) PDS ENVRN)
                (PUSH (EVAL X ENVRN) PDS)))
```

**37. Theorem** SUB1.LESSP1 (rewrite):

```
(IMPLIES (AND (NUMBERP X) (NOT (EQUAL X 0)))
         (LESSP (SUB1 X) X))
```

**38. Theorem** TRANSITIVITY.OF.LESSP (rewrite):

(IMPLIES (AND (NOT (LESSP X Z)) (LESSP X Z))
         (NOT (LESSP X Y)))

**39. Theorem** TRANSITIVITY.OF.LESSP2 (rewrite):

(IMPLIES (AND (NOT (LESSP X Z)) (LESSP X Y))
         (NOT (LESSP Y Z)))

**40. Theorem** TRANSITIVITY.OF.LESSP3 (rewrite):

(IMPLIES (AND (LESSP Y Z) (LESSP X Y))
         (LESSP X Z))

**41. Theorem** TRANSITIVITY.OF.NOT.LESSP (rewrite):

(IMPLIES (AND (LESSP X Z) (NOT (LESSP Y Z)))
         (LESSP X Y))

**42. Theorem** TRANSITIVITY.OF.NOT.LESSP2 (rewrite):

(IMPLIES (AND (LESSP X Z) (NOT (LESSP X Y)))
         (LESSP Y Z))

**43. Theorem** TRANSITIVITY.OF.NOT.LESSP3 (rewrite):

(IMPLIES (AND (NOT (LESSP Y Z))
              (NOT (LESSP X Y)))
         (NOT (LESSP X Z)))

**44. Theorem** LESSP.NOT.REFLEXIVE (rewrite):

(NOT (LESSP X X))

**45. Definition**

(EQP X Y)
=
(EQUAL (FIX X) (FIX Y))

**46. Theorem** LESSP.EQUAL (rewrite):

(IMPLIES (AND (NOT (EQP X Y))
              (NOT (LESSP Y X)))
         (LESSP X Y))

**47. Theorem** REVERSE.REVERSE (rewrite):

(IMPLIES (PLISTP X)
         (EQUAL (REVERSE (REVERSE X)) X))

**48. Definition**

(FLATTEN X)
=
(IF (LISTP X)
    (APPEND (FLATTEN (CAR X))
            (FLATTEN (CDR X)))
    (CONS X "NIL"))

## DEFINITIONS ACCEPTED AND THEOREMS PROVED BY OUR SYSTEM / 335

**49. Definition**

```
(MC.FLATTEN X Y)
   =
(IF (LISTP X)
    (MC.FLATTEN (CAR X)
                (MC.FLATTEN (CDR X) Y))
    (CONS X Y))
```

**50. Theorem** FLATTEN.MC.FLATTEN (rewrite):

```
(EQUAL (MC.FLATTEN X Y)
       (APPEND (FLATTEN X) Y))
```

**51. Definition**

```
(MEMBER X L)
   =
(IF (LISTP L)
    (IF (EQUAL X (CAR L))
        T
        (MEMBER X (CDR L)))
    F)
```

**52. Theorem** MEMBER.APPEND (rewrite):

```
(IMPLIES (MEMBER A B)
         (MEMBER A (APPEND B C)))
```

**53. Theorem** MEMBER.APPEND2 (rewrite):

```
(IMPLIES (MEMBER A C)
         (MEMBER A (APPEND B C)))
```

**54. Theorem** MEMBER.REVERSE (rewrite):

```
(EQUAL (MEMBER X (REVERSE Y))
       (MEMBER X Y))
```

**55. Definition**

```
(LENGTH X)
   =
(IF (LISTP X)
    (ADD1 (LENGTH (CDR X)))
    0)
```

**56. Theorem** LENGTH.REVERSE (rewrite):

```
(EQUAL (LENGTH (REVERSE X))
       (LENGTH X))
```

**57. Definition**

```
(INTERSECT X Y)
   =
(IF (LISTP X)
    (IF (MEMBER (CAR X) Y)
        (CONS (CAR X) (INTERSECT (CDR X) Y))
        (INTERSECT (CDR X) Y))
    "NIL")
```

**58. Theorem** MEMBER-INTERSECT:

```
(IMPLIES (AND (MEMBER A B) (MEMBER A C))
         (MEMBER A (INTERSECT B C)))
```

**59. Definition**

```
(UNION X Y)
  =
(IF (LISTP X)
    (IF (MEMBER (CAR X) Y)
        (UNION (CDR X) Y)
        (CONS (CAR X) (UNION (CDR X) Y)))
    Y)
```

**60. Theorem** MEMBER.UNION:

```
(IMPLIES (OR (MEMBER A B) (MEMBER A C))
         (MEMBER A (UNION B C)))
```

**61. Definition**

```
(SUBSETP X Y)
  =
(IF (LISTP X)
    (IF (MEMBER (CAR X) Y)
        (SUBSETP (CDR X) Y)
        F)
    T)
```

**62. Theorem** SUBSETP.UNION:

```
(IMPLIES (SUBSETP A B)
         (EQUAL (UNION A B) B))
```

**63. Theorem** SUBSETP.INTERSECT:

```
(IMPLIES (AND (PLISTP A) (SUBSETP A B))
         (EQUAL (INTERSECT A B) A))
```

**64. Definition**

```
(NTH X N)
  =
(IF (ZEROP N)
    X
    (NTH (CDR X) (SUB1 N)))
```

**65. Theorem** NTH.MEMBER:

```
(IMPLIES (LISTP (NTH X N))
         (MEMBER (CAR (NTH X N)) X))
```

**66. Definition**

```
(GREATERP X Y)
  =
(LESSP Y X)
```

### 67. Definition
```
(LESSEQP X Y)
   =
(NOT (LESSP Y X))
```

### 68. Definition
```
(GREATEREQP X Y)
   =
(NOT (LESSP X Y))
```

### 69. Theorem  TRANSITIVITY.OF.GREATERP:
```
(IMPLIES (AND (GREATERP X Y) (GREATERP Y Z))
         (GREATERP X Z))
```

### 70. Theorem  TRANSITIVITY.OF.LESSEQP:
```
(IMPLIES (AND (LESSEQP X Y) (LESSEQP Y Z))
         (LESSEQP X Z))
```

### 71. Theorem  TRICHOTOMY.OF.LESSP:
```
(OR (LESSP X Y)
    (OR (EQP X Y) (LESSP Y X)))
```

### 72. Definition
```
(ORDERED L)
   =
(IF (LISTP L)
    (IF (LISTP (CDR L))
        (IF (LESSP (CAR (CDR L)) (CAR L))
            F
            (ORDERED (CDR L)))
        T)
    T)
```

### 73. Definition
```
(ADDTOLIST X L)
   =
(IF (LISTP L)
    (IF (LESSP X (CAR L))
        (CONS X L)
        (CONS (CAR L) (ADDTOLIST X (CDR L))))
    (CONS X "NIL"))
```

### 74. Definition
```
(SORT L)
   =
(IF (LISTP L)
    (ADDTOLIST (CAR L) (SORT (CDR L)))
    "NIL")
```

### 75. Theorem  LESSEQP.PLUS (rewrite):
```
(NOT (LESSP (PLUS X Y) X))
```

76. **Theorem.** LESSEQP.PLUS2 (rewrite):

    (NOT (LESSP (PLUS X Y) Y))

77. **Theorem** COMMUTATIVITY.OF.APPEND.WRT.LENGTH:

    (EQUAL (LENGTH (APPEND A B))
           (LENGTH (APPEND B A)))

78. **Definition**

    (LAST L)
    =
    (IF (LISTP L)
        (IF (LISTP (CDR L)) (LAST (CDR L)) L)
        L)

79. **Definition**

    (ASSOC X Y)
    =
    (IF (LISTP Y)
        (IF (EQUAL X (CAR (CAR Y)))
            (CAR Y)
            (ASSOC X (CDR Y)))
        "NIL")

80. **Definition**

    (PAIRLIST X Y)
    =
    (IF (LISTP X)
        (IF (LISTP Y)
            (CONS (CONS (CAR X) (CAR Y))
                  (PAIRLIST (CDR X) (CDR Y)))
            "NIL")
        "NIL")

81. **Theorem** ASSOC.PAIRLIST (rewrite):

    (IMPLIES (AND (NOT (LESSP (LENGTH C) (LENGTH B)))
                  (MEMBER A B))
             (LISTP (ASSOC A (PAIRLIST B C))))

82. **Definition**

    (MAPCAR X FN)
    =
    (IF (LISTP X)
        (CONS (APPLY FN (CAR X) "NIL")
              (MAPCAR (CDR X) FN))
        "NIL")

83. **Theorem** MAPCAR.APPEND:

    (EQUAL (MAPCAR (APPEND A B) FN)
           (APPEND (MAPCAR A FN) (MAPCAR B FN)))

84. **Theorem** LENGTH.MAPCAR:

    (EQUAL (LENGTH (MAPCAR A FN))
           (LENGTH A))

### DEFINITIONS ACCEPTED AND THEOREMS PROVED BY OUR SYSTEM / 339

**85. Theorem** REVERSE.MAPCAR:

    (EQUAL (REVERSE (MAPCAR A FN))
           (MAPCAR (REVERSE A) FN))

**86. Definition**

    (LIT X Y FN)
    =
    (IF (LISTP X)
        (APPLY FN (CAR X) (LIT (CDR X) Y FN))
        Y)

**87. Theorem** LIT.APPEND (rewrite):

    (EQUAL (LIT (APPEND A B) C FN)
           (LIT A (LIT B C FN) FN))

**88. Definition**

    (BOOLEAN X)
    =
    (OR (EQUAL X T) (EQUAL X F))

**89. Definition**

    (IFF X Y)
    =
    (AND (IMPLIES X Y) (IMPLIES Y X))

**90. Theorem** IFF.EQUAL.EQUAL:

    (IMPLIES (AND (BOOLEAN P) (BOOLEAN Q))
             (EQUAL (IFF P Q) (EQUAL P Q)))

**91. Theorem** NTH.NIL (rewrite):

    (EQUAL (NTH "NIL" I) "NIL")

**92. Theorem** NTH.APPEND1:

    (EQUAL (NTH A (PLUS I J))
           (NTH (NTH A I) J))

**93. Theorem** COMMUTATIVITY.OF.EQUAL:

    (EQUAL (EQUAL A B) (EQUAL B A))

**94. Theorem** TRANSITIVITY.OF.EQUAL:

    (IMPLIES (AND (EQUAL A B) (EQUAL B C))
             (EQUAL A C))

**95. Theorem** ASSOCIATIVITY.OF.EQUAL:

    (IMPLIES (AND (BOOLEAN A)
                  (AND (BOOLEAN B) (BOOLEAN C)))
             (EQUAL (EQUAL (EQUAL A B) C)
                    (EQUAL A (EQUAL B C))))

## 96. Definition

```
(ODD X)
  =
(IF (ZEROP X)
    F
    (IF (ZEROP (SUB1 X))
        T
        (ODD (SUB1 (SUB1 X)))))
```

## 97. Definition

```
(EVEN1 X)
  =
(IF (ZEROP X) T (ODD (SUB1 X)))
```

## 98. Definition

```
(EVEN2 X)
  =
(IF (ZEROP X)
    T
    (IF (ZEROP (SUB1 X))
        F
        (EVEN2 (SUB1 (SUB1 X)))))
```

## 99. Definition

```
(DOUBLE I)
  =
(IF (ZEROP I)
    0
    (ADD1 (ADD1 (DOUBLE (SUB1 I)))))
```

## 100. Theorem EVEN1.DOUBLE:

```
(EVEN1 (DOUBLE I))
```

## 101. Definition

```
(HALF I)
  =
(IF (ZEROP I)
    0
    (IF (ZEROP (SUB1 I))
        0
        (ADD1 (HALF (SUB1 (SUB1 I))))))
```

## 102. Theorem HALF.DOUBLE:

```
(IMPLIES (NUMBERP I)
         (EQUAL (HALF (DOUBLE I)) I))
```

## 103. Theorem DOUBLE.HALF:

```
(IMPLIES (AND (NUMBERP I) (EVEN1 I))
         (EQUAL (DOUBLE (HALF I)) I))
```

## 104. Theorem DOUBLE.TIMES.2:

```
(EQUAL (DOUBLE I) (TIMES 2 I))
```

## DEFINITIONS ACCEPTED AND THEOREMS PROVED BY OUR SYSTEM / 341

**105. Theorem** SUBSETP.CONS (rewrite):

    (IMPLIES (SUBSETP X Y)
         (SUBSETP X (CONS Z Y)))

**106. Theorem** LAST.APPEND (rewrite):

    (EQUAL (LAST (APPEND A B))
       (IF (LISTP B)
         (LAST B)
         (IF (LISTP A)
           (CONS (CAR (LAST A)) B)
           B)))

**107. Theorem** LAST.REVERSE:

    (IMPLIES (LISTP A)
         (EQUAL (LAST (REVERSE A))
            (CONS (CAR A) "NIL")))

**108. Definition**

    (EXP I J)
      =
    (IF (ZEROP J)
        1
        (TIMES I (EXP I (SUB1 J))))

**109. Theorem** EXP.PLUS (rewrite):

    (EQUAL (EXP I (PLUS J K))
       (TIMES (EXP I J) (EXP I K)))

**110. Theorem** EXP.TIMES (rewrite):

    (EQUAL (EXP I (TIMES J K))
       (EXP (EXP I J) K))

**111. Theorem** EVEN1.EVEN2:

    (EQUAL (EVEN1 X) (EVEN2 X))

**112. Theorem** GREATERP.CONS:

    (GREATERP (LENGTH (CONS A B))
        (LENGTH B))

**113. Theorem** DUPLICITY.OF.LESSEQP:

    (OR (LESSEQP A B) (LESSEQP B A))

**114. Theorem** LESSEQP.NTH:

    (LESSEQP (LENGTH (NTH L I))
        (LENGTH L))

**115. Theorem** MEMBER.SORT:

    (EQUAL (MEMBER A (SORT B))
       (MEMBER A B))

**116. Theorem** LENGTH.SORT:

    (EQUAL (LENGTH (SORT A)) (LENGTH A))

**117. Definition**

```
(COUNT.LIST A L)
    =
(IF (LISTP L)
    (IF (EQUAL A (CAR L))
        (ADD1 (COUNT.LIST A (CDR L)))
        (COUNT.LIST A (CDR L)))
    0)
```

**118. Theorem** COUNT.LIST.SORT:

```
(EQUAL (COUNT.LIST A (SORT L))
       (COUNT.LIST A L))
```

**119. Theorem** ORDERED.APPEND:

```
(IMPLIES (ORDERED (APPEND A B))
         (ORDERED A))
```

**120. Theorem** LESSEQP.HALF:

```
(LESSEQP (HALF I) I)
```

**121. Definition**

```
(COPY A)
    =
(IF (LISTP A)
    (CONS (COPY (CAR A)) (COPY (CDR A)))
    A)
```

**122. Theorem** EQUAL.COPY:

```
(EQUAL (COPY A) A)
```

**123. Definition**

```
(EQUALP X Y)
    =
(IF (LISTP X)
    (IF (LISTP Y)
        (IF (EQUALP (CAR X) (CAR Y))
            (EQUALP (CDR X) (CDR Y))
            F)
        F)
    (EQUAL X Y))
```

**124. Definition**

```
(NUMBER.LISTP L)
    =
(IF (LISTP L)
    (IF (NUMBERP (CAR L))
        (NUMBER.LISTP (CDR L))
        F)
    (EQUAL L "NIL"))
```

**125. Theorem** ORDERED.SORT (rewrite):

```
(ORDERED (SORT X))
```

**126. Theorem**  SORT.OF.ORDERED.NUMBER.LIST (rewrite):

    (IMPLIES (AND (ORDERED X) (NUMBER.LISTP X))
             (EQUAL (SORT X) X))

**127. Defintion**

    (XOR P Q)
      =
    (IF Q (IF P F T) (EQUAL P T))

**128. Theorem**  SORT.ORDERED (rewrite):

    (IMPLIES (NUMBER.LISTP L)
             (EQUAL (EQUAL (SORT L) L)
                    (ORDERED L)))

**129. Definition**

    (SUBST X Y Z)
      =
    (IF (EQUAL Y Z)
        X
        (IF (LISTP Z)
            (CONS (SUBST X Y (CAR Z))
                  (SUBST X Y (CDR Z)))
            Z))

**130. Theorem**  SUBST.A.A:

    (EQUAL (SUBST A A B) B)

**131. Definition**

    (OCCUR X Y)
      =
    (IF (EQUAL X Y)
        T
        (IF (LISTP Y)
            (IF (OCCUR X (CAR Y))
                T
                (OCCUR X (CDR Y)))
            F))

**132. Theorem**  MEMBER.OCCUR:

    (IMPLIES (MEMBER A B) (OCCUR A B))

**133. Theorem**  OCCUR.SUBST:

    (IMPLIES (NOT (OCCUR A B))
             (EQUAL (SUBST C A B) B))

**134. Theorem**  COMMUTATIVITY.OF.EQUALP:

    (EQUAL (EQUALP A B) (EQUALP B A))

**135. Theorem**  TRANSITIVITY.OF.EQUALP:

    (IMPLIES (AND (EQUALP A B) (EQUALP B C))
             (EQUALP A C))

**136. Theorem**  EQUAL.EQUALP (rewrite):
> (EQUAL (EQUALP A B) (EQUAL A B))

**137. Definition**
> (SWAPTREE X)
> =
> (IF (LISTP X)
>     (CONS (SWAPTREE (CDR X))
>           (SWAPTREE (CAR X)))
>     X)

**138. Theorem**  SWAPTREE.SWAPTREE (rewrite):
> (EQUAL (SWAPTREE (SWAPTREE X)) X)

**139. Theorem**  FLATTEN.SWAPTREE (rewrite):
> (EQUAL (FLATTEN (SWAPTREE A))
>        (REVERSE (FLATTEN A)))

**140. Definition**
> (TIPCOUNT X)
> =
> (IF (LISTP X)
>     (PLUS (TIPCOUNT (CAR X))
>           (TIPCOUNT (CDR X)))
>     1)

**141. Theorem**  LENGTH.TIPCOUNT:
> (EQUAL (LENGTH (FLATTEN A))
>        (TIPCOUNT A))

**142. Definition**
> (COUNTPS-LOOP L PRED ANS)
> =
> (IF (LISTP L)
>     (IF (APPLY PRED (CAR L) "NIL")
>         (COUNTPS-LOOP (CDR L) PRED (ADD1 ANS))
>         (COUNTPS-LOOP (CDR L) PRED ANS))
>     ANS)

**143. Definition**
> (COUNTPS- L PRED)
> =
> (COUNTPS-LOOP L PRED 0)

**144. Definition**
> (COUNTPS L PRED)
> =
> (IF (LISTP L)
>     (IF (APPLY PRED (CAR L) "NIL")
>         (ADD1 (COUNTPS (CDR L) PRED))
>         (COUNTPS (CDR L) PRED))
>     0)

**145. Theorem**  COUNTPS–COUNTPS (rewrite):

        (IMPLIES (NUMBERP N)
                 (EQUAL (COUNTPS-LOOP L PRED N)
                        (PLUS N (COUNTPS L PRED)))))

**146. Definition**

        (FACT I)
           =
        (IF (ZEROP I)
            1
            (TIMES I (FACT (SUB1 I))))

**147. Definition**

        (FACT-LOOP I ANS)
           =
        (IF (ZEROP I)
            ANS
            (FACT-LOOP (SUB1 I) (TIMES I ANS)))

**148. Definition**

        (FACT- I)
           =
        (FACT-LOOP I 1)

**149. Theorem**  FACT–LOOP.FACT (rewrite):

        (IMPLIES (NUMBERP I)
                 (EQUAL (FACT-LOOP J I)
                        (TIMES I (FACT J)))))

**150. Theorem**  FACT–FACT:

        (EQUAL (FACT- I) (FACT I))

**151. Definition**

        (REVERSE-LOOP X ANS)
           =
        (IF (LISTP X)
            (REVERSE-LOOP (CDR X)
                          (CONS (CAR X) ANS))
            ANS)

**152. Definition**

        (REVERSE- X)
           =
        (REVERSE-LOOP X "NIL")

**153. Theorem**  REVERSE–LOOP.APPEND.REVERSE (rewrite):

        (EQUAL (REVERSE-LOOP X Y)
               (APPEND (REVERSE X) Y))

**154. Theorem**  REVERSE–REVERSE (rewrite):

        (EQUAL (REVERSE-LOOP X "NIL")
               (REVERSE X))

346 / APPENDIX A

**155. Theorem** REVERSE-APPEND:
```
(EQUAL (REVERSE- (APPEND A B))
       (APPEND (REVERSE- B) (REVERSE- A)))
```

**156. Theorem** REVERSE-REVERSE-:
```
(IMPLIES (PLISTP X)
         (EQUAL (REVERSE- (REVERSE- X)) X))
```

**157. Definition**
```
(PLUS- I J)
=
(IF (ZEROP I)
    J
    (PLUS- (SUB1 I) (ADD1 J)))
```

**158. Theorem** PLUS-PLUS:
```
(IMPLIES (NUMBERP J)
         (EQUAL (PLUS- I J) (PLUS I J)))
```

**159. Definition**
```
(UNION- X Y)
=
(IF (LISTP X)
    (IF (MEMBER (CAR X) Y)
        (UNION- (CDR X) Y)
        (UNION- (CDR X) (CONS (CAR X) Y)))
    Y)
```

**160. Definition**
```
(SORT-LP X Y)
=
(IF (LISTP X)
    (SORT-LP (CDR X)
             (ADDTOLIST (CAR X) Y))
    Y)
```

**161. Definition**
```
(SORT- X)
=
(SORT-LP X "NIL")
```

**162. Theorem** MEMBER.UNION-:
```
(IMPLIES (MEMBER A C)
         (MEMBER A (UNION- B C)))
```

**163. Theorem** ORDERED.ADDTOLIST (rewrite):
```
(IMPLIES (ORDERED Y)
         (ORDERED (ADDTOLIST X Y)))
```

**164. Theorem** ORDERED.SORT-LP (rewrite):
```
(IMPLIES (ORDERED Y)
         (ORDERED (SORT-LP X Y)))
```

### DEFINITIONS ACCEPTED AND THEOREMS PROVED BY OUR SYSTEM / 347

**165. Theorem**  COUNT.SORT-LP (rewrite):

```
(EQUAL (COUNT.LIST Z (SORT-LP X Y))
       (PLUS (COUNT.LIST Z X)
             (COUNT.LIST Z Y)))
```

**166. Theorem**  APPEND.CANCELLATION (rewrite):

```
(EQUAL (EQUAL (APPEND A B) (APPEND A C))
       (EQUAL B C))
```

**167. Definition**

```
(DIFFERENCE I J)
    =
(IF (ZEROP I)
    0
    (IF (ZEROP J)
        I
        (DIFFERENCE (SUB1 I) (SUB1 J))))
```

**168. Theorem**  COUNTING.UP.BY.1 (induction):

```
(IMPLIES (LESSP X Y)
         (LESSP (DIFFERENCE Y (ADD1 X))
                (DIFFERENCE Y X)))
```

**169. Theorem**  COUNTING.DOWN.BY.N+1 (rewrite):

```
(EQUAL (LESSP (DIFFERENCE I N) I)
       (AND (NOT (ZEROP I)) (NOT (ZEROP N))))
```

**170. Theorem**  RECURSION.BY.DIFFERENCE (induction):

```
(IMPLIES (AND (NUMBERP I)
              (NUMBERP N)
              (NOT (EQUAL I 0))
              (NOT (EQUAL N 0)))
         (LESSP (DIFFERENCE I N) I))
```

**171. Definition**

```
(QUOTIENT I J)
    =
(IF (ZEROP J)
    0
    (IF (LESSP I J)
        0
        (ADD1 (QUOTIENT (DIFFERENCE I J) J))))
```

**172. Definition**

```
(REMAINDER I J)
    =
(IF (ZEROP J)
    (FIX I)
    (IF (LESSP I J)
        (FIX I)
        (REMAINDER (DIFFERENCE I J) J)))
```

348 / APPENDIX A

**173. Definition**

```
(POWER.EVAL L BASE)
   =
(IF (LISTP L)
    (PLUS (CAR L)
          (TIMES
             BASE
             (POWER.EVAL (CDR L) BASE)))
    0)
```

**174. Definition**

```
(BIG.PLUS1 L I BASE)
   =
(IF
 (LISTP L)
 (IF (ZEROP I)
     L
     (CONS (REMAINDER (PLUS (CAR L) I) BASE)
           (BIG.PLUS1 (CDR L)
                      (QUOTIENT (PLUS (CAR L) I) BASE)
                      BASE)))
 (CONS I "NIL"))
```

**175. Theorem** REMAINDER.QUOTIENT (rewrite):

```
(EQUAL (PLUS (REMAINDER X Y)
             (TIMES Y (QUOTIENT X Y)))
       (FIX X))
```

**176. Theorem** POWER.EVAL.BIG.PLUS1 (rewrite):

```
(EQUAL (POWER.EVAL (BIG.PLUS1 L I BASE) BASE)
       (PLUS (POWER.EVAL L BASE) I))
```

**177. Definition**

```
(BIG.PLUS X Y I BASE)
   =
(IF
 (LISTP X)
 (IF
  (LISTP Y)
  (CONS
    (REMAINDER (PLUS I (PLUS (CAR X) (CAR Y)))
               BASE)
    (BIG.PLUS (CDR X)
              (CDR Y)
              (QUOTIENT (PLUS I (PLUS (CAR X) (CAR Y)))
                        BASE)
              BASE))
  (BIG.PLUS1 X I BASE))
 (BIG.PLUS1 Y I BASE))
```

**178. Theorem** POWER.EVAL.BIG.PLUS (rewrite):

```
(EQUAL (POWER.EVAL (BIG.PLUS X Y I BASE)
                   BASE)
```

```
           (PLUS I
              (PLUS (POWER.EVAL X BASE)
                    (POWER.EVAL Y BASE)))))
```

**179. Theorem** LESSP.DIFFERENCE1 (rewrite):

```
(NOT (LESSP X (DIFFERENCE X Y)))
```

**180. Theorem** REMAINDER.WRT.1 (rewrite):

```
(EQUAL (REMAINDER Y 1) 0)
```

**181. Theorem** REMAINDER.WRT.12 (rewrite):

```
(IMPLIES (NOT (NUMBERP X))
         (EQUAL (REMAINDER Y X) (FIX Y)))
```

**182. Theorem** LESSP.REMAINDER2 (rewrite and generalize):

```
(EQUAL (LESSP (REMAINDER X Y) Y)
       (NOT (ZEROP Y)))
```

**183. Theorem** LESSP.DIFFERENCE (rewrite):

```
(IMPLIES (NOT (LESSP Y X))
         (EQUAL (DIFFERENCE X Y) 0))
```

**184. Theorem** REMAINDER.X.X (rewrite):

```
(EQUAL (REMAINDER X X) 0)
```

**185. Theorem** REMAINDER.QUOTIENT.ELIM (elimination):

```
(IMPLIES (AND (NOT (ZEROP Y)) (NUMBERP X))
         (EQUAL (PLUS (REMAINDER X Y)
                      (TIMES Y (QUOTIENT X Y)))
                X))
```

**186. Theorem** PLUS.EQUAL.0 (rewrite):

```
(EQUAL (EQUAL (PLUS X Y) 0)
       (AND (ZEROP X) (ZEROP Y)))
```

**187. Theorem** PLUS.CANCELLATION1 (rewrite):

```
(EQUAL (EQUAL (PLUS X Y) X)
       (AND (NUMBERP X) (ZEROP Y)))
```

**188. Theorem** LESSP.PLUS.TIMES (rewrite):

```
(IMPLIES (AND (NOT (ZEROP Z)) (NOT (ZEROP J)))
         (LESSP X (PLUS Z (TIMES J X))))
```

**189. Theorem** LESSP.QUOTIENT1 (rewrite):

```
(EQUAL (LESSP (QUOTIENT I J) I)
       (AND (NOT (ZEROP I))
            (OR (ZEROP J) (NOT (EQUAL J 1)))))
```

**190. Theorem** LESSP.REMAINDER1 (rewrite):

```
(EQUAL (LESSP (REMAINDER X Y) X)
       (AND (NOT (ZEROP Y))
            (NOT (ZEROP X))
            (NOT (LESSP X Y))))
```

**191. Theorem** RECURSION.BY.QUOTIENT (induction):
```
(IMPLIES (AND (NUMBERP I)
              (NOT (EQUAL I 0))
              (NUMBERP J)
              (NOT (EQUAL J 0))
              (NOT (EQUAL J 1)))
         (LESSP (QUOTIENT I J) I))
```

**192. Definition**
```
(POWER.REP I BASE)
  =
(IF
  (ZEROP I)
  "NIL"
  (IF (ZEROP BASE)
      (CONS I "NIL")
      (IF (EQUAL BASE 1)
          (CONS I "NIL")
          (CONS (REMAINDER I BASE)
                (POWER.REP (QUOTIENT I BASE) BASE)))))
```

**193. Theorem** POWER.EVAL.POWER.REP (rewrite):
```
(EQUAL (POWER.EVAL (POWER.REP I BASE) BASE)
       (FIX I))
```

**194. Theorem** CORRECTNESS.OF.BIG.PLUS (rewrite):
```
(EQUAL (POWER.EVAL (BIG.PLUS (POWER.REP I BASE)
                             (POWER.REP J BASE)
                             0 BASE)
                   BASE)
       (PLUS I J))
```

**195. Definition**
```
(SGCD X Y I)
  =
(IF (ZEROP I)
    0
    (IF (ZEROP (REMAINDER X I))
        (IF (ZEROP (REMAINDER Y I))
            I
            (SGCD X Y (SUB1 I)))
        (SGCD X Y (SUB1 I))))
```

**196. Definition**
```
(GCD X Y)
  =
(IF (ZEROP X)
    (FIX Y)
    (IF (ZEROP Y)
        X
        (IF (LESSP X Y)
            (GCD X (DIFFERENCE Y X))
            (GCD (DIFFERENCE X Y) Y))))
```

### DEFINITIONS ACCEPTED AND THEOREMS PROVED BY OUR SYSTEM / 351

**197. Theorem** LESSP.NOT.COMMUTATIVE (rewrite):

    (IMPLIES (LESSP Y X)
         (NOT (LESSP X Y)))

**198. Theorem** COMMUTATIVITY.OF.GCD (rewrite):

    (EQUAL (GCD X Y) (GCD Y X))

**199. Theorem** SGCD.X.O.X (rewrite):

    (EQUAL (SGCD X O X) (FIX X))

**200. Theorem** SGCD.X.X.X (rewrite):

    (EQUAL (SGCD X X X) (FIX X))

**201. Theorem** COMMUTATIVITY.OF.SGCD (rewrite):

    (EQUAL (SGCD X Y Z) (SGCD Y X Z))

**202. Theorem** RECURSION.BY.REMAINDER (induction):

    (IMPLIES (AND (NUMBERP Y)
             (NOT (EQUAL Y 0))
             (LESSP Y X))
        (LESSP (REMAINDER X Y) X))

**203. Definition**

    (NTHCHAR N STR)
    =
    (CAR (NTH STR N))

**204. Theorem** ELEMENT.APPEND:

    (EQUAL (NTH A I)
       (NTH (APPEND C A)
          (PLUS I (LENGTH C)))))

**205. Theorem** NTH.APPEND (rewrite):

    (EQUAL (NTH (APPEND A B) I)
       (APPEND (NTH A I)
          (NTH B (DIFFERENCE I (LENGTH A))))))

**206. Theorem** PLUS.NEQUAL.X (rewrite):

    (NOT (EQUAL (ADD1 (PLUS X Y)) X))

**207. Theorem** NEQUAL.PLUS.ADD1 (rewrite):

    (NOT (EQUAL (ADD1 (PLUS Y X)) X))

**208. Theorem** DIFFERENCE.PLUS (rewrite):

    (EQUAL (DIFFERENCE (PLUS X Y) X)
       (FIX Y))

**209. Theorem** PLUS.DIFFERENCE3 (rewrite):

    (EQUAL (DIFFERENCE (PLUS X Y) (PLUS X Z))
       (DIFFERENCE Y Z))

352 / APPENDIX A

**210. Theorem** TIMES.DIFFERENCE (rewrite):
```
(EQUAL (DIFFERENCE (TIMES C X) (TIMES W X))
       (TIMES X (DIFFERENCE C W)))
```

**211. Definition**
```
(DIVIDES X Y)
   =
(ZEROP (REMAINDER Y X))
```

**212. Theorem** DIVIDES.TIMES (rewrite):
```
(EQUAL (REMAINDER (TIMES X Z) Z) 0)
```

**213. Theorem** DIFFERENCE.PLUS2 (rewrite):
```
(EQUAL (DIFFERENCE (PLUS B (PLUS A C)) A)
       (PLUS B C))
```

**214. Theorem** EQUAL.DIFFERENCE (rewrite):
```
(EQUAL (EQUAL X (DIFFERENCE Y K))
       (IF (LESSP K Y)
           (IF (NUMBERP X)
               (IF (NUMBERP Y)
                   (EQUAL Y (PLUS X K))
                   (EQUAL X 0))
               F)
           (EQUAL X 0)))
```

**215. Theorem** DIFFERENCE.ADD1.CANCELLATION (rewrite):
```
(EQUAL (DIFFERENCE (ADD1 (PLUS Y Z)) Z)
       (ADD1 Y))
```

**216. Theorem** REMAINDER.ADD1 (rewrite):
```
(IMPLIES (AND (NOT (ZEROP Y))
              (NOT (EQUAL Y 1)))
         (NOT (EQUAL (REMAINDER (ADD1 (TIMES X Y)) Y)
                     0)))
```

**217. Theorem** DIVIDES.PLUS.REWRITE1 (rewrite):
```
(IMPLIES (AND (EQUAL (REMAINDER X Z) 0)
              (EQUAL (REMAINDER Y Z) 0))
         (EQUAL (REMAINDER (PLUS X Y) Z) 0))
```

**218. Theorem** DIVIDES.PLUS.REWRITE2 (rewrite):
```
(IMPLIES (AND (EQUAL (REMAINDER X Z) 0)
              (NOT (EQUAL (REMAINDER Y Z) 0)))
         (NOT (EQUAL (REMAINDER (PLUS X Y) Z) 0)))
```

**219. Theorem** DIVIDES.PLUS.REWRITE (rewrite):
```
(IMPLIES (EQUAL (REMAINDER X Z) 0)
         (EQUAL (EQUAL (REMAINDER (PLUS X Y) Z) 0)
                (EQUAL (REMAINDER Y Z) 0)))
```

**220. Theorem** DIVIDES.PLUS.REWRITE.COMMUTED (rewrite):

    (IMPLIES (EQUAL (REMAINDER X Z) 0)
         (EQUAL (EQUAL (REMAINDER (PLUS Y X) Z) 0)
              (EQUAL (REMAINDER Y Z) 0)))

**221. Theorem** LESSP.DIFFERENCE2 (rewrite):

    (EQUAL (EQUAL (DIFFERENCE X Y) 0)
       (NOT (LESSP Y X)))

**222. Theorem** DIFFERENCE.ELIM (elimination):

    (IMPLIES (AND (NUMBERP Y) (LESSEQP X Y))
         (EQUAL (PLUS X (DIFFERENCE Y X)) Y))

**223. Theorem** LESSP.PLUS.CANCELLATION (rewrite):

    (EQUAL (LESSP (PLUS X Y) (PLUS X Z))
       (LESSP Y Z))

**224. Theorem** LESSP.PLUS.CANCELLATION2 (rewrite):

    (EQUAL (LESSP (PLUS X Y) (PLUS Z X))
       (LESSP Y Z))

**225. Theorem** EUCLID (rewrite):

    (IMPLIES (EQUAL (REMAINDER X Z) 0)
         (EQUAL (EQUAL (REMAINDER (DIFFERENCE Y X) Z)
                0)
           (IF (LESSP X Y)
              (EQUAL (REMAINDER Y Z) 0)
              T)))

**226. Theorem** LESSP.TIMES.CANCELLATION (rewrite):

    (EQUAL (LESSP (TIMES X Z) (TIMES Y Z))
       (AND (NOT (ZEROP Z)) (LESSP X Y)))

**227. Theorem** DISTRIBUTIVITY.OF.TIMES.OVER.GCD (rewrite):

    (EQUAL (GCD (TIMES X Z) (TIMES Y Z))
       (TIMES Z (GCD X Y)))

**228. Theorem** GCD.DIVIDES.BOTH (rewrite):

    (AND (EQUAL (REMAINDER X (GCD X Y)) 0)
       (EQUAL (REMAINDER Y (GCD X Y)) 0))

**229. Theorem** GCD.IS.THE.GREATEST:

    (IMPLIES (AND (NOT (ZEROP X))
            (NOT (ZEROP Y))
            (DIVIDES Z X)
            (DIVIDES Z Y))
        (LESSEQP Z (GCD X Y)))

## 354 / APPENDIX A

**230. Shell Definition.**

Add the shell CONS.IF of three arguments
with recognizer IF.EXPRP,
accessors TEST, LEFT.BRANCH, and RIGHT.BRANCH,
default values "NIL", "NIL", and "NIL", and
well-founded relation TEST.LEFT.BRANCH.RIGHT.BRANCHP.

**231. Definition**

```
(ASSIGNMENT VAR ALIST)
    =
(IF (EQUAL VAR T)
    T
    (IF (EQUAL VAR F)
        F
        (IF (NLISTP ALIST)
            F
            (IF (EQUAL VAR (CAAR ALIST))
                (CDAR ALIST)
                (ASSIGNMENT VAR (CDR ALIST))))))
```

**232. Definition**

```
(VALUE X ALIST)
    =
(IF (IF.EXPRP X)
    (IF (VALUE (TEST X) ALIST)
        (VALUE (LEFT.BRANCH X) ALIST)
        (VALUE (RIGHT.BRANCH X) ALIST))
    (ASSIGNMENT X ALIST))
```

**233. Definition**

```
(IF.DEPTH X)
    =
(IF (IF.EXPRP X)
    (ADD1 (IF.DEPTH (TEST X)))
    0)
```

**234. Theorem** IF.DEPTH.GOES.DOWN (induction):

```
(IMPLIES
    (AND (IF.EXPRP X) (IF.EXPRP (TEST X)))
    (LESSP (IF.DEPTH (CONS.IF (TEST (TEST X)) Y Z))
           (IF.DEPTH X)))
```

**235. Definition**

```
(IF.COMPLEXITY X)
    =
(IF (IF.EXPRP X)
    (TIMES (IF.COMPLEXITY (TEST X))
           (PLUS (IF.COMPLEXITY (LEFT.BRANCH X))
                 (IF.COMPLEXITY (RIGHT.BRANCH X))))
    1)
```

**236. Theorem** IF.COMPLEXITY.NOT.0 (rewrite):

```
(NOT (EQUAL (IF.COMPLEXITY X) 0))
```

**237. Theorem**  LESSP.D.V (rewrite):

```
(IMPLIES (AND (NOT (ZEROP D))
              (NOT (ZEROP V))
              (NOT (ZEROP Z)))
         (LESSP V
                (PLUS (TIMES D V) (TIMES D Z))))
```

**238. Theorem**  IF.COMPLEXITY.GOES.DOWN1 (induction):

```
(IMPLIES (IF.EXPRP X)
         (LESSP (IF.COMPLEXITY (LEFT.BRANCH X))
                (IF.COMPLEXITY X)))
```

**239. Theorem**  IF.COMPLEXITY.GOES.DOWN2 (induction):

```
(IMPLIES (IF.EXPRP X)
         (LESSP (IF.COMPLEXITY (RIGHT.BRANCH X))
                (IF.COMPLEXITY X)))
```

**240. Theorem**  IF.COMPLEXITY.STAYS.EVEN (induction):

```
(IMPLIES
 (AND (IF.EXPRP X) (IF.EXPRP (TEST X)))
 (EQUAL
  (IF.COMPLEXITY
             (CONS.IF (TEST (TEST X))
                      (CONS.IF (LEFT.BRANCH (TEST X))
                               (LEFT.BRANCH X)
                               (RIGHT.BRANCH X))
                      (CONS.IF (RIGHT.BRANCH (TEST X))
                               (LEFT.BRANCH X)
                               (RIGHT.BRANCH X))))
  (IF.COMPLEXITY X)))
```

**241. Definition**

```
(NORMALIZE X)
   =
(IF
 (IF.EXPRP X)
 (IF
  (IF.EXPRP (TEST X))
  (NORMALIZE (CONS.IF (TEST (TEST X))
                      (CONS.IF (LEFT.BRANCH (TEST X))
                               (LEFT.BRANCH X)
                               (RIGHT.BRANCH X))
                      (CONS.IF (RIGHT.BRANCH (TEST X))
                               (LEFT.BRANCH X)
                               (RIGHT.BRANCH X))))
  (CONS.IF (TEST X)
           (NORMALIZE (LEFT.BRANCH X))
           (NORMALIZE (RIGHT.BRANCH X))))
 X)
```

## 356 / APPENDIX A

**242. Definition**

```
(NORMALIZED.IF.EXPRP X)
    =
(IF (IF.EXPRP X)
    (AND (NOT (IF.EXPRP (TEST X)))
         (NORMALIZED.IF.EXPRP (LEFT.BRANCH X))
         (NORMALIZED.IF.EXPRP (RIGHT.BRANCH X)))
    T)
```

**243. Definition**

```
(ASSIGNEDP VAR ALIST)
    =
(IF (EQUAL VAR T)
    T
    (IF (EQUAL VAR F)
        T
        (IF (NLISTP ALIST)
            F
            (IF (EQUAL VAR (CAAR ALIST))
                T
                (ASSIGNEDP VAR (CDR ALIST))))))
```

**244. Definition**

```
(ASSUME.TRUE VAR ALIST)
    =
(CONS (CONS VAR T) ALIST)
```

**245. Definition**

```
(ASSUME.FALSE VAR ALIST)
    =
(CONS (CONS VAR F) ALIST)
```

**246. Definition**

```
(TAUTOLOGYP X ALIST)
    =
(IF
 (IF.EXPRP X)
 (IF (ASSIGNEDP (TEST X) ALIST)
     (IF (ASSIGNMENT (TEST X) ALIST)
         (TAUTOLOGYP (LEFT.BRANCH X) ALIST)
         (TAUTOLOGYP (RIGHT.BRANCH X) ALIST))
     (AND (TAUTOLOGYP (LEFT.BRANCH X)
                      (ASSUME.TRUE (TEST X) ALIST))
          (TAUTOLOGYP (RIGHT.BRANCH X)
                      (ASSUME.FALSE (TEST X) ALIST))))
 (ASSIGNMENT X ALIST))
```

**247. Theorem** ASSIGNMENT.APPEND (rewrite):

```
(EQUAL (ASSIGNMENT X (APPEND A B))
       (IF (ASSIGNEDP X A)
           (ASSIGNMENT X A)
           (ASSIGNMENT X B)))
```

**248. Theorem** VALUE.CAN.IGNORE.REDUNDANT.ASSIGNMENTS (rewrite):

```
(AND
    (IMPLIES (AND (IFF VAL (ASSIGNMENT VAR A))
                  (VALUE X A))
             (VALUE X (CONS (CONS VAR VAL) A)))
    (IMPLIES (AND (IFF VAL (ASSIGNMENT VAR A))
                  (NOT (VALUE X A)))
             (NOT (VALUE X (CONS (CONS VAR VAL) A)))))
```

**249. Theorem** VALUE.SHORT.CUT (rewrite):

```
(IMPLIES (AND (IF.EXPRP X)
              (NORMALIZED.IF.EXPRP X))
         (EQUAL (VALUE (TEST X) A)
                (ASSIGNMENT (TEST X) A)))
```

**250. Theorem** ASSIGNMENT.IMPLIES.ASSIGNEDP (rewrite):

```
(IMPLIES (ASSIGNMENT X A)
         (ASSIGNEDP X A))
```

**251. Theorem** TAUTOLOGYP.IS.SOUND (rewrite):

```
(IMPLIES (AND (NORMALIZED.IF.EXPRP X)
              (TAUTOLOGYP X A1))
         (VALUE X (APPEND A1 A2)))
```

**252. Definition**

```
(TAUTOLOGY.CHECKER X)
=
(TAUTOLOGYP (NORMALIZE X) "NIL")
```

**253. Definition**

```
(FALSIFY1 X ALIST)
=
(IF (IF.EXPRP X)
    (IF (ASSIGNEDP (TEST X) ALIST)
        (IF (ASSIGNMENT (TEST X) ALIST)
            (FALSIFY1 (LEFT.BRANCH X) ALIST)
            (FALSIFY1 (RIGHT.BRANCH X) ALIST))
        (IF (FALSIFY1 (LEFT.BRANCH X)
                      (ASSUME.TRUE (TEST X) ALIST))
            (FALSIFY1 (LEFT.BRANCH X)
                      (ASSUME.TRUE (TEST X) ALIST))
            (FALSIFY1 (RIGHT.BRANCH X)
                      (ASSUME.FALSE (TEST X) ALIST))))
    (IF (ASSIGNEDP X ALIST)
        (IF (ASSIGNMENT X ALIST) F ALIST)
        (CONS (CONS X F) ALIST)))
```

**254. Definition**

```
(FALSIFY X)
=
(FALSIFY1 (NORMALIZE X) "NIL")
```

## 358 / APPENDIX A

**255. Theorem** FALSIFY1.EXTENDS.MODELS (rewrite):
```
(IMPLIES (ASSIGNEDP X A)
         (EQUAL (ASSIGNMENT X (FALSIFY1 Y A))
                (IF (FALSIFY1 Y A)
                    (ASSIGNMENT X A)
                    (EQUAL X T))))
```

**256. Theorem** FALSIFY1.FALSIFIES (rewrite):
```
(IMPLIES (AND (NORMALIZED.IF.EXPRP X)
              (FALSIFY1 X A))
         (EQUAL (VALUE X (FALSIFY1 X A)) F))
```

**257. Theorem** TAUTOLOGYP.FAILS.MEANS.FALSIFY1.WINS (rewrite):
```
(IMPLIES (AND (NORMALIZED.IF.EXPRP X)
              (NOT (TAUTOLOGYP X A))
              A)
         (FALSIFY1 X A))
```

**258. Theorem** NORMALIZE.IS.SOUND (rewrite):
```
(EQUAL (VALUE (NORMALIZE X) A)
       (VALUE X A))
```

**259. Theorem** NORMALIZE.NORMALIZES (rewrite):
```
(NORMALIZED.IF.EXPRP (NORMALIZE X))
```

**260. Theorem** TAUTOLOGY.CHECKER.COMPLETENESS.BRIDGE (rewrite):
```
(IMPLIES (AND (EQUAL (VALUE Y (FALSIFY1 X A))
                     (VALUE X (FALSIFY1 X A)))
              (FALSIFY1 X A)
              (NORMALIZED.IF.EXPRP X))
         (EQUAL (VALUE Y (FALSIFY1 X A)) F))
```

**261. Theorem** TAUTOLOGY.CHECKER.IS.COMPLETE:
```
(IMPLIES (NOT (TAUTOLOGY.CHECKER X))
         (EQUAL (VALUE X (FALSIFY X)) F))
```

**262. Theorem** TAUTOLOGY.CHECKER.SOUNDNESS.BRIDGE (rewrite):
```
(IMPLIES (AND (TAUTOLOGYP Y A1)
              (NORMALIZED.IF.EXPRP Y)
              (EQUAL (VALUE X A2)
                     (VALUE Y (APPEND A1 A2))))
         (VALUE X A2))
```

**263. Theorem** TAUTOLOGY.CHECKER.IS.SOUND:
```
(IMPLIES (TAUTOLOGY.CHECKER X)
         (VALUE X A))
```

**264. Theorem** FLATTEN.SINGLETON (rewrite):
```
(EQUAL (EQUAL (FLATTEN X) (CONS Y "NIL"))
       (AND (NLISTP X) (EQUAL X Y)))
```

## DEFINITIONS ACCEPTED AND THEOREMS PROVED BY OUR SYSTEM / 359

**265. Definition**

```
(LEFTCOUNT X)
    =
(IF (NLISTP X)
    0
    (ADD1 (LEFTCOUNT (CAR X))))
```

**266. Theorem** LEFTCOUNT.GOES.DOWN (induction):

```
(IMPLIES
    (AND (LISTP X) (LISTP (CAR X)))
    (LESSP (LEFTCOUNT (CONS (CAAR X)
                            (CONS (CDAR X) (CDR X))))
           (LEFTCOUNT X)))
```

**267. Definition**

```
(GOPHER X)
    =
(IF (OR (NLISTP X) (NLISTP (CAR X)))
    X
    (GOPHER (CONS (CAAR X)
                  (CONS (CDAR X) (CDR X)))))
```

**268. Theorem** GOPHER.PRESERVES.COUNT (induction):

```
(EQUAL (COUNT (GOPHER X)) (COUNT X))
```

**269. Definition**

```
(SAMEFRINGE X Y)
    =
(IF (OR (NLISTP X) (NLISTP Y))
    (EQUAL X Y)
    (AND (EQUAL (CAR (GOPHER X))
                (CAR (GOPHER Y)))
         (SAMEFRINGE (CDR (GOPHER X))
                     (CDR (GOPHER Y)))))
```

**270. Theorem** LISTP.GOPHER (rewrite):

```
(EQUAL (LISTP (GOPHER X)) (LISTP X))
```

**271. Theorem** GOPHER.RETURNS.LEFTMOST.ATOM (rewrite):

```
(EQUAL (CAR (GOPHER X))
       (IF (LISTP X)
           (CAR (FLATTEN X))
           "NIL"))
```

**272. Theorem** GOPHER.RETURNS.CORRECT.STATE (rewrite):

```
(EQUAL (FLATTEN (CDR (GOPHER X)))
       (IF (LISTP X)
           (CDR (FLATTEN X))
           (CONS "NIL" "NIL")))
```

**273. Theorem** CORRECTNESS.OF.SAMEFRINGE (rewrite):

```
(EQUAL (SAMEFRINGE X Y)
       (EQUAL (FLATTEN X) (FLATTEN Y)))
```

**274. Definition**

```
(PRIME1 X Y)
    =
(IF (ZEROP Y)
    F
    (IF (EQUAL Y 1)
        T
        (AND (NOT (DIVIDES Y X))
             (PRIME1 X (SUB1 X)))))
```

**275. Definition**

```
(PRIME X)
    =
(AND (NOT (ZEROP X))
     (NOT (EQUAL X 1))
     (PRIME1 X (SUB1 X)))
```

**276. Definition**

```
(GREATEST.FACTOR X Y)
    =
(IF (OR (ZEROP Y) (EQUAL Y 1))
    X
    (IF (DIVIDES Y X)
        Y
        (GREATEST.FACTOR X (SUB1 Y))))
```

**277. Definition**

```
(ID X)
    =
(IF (ZEROP X) X (ADD1 (ID (SUB1 X))))
```

**278. Theorem** LESSP.ID2 (rewrite):

```
(IMPLIES (NOT (LESSP X Y))
         (NOT (LESSP X (ID Y))))
```

**279. Theorem** GREATEST.FACTOR.LESSP (rewrite):

```
(IMPLIES (AND (LESSP Y (ID X))
              (NOT (PRIME1 X Y))
              (NOT (ZEROP X))
              (NOT (EQUAL X 1))
              (NOT (ZEROP Y)))
         (LESSP (GREATEST.FACTOR X Y) X))
```

## DEFINITIONS ACCEPTED AND THEOREMS PROVED BY OUR SYSTEM / 361

**280. Theorem** GREATEST.FACTOR.DIVIDES (rewrite):

```
(IMPLIES (AND (LESSP Y (ID X))
              (NOT (PRIME1 X Y))
              (NOT (ZEROP X))
              (NOT (EQUAL X 1))
              (NOT (ZEROP Y)))
         (EQUAL (REMAINDER X (GREATEST.FACTOR X Y))
                0))
```

**281. Theorem** LESSP.ID3 (rewrite):

```
(IMPLIES (LESSP X Y) (LESSP X (ID Y)))
```

**282. Theorem** GREATEST.FACTOR.LESSP.IND (induction):

```
(IMPLIES (AND (LESSP Y X)
              (NOT (PRIME1 X Y))
              (NOT (ZEROP X))
              (NOT (EQUAL X 1))
              (NOT (ZEROP Y)))
         (LESSP (COUNT (GREATEST.FACTOR X Y))
                (COUNT X)))
```

**283. Theorem** GREATEST.FACTOR.0 (rewrite):

```
(EQUAL (EQUAL (GREATEST.FACTOR X Y) 0)
       (AND (OR (ZEROP Y) (EQUAL Y 1))
            (EQUAL X 0)))
```

**284. Theorem** GREATEST.FACTOR.1 (rewrite):

```
(EQUAL (EQUAL (GREATEST.FACTOR X Y) 1)
       (EQUAL X 1))
```

**285. Theorem** NUMBERP.GREATEST.FACTOR (rewrite):

```
(EQUAL (NUMBERP (GREATEST.FACTOR X Y))
       (NOT (AND (OR (ZEROP Y) (EQUAL Y 1))
                 (NOT (NUMBERP X)))))
```

**286. Definition**

```
(PRIME.FACTORS X)
    =
(IF
 (OR (ZEROP X) (EQUAL (SUB1 X) 0))
 "NIL"
 (IF
  (PRIME1 X (SUB1 X))
  (CONS X "NIL")
  (APPEND
   (PRIME.FACTORS (GREATEST.FACTOR X (SUB1 X)))
   (PRIME.FACTORS
           (QUOTIENT X
                     (GREATEST.FACTOR X (SUB1 X)))))))
```

**287. Definition**

        (PRIME.LIST L)
        =
        (IF (NLISTP L)
            T
            (AND (PRIME (CAR L))
                 (PRIME.LIST (CDR L))))

**288. Definition**

        (TIMES.LIST L)
        =
        (IF (NLISTP L)
            1
            (TIMES (CAR L) (TIMES.LIST (CDR L))))

**289. Theorem** TIMES.LIST.APPEND (rewrite):

        (EQUAL (TIMES.LIST (APPEND X Y))
               (TIMES (TIMES.LIST X) (TIMES.LIST Y)))

**290. Theorem** PRIME.LIST.APPEND (rewrite):

        (EQUAL (PRIME.LIST (APPEND X Y))
               (AND (PRIME.LIST X) (PRIME.LIST Y)))

**291. Theorem** PRIME.LIST.PRIME.FACTORS (rewrite):

        (PRIME.LIST (PRIME.FACTORS X))

**292. Theorem** QUOTIENT.TIMES1 (rewrite):

        (IMPLIES (AND (NUMBERP Y)
                      (NUMBERP X)
                      (NOT (EQUAL X 0))
                      (DIVIDES X Y))
                 (EQUAL (TIMES X (QUOTIENT Y X)) Y))

**293. Theorem** QUOTIENT.LESSP (rewrite):

        (IMPLIES (AND (NOT (ZEROP X)) (LESSP X Y))
                 (NOT (EQUAL (QUOTIENT Y X) 0)))

**294. Theorem** ID.ADD1 (rewrite):

        (EQUAL (ID (ADD1 X)) (ADD1 (ID X)))

**295. Theorem** LESSP.ADD1.ID (rewrite):

        (LESSP X (ADD1 (ID X)))

**296. Theorem** ENOUGH.FACTORS (rewrite):

        (IMPLIES (NOT (ZEROP X))
                 (EQUAL (TIMES.LIST (PRIME.FACTORS X))
                        X))

**297. Theorem** PRIME.FACTORIZATION.EXISTENCE:

        (IMPLIES (NOT (ZEROP X))
                 (AND (EQUAL (TIMES.LIST (PRIME.FACTORS X))
                             X)
                      (PRIME.LIST (PRIME.FACTORS X))))

### DEFINITIONS ACCEPTED AND THEOREMS PROVED BY OUR SYSTEM / 363

**298. Theorem** PRIME.KRUTCH (rewrite):

```
(IMPLIES (AND (LESSP X Z)
              (NOT (EQUAL Z (ADD1 X))))
         (NOT (LESSP Z (ADD1 X))))
```

**299. Theorem** PRIME.BRIDGE (rewrite):

```
(IMPLIES
         (AND (EQUAL (REMAINDER (ADD1 X) Z) 0)
              (NOT (EQUAL Z (ADD1 X)))
              (NOT (PRIME1 (ADD1 X)
                           (PLUS (DIFFERENCE X Z) Z))))
         (NOT (PRIME1 (ADD1 X) X)))
```

**300. Theorem** PRIME1.BASIC (rewrite):

```
(IMPLIES (AND (NOT (EQUAL Z 1))
              (NOT (EQUAL Z (ADD1 X)))
              (EQUAL (REMAINDER (ADD1 X) Z) 0))
         (NOT (PRIME1 (ADD1 X) (PLUS Z L))))
```

**301. Theorem** PRIME.BASIC (rewrite):

```
(IMPLIES (AND (NOT (EQUAL Z 1))
              (NOT (EQUAL Z X))
              (DIVIDES Z X))
         (NOT (PRIME1 X (SUB1 X))))
```

**302. Theorem** REMAINDER.GCD (rewrite):

```
(IMPLIES (EQUAL (GCD B X) Y)
         (EQUAL (REMAINDER B Y) 0))
```

**303. Theorem** REMAINDER.GCD.1 (rewrite):

```
(IMPLIES (NOT (EQUAL (REMAINDER B X) 0))
         (NOT (EQUAL (GCD B X) X)))
```

**304. Theorem** DIVIDES.TIMES1 (rewrite):

```
(IMPLIES (EQUAL A (TIMES Z Y))
         (EQUAL (REMAINDER A Z) 0))
```

**305. Theorem** TIMES.IDENTITY (rewrite):

```
(IMPLIES (EQUAL Y 1)
         (EQUAL (EQUAL X (TIMES X Y))
                (NUMBERP X)))
```

**306. Theorem** KLUDGE.BRIDGE (rewrite):

```
(IMPLIES (EQUAL Y (TIMES K X))
         (EQUAL (GCD Y (TIMES A X))
                (TIMES X (GCD A K))))
```

**307. Theorem** HACK1 (rewrite):

```
(IMPLIES (AND (NOT (DIVIDES X A))
              (EQUAL A
                     (GCD (TIMES X A) (TIMES B A))))
         (NOT (EQUAL (TIMES K X) (TIMES B A))))
```

**308. Theorem** PRIME.GCD (rewrite):

```
(IMPLIES (AND (NOT (DIVIDES X B))
              (PRIME1 X (SUB1 X)))
         (EQUAL (EQUAL (GCD B X) 1) T))
```

**309. Theorem** PRIME.KEY (rewrite):

```
(IMPLIES (AND (NUMBERP Z)
              (PRIME X)
              (NOT (DIVIDES X Z))
              (NOT (DIVIDES X B)))
         (NOT (EQUAL (TIMES X K) (TIMES B Z))))
```

**310. Theorem** QUOTIENT.DIVIDES (rewrite):

```
(IMPLIES (AND (NUMBERP Y)
              (NOT (EQUAL (TIMES X (QUOTIENT Y X)) Y)))
         (NOT (EQUAL (REMAINDER Y X) 0)))
```

**311. Theorem** LITTLE.STEP (rewrite):

```
(IMPLIES (AND (PRIME X)
              (NOT (EQUAL Y 1))
              (NOT (EQUAL X Y)))
         (NOT (EQUAL (REMAINDER X Y) 0)))
```

**312. Definition**

```
(DELETE X L)
  =
(IF (NLISTP L)
    L
    (IF (EQUAL X (CAR L))
        (CDR L)
        (CONS (CAR L) (DELETE X (CDR L)))))
```

**313. Definition**

```
(PERM A B)
  =
(IF (NLISTP A)
    (NLISTP B)
    (AND (MEMBER (CAR A) B)
         (PERM (CDR A) (DELETE (CAR A) B))))
```

**314. Theorem** REMAINDER.TIMES (rewrite):

```
(EQUAL (REMAINDER (TIMES Y X) Y) 0)
```

**315. Theorem** PRIME.LIST.DELETE (rewrite):

```
(IMPLIES (PRIME.LIST L2)
         (PRIME.LIST (DELETE X L2)))
```

**316. Theorem** DIVIDES.TIMES.LIST (rewrite):

```
(IMPLIES (AND (NOT (ZEROP C)) (MEMBER C L))
         (EQUAL (REMAINDER (TIMES.LIST L) C)
                0))
```

**317. Theorem** QUOTIENT.TIMES (rewrite):

    (EQUAL (QUOTIENT (TIMES Y X) Y)
        (IF (ZEROP Y) 0 (FIX X)))

**318. Theorem** DISTRIBUTIVITY.OF.DIVIDES (rewrite):

    (IMPLIES (AND (NOT (ZEROP A)) (DIVIDES A W))
        (EQUAL (TIMES C (QUOTIENT W A))
            (QUOTIENT (TIMES C W) A)))

**319. Theorem** TIMES.LIST.DELETE (rewrite):

    (IMPLIES (AND (NOT (ZEROP C)) (MEMBER C L))
        (EQUAL (TIMES.LIST (DELETE C L))
            (QUOTIENT (TIMES.LIST L) C)))

**320. Theorem** PRIME.LIST.TIMES.LIST (rewrite):

    (IMPLIES (AND (PRIME C)
            (PRIME.LIST L2)
            (NOT (MEMBER C L2)))
        (NOT (EQUAL (REMAINDER (TIMES.LIST L2) C)
                0)))

**321. Theorem** IF.TIMES.THEN.DIVIDES (rewrite):

    (IMPLIES (AND (NOT (ZEROP C))
           (NOT (DIVIDES C X)))
        (NOT (EQUAL (TIMES C Y) X)))

**322. Theorem** PRIME.MEMBER (rewrite):

    (IMPLIES (AND (PRIME C)
            (PRIME.LIST L2)
            (EQUAL (TIMES C (TIMES.LIST L1))
                (TIMES.LIST L2)))
        (MEMBER C L2))

**323. Theorem** DIVIDES.IMPLIES.TIMES (rewrite):

    (IMPLIES (AND (NOT (ZEROP A))
            (NUMBERP C)
            (EQUAL (TIMES A C) B))
        (EQUAL (EQUAL C (QUOTIENT B A)) T))

**324. Theorem** TIMES.EQUAL.1 (rewrite):

    (EQUAL (EQUAL (TIMES A B) 1)
        (AND (NOT (EQUAL A 0))
            (NOT (EQUAL B 0))
            (NUMBERP A)
            (NUMBERP B)
            (EQUAL (SUB1 A) 0)
            (EQUAL (SUB1 B) 0)))

**325. Theorem** PRIME.FACTORIZATION.UNIQUENESS:

```
(IMPLIES (AND (PRIME.LIST L1)
              (PRIME.LIST L2)
              (EQUAL (TIMES.LIST L1)
                     (TIMES.LIST L2)))
         (PERM L1 L2))
```

**326. Definition**

```
(MAXIMUM L)
    =
(IF (NLISTP L)
    0
    (IF (LESSP (CAR L) (MAXIMUM (CDR L)))
        (MAXIMUM (CDR L))
        (CAR L)))
```

**327. Theorem** MEMBER.MAXIMUM (rewrite):

```
(IMPLIES (LISTP X)
         (MEMBER (MAXIMUM X) X))
```

**328. Theorem** LESSP.DELETE.REWRITE (rewrite):

```
(EQUAL (LESSP (LENGTH (DELETE X L))
              (LENGTH L))
       (MEMBER X L))
```

**329. Theorem** LESSP.LENGTH (induction):

```
(IMPLIES (LISTP L)
         (LESSP (LENGTH (DELETE (MAXIMUM L) L))
                (LENGTH L)))
```

**330. Definition**

```
(ORDERED2 L)
    =
(IF (LISTP L)
    (IF (LISTP (CDR L))
        (IF (LESSP (CAR L) (CADR L))
            F
            (ORDERED2 (CDR L)))
        T)
    T)
```

**331. Definition**

```
(DSORT L)
    =
(IF (NLISTP L)
    "NIL"
    (CONS (MAXIMUM L)
          (DSORT (DELETE (MAXIMUM L) L))))
```

## 332. Definition

```
(ADDTOLIST2 X L)
  =
(IF (LISTP L)
    (IF (LESSP X (CAR L))
        (CONS (CAR L) (ADDTOLIST2 X (CDR L)))
        (CONS X L))
    (CONS X "NIL"))
```

## 333. Definition

```
(SORT2 L)
  =
(IF (NLISTP L)
    "NIL"
    (ADDTOLIST2 (CAR L) (SORT2 (CDR L))))
```

## 334. Theorem  SORT2.GEN.1 (rewrite):

```
(PLISTP (SORT2 X))
```

## 335. Theorem  SORT2.GEN.2 (rewrite):

```
(ORDERED2 (SORT2 X))
```

## 336. Theorem  SORT2.GEN (generalize):

```
(AND (PLISTP (SORT2 X))
     (ORDERED2 (SORT2 X)))
```

## 337. Theorem  ADDTOLIST2.DELETE (rewrite):

```
(IMPLIES (AND (PLISTP Y)
              (ORDERED2 Y)
              (NOT (EQUAL X V)))
         (EQUAL (ADDTOLIST2 V (DELETE X Y))
                (DELETE X (ADDTOLIST2 V Y))))
```

## 338. Theorem  DELETE.ADDTOLIST2 (rewrite):

```
(IMPLIES (PLISTP Y)
         (EQUAL (DELETE V (ADDTOLIST2 V Y)) Y))
```

## 339. Theorem  ADDTOLIST2.KLUDGE (rewrite):

```
(IMPLIES (AND (NOT (LESSP V W))
              (EQUAL (ADDTOLIST2 V Y) (CONS V Y)))
         (EQUAL (ADDTOLIST2 V (ADDTOLIST2 W Y))
                (CONS V (ADDTOLIST2 W Y))))
```

## 340. Theorem  LESSP.MAXIMUM.ADDTOLIST2 (rewrite):

```
(IMPLIES (NOT (LESSP V (MAXIMUM Z)))
         (EQUAL (ADDTOLIST2 V (SORT2 Z))
                (CONS V (SORT2 Z))))
```

## 341. Theorem  SORT2.DELETE.CONS (rewrite):

```
(IMPLIES (LISTP X)
         (EQUAL (CONS (MAXIMUM X)
                      (DELETE (MAXIMUM X) (SORT2 X)))
                (SORT2 X)))
```

**342. Theorem** SORT2.DELETE (rewrite):
```
(EQUAL (SORT2 (DELETE X L))
       (DELETE X (SORT2 L)))
```

**343. Theorem** DSORT.SORT2 (rewrite):
```
(EQUAL (DSORT X) (SORT2 X))
```

**344. Theorem** COUNT.LIST.SORT2:
```
(EQUAL (COUNT.LIST A (SORT2 L))
       (COUNT.LIST A L))
```

**345. Theorem** LESSP.PLUS.SUB1 (rewrite):
```
(NOT (LESSP (PLUS Y Z) (SUB1 Z)))
```

**346. Undefined Function**
```
(DECREMENTP X)
```

**347. Undefined Function**
```
(MEASURE X)
```

**348. Undefined Function**
```
(DECREMENT X)
```

**349. Axiom** PK (induction):
```
(IMPLIES (NOT (DECREMENTP X))
         (LESSP (MEASURE (DECREMENT X))
                (MEASURE X)))
```

**350. Undefined Function**
```
(FIDDLE X)
```

**351. Definition**
```
(FIDDLE.DOWN X Y)
    =
(IF (DECREMENTP X)
    Y
    (FIDDLE (FIDDLE.DOWN (DECREMENT X) Y)))
```

**352. Definition**
```
(FIDDLE.DOWN.2 X Y)
    =
(IF (DECREMENTP X)
    Y
    (FIDDLE.DOWN.2 (DECREMENT X)
                   (FIDDLE Y)))
```

## 353. Theorem FIDDLE.EQUAL:

```
(EQUAL (FIDDLE.DOWN X Y)
       (FIDDLE.DOWN.2 X Y))
```

## 354. Theorem PLUS.CANCELLATION (rewrite):

```
(EQUAL (EQUAL (PLUS X Y) (PLUS X Z))
       (EQP Y Z))
```

## 355. Definition

```
(MATCH PAT STR)
   =
(IF (LISTP PAT)
    (IF (LISTP STR)
        (IF (EQUAL (CAR PAT) (CAR STR))
            (MATCH (CDR PAT) (CDR STR))
            F)
        F)
    T)
```

## 356. Definition

```
(STRPOS PAT STR)
   =
(IF (MATCH PAT STR)
    0
    (IF (LISTP STR)
        (ADD1 (STRPOS PAT (CDR STR)))
        0))
```

## 357. Definition

```
(DELTA1 CHAR PAT)
   =
(STRPOS (CONS CHAR "NIL")
        (REVERSE PAT))
```

## 358. Definition

```
(TOP.ASSERT PAT STR I PATLEN STRLEN PAT* STR*)
   =
(AND (EQUAL PAT PAT*)
     (EQUAL STR STR*)
     (EQUAL PATLEN (LENGTH PAT))
     (LISTP PAT)
     (EQUAL STRLEN (LENGTH STR))
     (NUMBERP I)
     (LESSEQP (SUB1 PATLEN) I)
     (LESSP I
            (PLUS PATLEN (STRPOS PAT STR))))
```

### 359. Definition

    (LOOP.ASSERT PAT STR I J PATLEN STRLEN NEXTI PAT* STR*)
    =
    (AND (TOP.ASSERT PAT STR
                     (SUB1 NEXTI)
                     PATLEN STRLEN PAT* STR*)
         (NUMBERP I)
         (NUMBERP J)
         (NUMBERP NEXTI)
         (LESSP J PATLEN)
         (LESSP I STRLEN)
         (EQUAL NEXTI
                (PLUS PATLEN (DIFFERENCE I J)))
         (LESSEQP NEXTI STRLEN)
         (LESSEQP J I)
         (MATCH (NTH PAT (ADD1 J))
                (NTH STR (ADD1 I)))))

### 360. Theorem ZEROP.LENGTH (rewrite):

    (EQUAL (EQUAL (LENGTH X) 0)
           (NOT (LISTP X)))

### 361. Theorem FSTRPOS.VC1:

    (IMPLIES (EQUAL (LENGTH PAT*) 0)
             (EQUAL 0 (STRPOS PAT* STR*)))

### 362. Theorem SUB1.LESSP.PLUS (rewrite):

    (EQUAL (LESSP (SUB1 X) (PLUS X Y))
           (IF (ZEROP X) (NOT (ZEROP Y)) T))

### 363. Theorem FSTRPOS.VC2:

    (IMPLIES (NOT (EQUAL (LENGTH PAT*) 0))
             (TOP.ASSERT PAT* STR*
                         (SUB1 (LENGTH PAT*))
                         (LENGTH PAT*)
                         (LENGTH STR*)
                         PAT* STR*))

### 364. Theorem MATCH.LENGTHS (rewrite):

    (IMPLIES (MATCH X Y)
             (NOT (LESSP (LENGTH Y) (LENGTH X))))

### 365. Theorem MATCH.LENGTHS1 (rewrite):

    (IMPLIES (LESSP (LENGTH Y) (LENGTH X))
             (NOT (MATCH X Y)))

### 366. Theorem STRPOS.BOUNDARY.CONDITION (rewrite):

    (IMPLIES (NOT (EQUAL (STRPOS PAT STR) (LENGTH STR)))
             (NOT (LESSP (LENGTH STR)
                         (PLUS (LENGTH PAT)
                               (STRPOS PAT STR)))))

### DEFINITIONS ACCEPTED AND THEOREMS PROVED BY OUR SYSTEM / 371

**367. Theorem** FSTRPOS.VC3:

```
(IMPLIES (AND (GREATEREQP I STRLEN)
              (TOP.ASSERT PAT STR I PATLEN STRLEN
                          PAT* STR*))
         (EQUAL STRLEN (STRPOS PAT* STR*)))
```

**368. Theorem** PLUS.DIFFERENCE.SUB1.REWRITE (rewrite):

```
(EQUAL (PLUS X (DIFFERENCE Y (SUB1 X)))
       (IF (ZEROP X)
           (FIX Y)
           (IF (LESSP Y (SUB1 X))
               (FIX X)
               (ADD1 Y))))
```

**369. Theorem** LISTP.NTH (rewrite):

```
(EQUAL (LISTP (NTH X I))
       (LESSP I (LENGTH X)))
```

**370. Theorem** CDR.NTH (rewrite):

```
(EQUAL (CDR (NTH X Y))
       (NTH (CDR X) Y))
```

**371. Theorem** STRPOS.EQUAL (rewrite):

```
(IMPLIES (AND (LESSP I (LENGTH STR))
              (NOT (LESSP (STRPOS PAT STR) I))
              (NUMBERP I)
              (MATCH PAT (NTH STR I)))
         (EQUAL (STRPOS PAT STR) I))
```

**372. Theorem** VC4.HACK.1 (rewrite):

```
(IMPLIES (LESSP I (LENGTH STR))
         (NOT (LESSP (SUB1 (LENGTH STR)) I)))
```

**373. Theorem** FSTRPOS.VC4:

```
(IMPLIES (AND (NOT (GREATEREQP I STRLEN))
              (TOP.ASSERT PAT STR I PATLEN STRLEN
                          PAT* STR*))
         (LOOP.ASSERT PAT STR I
                      (SUB1 PATLEN)
                      PATLEN STRLEN
                      (ADD1 I)
                      PAT* STR*))
```

**374. Theorem** SWAPPED.PLUS.CANCELLATION (rewrite):

```
(EQUAL (LESSP (PLUS B A) (PLUS A C))
       (LESSP B C))
```

**375. Theorem** LESSP.SUB1.PLUS.CANCELLATION (rewrite):

```
(EQUAL (LESSP (SUB1 (PLUS Y X)) (PLUS X Z))
       (IF (ZEROP Y)
           (IF (ZEROP X) (NOT (ZEROP Z)) T)
           (LESSP (SUB1 Y) Z)))
```

372 / APPENDIX A

**376. Theorem** VC5.HACK1 (rewrite):
```
(IMPLIES (LESSP (SUB1 I) (STRPOS PAT STR))
         (NOT (LESSP (STRPOS PAT STR) I)))
```

**377. Theorem** FSTRPOS.VC5:
```
(IMPLIES (AND (EQUAL J 0)
              (EQUAL (NTHCHAR I STR)
                     (NTHCHAR J PAT))
              (LOOP.ASSERT PAT STR I J PATLEN STRLEN
                           NEXTI PAT* STR*))
         (EQUAL I (STRPOS PAT* STR*)))
```

**378. Theorem** FSTRPOS.VC6:
```
(IMPLIES (AND (NOT (EQUAL J 0))
              (EQUAL (NTHCHAR I STR)
                     (NTHCHAR J PAT))
              (LOOP.ASSERT PAT STR I J PATLEN STRLEN
                           NEXTI PAT* STR*))
         (LOOP.ASSERT PAT STR
                      (SUB1 I)
                      (SUB1 J)
                      PATLEN STRLEN NEXTI PAT* STR*))
```

**379. Theorem** STRPOS.LESSEQP.STRLEN (rewrite):
```
(NOT (LESSP (LENGTH STR) (STRPOS PAT STR)))
```

**380. Theorem** LESSP.KLUDGE1 (rewrite):
```
(IMPLIES (NOT (LESSP B A))
         (EQUAL (LESSP A (PLUS B C))
                (IF (ZEROP C) (LESSP A B) T)))
```

**381. Theorem** STRPOS.LIST.APPEND (rewrite):
```
(EQUAL (STRPOS (CONS C "NIL") (APPEND A B))
       (IF (MEMBER C A)
           (STRPOS (CONS C "NIL") A)
           (PLUS (LENGTH A)
                 (STRPOS (CONS C "NIL") B))))
```

**382. Theorem** STRPOS.LESSEQP.CRUTCH (rewrite):
```
(IMPLIES (NOT (LESSP (LENGTH Q) (LENGTH P)))
         (NOT (LESSP (LENGTH Q) (STRPOS PAT P))))
```

**383. Theorem** STRPOS.EQUAL.0 (rewrite):
```
(EQUAL (EQUAL (STRPOS PAT STR) 0)
       (OR (NLISTP STR) (MATCH PAT STR)))
```

**384. Theorem** LESSP.KLUDGE2 (rewrite):
```
(IMPLIES (LESSP I (LENGTH PAT))
         (LESSP (SUB1 I)
                (PLUS (LENGTH PAT) Z)))
```

# DEFINITIONS ACCEPTED AND THEOREMS PROVED BY OUR SYSTEM / 373

**385. Theorem** MATCH.IMPLIES.CAR.MEMBER (rewrite):

```
(IMPLIES (AND (LISTP PAT)
              (NOT (MEMBER (CAR STR) PAT)))
         (NOT (MATCH PAT STR)))
```

**386. Theorem** MATCH.IMPLIES.CAR.MEMBER1 (rewrite):

```
(IMPLIES (AND (LISTP PAT) (MATCH PAT STR))
         (MEMBER (CAR STR) PAT))
```

**387. Theorem** MATCH.IMPLIES.MEMBER (rewrite):

```
(IMPLIES (AND (LESSP I (LENGTH PAT))
              (MATCH PAT STR))
         (MEMBER (CAR (NTH STR I)) PAT))
```

**388. Theorem** DELTA1.LESSP.IFF.MEMBER (rewrite):

```
(EQUAL (LESSP (STRPOS (CONS CHAR "NIL")
                      (REVERSE PAT))
              (LENGTH PAT))
       (MEMBER CHAR PAT))
```

**389. Theorem** LESSP.PLUS (rewrite):

```
(IMPLIES (NOT (LESSP X Z))
         (NOT (LESSP (PLUS X Y) Z)))
```

**390. Theorem** LESSP.PLUS1 (rewrite):

```
(IMPLIES (NOT (LESSP Y Z))
         (NOT (LESSP (PLUS X Y) Z)))
```

**391. Theorem** MATCH.IMPLIES.DELTA1.OK (rewrite):

```
(IMPLIES
   (AND (MATCH PAT STR)
        (LESSP I (LENGTH PAT)))
   (LESSP (PLUS I
                (STRPOS (CONS (CAR (NTH STR I)) "NIL")
                        (REVERSE PAT)))
          (LENGTH PAT)))
```

**392. Theorem** SUB1.LENGTH (rewrite):

```
(EQUAL (SUB1 (LENGTH X))
       (LENGTH (CDR X)))
```

**393. Theorem** DELTA1.LEMMA (rewrite):

```
(IMPLIES
   (AND (LISTP PAT)
        (LESSP I (LENGTH STR))
        (LESSP I
               (PLUS (LENGTH PAT) (STRPOS PAT STR))))
   (LESSP (PLUS I
                (STRPOS (CONS (CAR (NTH STR I)) "NIL")
                        (REVERSE PAT)))
          (PLUS (LENGTH PAT) (STRPOS PAT STR))))
```

**394. Theorem**  MATCH.EPSILON (rewrite):

```
(IMPLIES (AND (LESSP J (LENGTH PAT))
              (MATCH PAT STR))
         (EQUAL (CAR (NTH PAT J))
                (CAR (NTH STR J))))
```

**395. Theorem**  STRPOS.EPSILON (rewrite):

```
(IMPLIES
     (AND (LESSP J (LENGTH PAT))
          (LESSP (PLUS J (STRPOS PAT STR))
                 (LENGTH STR)))
     (EQUAL (CAR (NTH STR (PLUS J (STRPOS PAT STR))))
            (CAR (NTH PAT J))))
```

**396. Theorem**  EQ.CHARS.AT.STRPOS (rewrite):

```
(IMPLIES (AND (NOT (LESSP I J))
              (NOT (EQUAL (CAR (NTH STR I))
                          (CAR (NTH PAT J))))
              (LESSP I (LENGTH STR))
              (LESSP J (LENGTH PAT)))
         (NOT (EQUAL (STRPOS PAT STR)
                     (DIFFERENCE I J))))
```

**397. Theorem**  LESSP.DIFFERENCE.1 (rewrite):

```
(IMPLIES (LESSP I J)
         (EQUAL (DIFFERENCE I J) 0))
```

**398. Theorem**  LESSP.SUB1.SUB1 (rewrite):

```
(EQUAL (LESSP (SUB1 (SUB1 X)) X)
       (NOT (ZEROP X)))
```

**399. Theorem**  PLUS.2.NOT (rewrite):

```
(NOT (EQUAL (ADD1 (ADD1 (PLUS V Z))) V))
```

**400. Theorem**  LESSP.SUB1.SUB1.PLUS (rewrite):

```
(EQUAL (LESSP (SUB1 (SUB1 Y)) (PLUS Z Y))
       (OR (NOT (ZEROP Z)) (NOT (ZEROP Y))))
```

**401. Theorem**  LESSP.KLUDGE3 (rewrite):

```
(EQUAL (LESSP (SUB1 (PLUS L (DIFFERENCE I J)))
              I)
       (IF (LESSP I J)
           (LESSP (SUB1 L) I)
           (IF (ZEROP L)
               (NOT (ZEROP I))
               (NOT (LESSP J L)))))
```

**402. Theorem**  GT.SUB1 (rewrite):

```
(NOT (LESSP (PLUS X Y) (SUB1 X)))
```

**403. Theorem** LESSP.SUB1.HACK1 (rewrite):

```
(IMPLIES (AND (NOT (LESSP I J))
              (NOT (EQUAL I J))
              (NUMBERP J)
              (NUMBERP I))
         (NOT (LESSP (SUB1 I) J)))
```

**404. Theorem** FSTRPOS.VC7:

```
(IMPLIES
 (AND (NOT (EQUAL (NTHCHAR I STR)
                  (NTHCHAR J PAT)))
      (LOOP.ASSERT PAT STR I J PATLEN STRLEN NEXTI
                   PAT* STR*))
 (TOP.ASSERT PAT STR
       (IF (LESSP (PLUS I (DELTA1 (NTHCHAR I STR) PAT))
                  NEXTI)
           NEXTI
           (PLUS I (DELTA1 (NTHCHAR I STR) PAT)))
       PATLEN STRLEN PAT* STR*))
```

# Appendix B

## The Implementation of the Shell Principle

Below we give the axioms added by our theorem-prover in response to the user command:

Add the shell `const`, of n arguments,
with (optionally, bottom object (`btm`),)
recognizer r,
accessors $ac_1$, ..., $ac_n$,
type restrictions $tr_1$, ..., $tr_n$, and
default values $dv_1$, ..., $dv_n$.

Our implementation of the shell principle differs from the formal presentation in Chapter III in two ways. First, we do not actually require (or allow) the user to specify the name of a new well-founded relation to be defined for the class. Instead, we axiomatize COUNT for the class and add the appropriate induction lemmas, using COUNT and LESSP. In Chapter III we justified the introduction of COUNT and LESSP. The second difference between the formal description and our implementation is that our implementation adds lemmas that are more useful to the theorem-prover than would be the axioms noted in Chapter III. For example, certain of the formal axioms are reformulated so that they are more useful as rewrite rules. In all cases, the lemmas added by our implementation follow immediately from the axioms given in Chapter III.

Most of the axioms have names. We indicate the names schematically below. For example, when the CONS shell is added, the schematic name $ac_1$.const denotes CAR.CONS.

After the name, we indicate in parentheses the "type" of the axiom (e.g., rewrite, elimination, or induction). No generalization lemmas, per se, are added by the shell mechanism; however, rewrite lemmas encoded as type prescriptions may restrict generalizations of shell functions since the generalization heuristic employs type sets.

**Axiom** r.const (rewrite):

  (r (const X1 ... Xn)).

**Axiom** r.btm (rewrite):

  (r (btm)).

**Axiom** BOOLEAN.r (rewrite):

  (OR (EQUAL (r X) T) (EQUAL (r X) F)).

For each i from 1 to n, let $tr_i'$ be $tr_i$ with all occurrences of Xi replaced by ($ac_i$ X), and add:

**Axiom** TYPE.OF.$ac_i$ (rewrite):

  $tr_i'$

(Observe that all of the above axioms are stored as type prescriptions.)

For each i from 1 to n, add the following five axioms:

**Axiom** $ac_i$.const (rewrite):

  (EQUAL ($ac_i$ (const X1 ... Xn))
    (IF $tr_i$ Xi $dv_i$)).

**Axiom** $ac_i$.Nr (rewrite):

  (IMPLIES (NOT (r X))
    (EQUAL ($ac_i$ X) $dv_i$)).

**Axiom** $ac_i$.TYPE.RESTRICTION (rewrite):

  (IMPLIES (NOT $tr_i$)
    (EQUAL (const X1 ... Xi ... Xn)
      (const X1 ... $dv_i$ ... Xn))).

**Axiom** $ac_i$.btm (rewrite):

  (EQUAL ($ac_i$ (btm)) $dv_i$).

**Axiom** $ac_i$.LESSP (induction):

```
(IMPLIES (AND (r X)
              (NOT (EQUAL X (btm))))
         (LESSP (COUNT (ac_i X)) (COUNT X))).
```

Let s be the substitution replacing X1 by Y1, ... and Xn by Yn, and let $tr_i/s$ denote the result of applying s to $tr_i$. Add:

**Axiom** const.EQUAL (rewrite):

```
(EQUAL (EQUAL (const X1 ... Xn)
              (const Y1 ... Yn))
       (AND (IF tr_1
                (IF tr_1/s
                    (EQUAL X1 Y1)
                    (EQUAL X1 dv_1))
                (IF tr_1/s
                    (EQUAL dv_1 Y1)
                    T))
            ...
            (IF tr_n
                (IF tr_n/s
                    (EQUAL Xn Yn)
                    (EQUAL Xn dv_n))
                (IF tr_n/s
                    (EQUAL dv_n Yn)
                    T)))).
```

**Axiom** $const.ac_1. \ldots .ac_n$ (rewrite):

```
(EQUAL (const (ac_1 X) ... (ac_n X))
       (IF (AND (r X)
                (NOT (EQUAL X (btm))))
           X
           (const dv_1 ... dv_n))).
```

**Axiom** $ac_1/ \ldots /ac_n$.ELIM (elimination):

```
(IMPLIES (AND (r X)
              (NOT (EQUAL X (btm))))
         (EQUAL (const (ac_1 X) ... (ac_n X))
                X)).
```

**Axiom** COUNT.const (rewrite):

```
(EQUAL (COUNT (const X1 ... Xn))
```

$$(\text{ADD1 (PLUS (IF tr}_1 \text{ (COUNT X1) 0)}$$
$$\ldots$$
$$(\text{IF tr}_n \text{ (COUNT Xn) 0)}))).$$

**Axiom** COUNT.btm (rewrite):

(EQUAL (COUNT (btm)) 0).

The handling of the special case in which no bottom object is supplied should be obvious.

We simplify the right-hand sides of concluding equalities in rewrite axioms by expanding nonrecursive functions (e.g., AND), putting IF-expressions into IF-normal form, and simplifying IF-expressions with explicit value tests (e.g., (IF T x y) is replaced by x).

The following axioms resulting from an application of the shell principle are "wired-in" to the theorem-prover.

(NOT (EQUAL (const X1 ... Xn) (btm))),

(NOT (r T)),

(NOT (r F)),

and

(IMPLIES (r X) (NOT (r' X))).

The first of the above axioms is built into the rules for rewriting an EQUAL expression given in Chapter IX. The other axioms above are built into the type set mechanism (see Chapter VI).

# Appendix C

## Clauses for Our Theory

Readers familiar with other mechanical theorem-provers might be interested in seeing our theory cast in the more usual clausal form. We do not formulate our shell principle, induction principle, or definition principle in clausal form.[48] However, we do give the clauses generated by these principles on specific examples.

Here we give, in clausal form, axioms for T, F, IF, EQUAL, numbers, literal atoms, and ordered pairs. These axioms, together with our induction principle, definition principle, and the well-foundedness of LESSP and induced lexicographic relations, are equivalent to the theory described in Chapter III. We also exhibit the axioms of definition for the functions used in the MC.FLATTEN example in Chapter II, and we exhibit a clausal formulation of the first induction step used in that example.

In the following clauses, we use a common notation for function application and clauses [49]. T, F, O, and NIL are constants. X, Y, Z, P, Q, X1, X2, Y1, and Y2 are variables.

---

[48] It is interesting that the set theory of von Neumann, Bernays, and Gödel [21] can be stated in a finite number of clauses. Thus, in principle, one could use a finitely axiomatized set theory with a resolution theorem-prover to investigate problems normally requiring the axiom schemes (i.e., infinite axioms) of induction and comprehension.

## 1. LOGICAL DEFINITIONS

L1. $\{T \neq F\}$
L2. $\{X \neq Y \quad \text{EQUAL}(X,Y)=T\}$
L3. $\{X=Y \quad \text{EQUAL}(X,Y)=F\}$
L4. $\{X \neq F \quad \text{IF}(X,Y,Z)=Z\}$
L5. $\{X=F \quad \text{IF}(X,Y,Z)=Y\}$
L6. $\{\text{NOT}(P)=\text{IF}(P,F,T)\}$
L7. $\{\text{AND}(P,Q)=\text{IF}(P,\text{IF}(Q,T,F),F))\}$
L8. $\{\text{OR}(P,Q)=\text{IF}(P,T,\text{IF}(Q,T,F))\}$
L9. $\{\text{IMPLIES}(P,Q)=\text{IF}(P,\text{IF}(Q,T,F),T)\}$

## 2. AXIOMS FOR NATURAL NUMBERS

A1. $\{\text{NUMBERP}(\text{ADD1}(X1))=T\}$
A2. $\{\text{NUMBERP}(0)=T\}$
A3. $\{\text{NUMBERP}(X)=T \quad \text{NUMBERP}(X)=F\}$
A4. $\{\text{NUMBERP}(\text{SUB1}(X))=T\}$
A5. $\{\text{SUB1}(\text{ADD1}(X1))=\text{IF}(\text{NUMBERP}(X1),X1,0)\}$
A6. $\{\text{NUMBERP}(X)=T \quad \text{SUB1}(X)=0\}$
A7. $\{\text{NUMBERP}(X1)=T \quad \text{ADD1}(X1)=\text{ADD1}(0)\}$
A8. $\{\text{SUB1}(0)=0\}$
A9. $\{\text{NUMBERP}(X)=F \quad X=0 \quad \text{LESSP}(\text{COUNT}(\text{SUB1}(X)),$
$\qquad\qquad\qquad\qquad \text{COUNT}(X))=T\}$
A10. $\{\text{EQUAL}(\text{ADD1}(X1),\text{ADD1}(Y1))=$
$\quad \text{IF}(\text{NUMBERP}(X1),$
$\qquad \text{IF}(\text{NUMBERP}(Y1),\text{EQUAL}(X1,Y1),\text{EQUAL}(X1,0)),$
$\qquad \text{IF}(\text{NUMBERP}(Y1),\text{EQUAL}(0,Y1),T))\}$
A11. $\{\text{ADD1}(\text{SUB1}(X))=\text{IF}(\text{AND}(\text{NUMBERP}(X),$
$\qquad\qquad\qquad \text{NOT}(\text{EQUAL}(X,0))),$
$\qquad\qquad\qquad X,\text{ADD1}(0))\}$
A12. $\{\text{COUNT}(\text{ADD1}(X1))=\text{ADD1}(\text{IF}(\text{NUMBERP}(X1),$
$\qquad\qquad\qquad\qquad \text{COUNT}(X1),0))\}$
A13. $\{\text{COUNT}(0)=0\}$
A14. $\{0 \neq \text{ADD1}(X1)\}$
A15. $\{\text{NUMBERP}(T)=F\}$
A16. $\{\text{NUMBERP}(F)=F\}$
A17. $\{\text{ZEROP}(X)=\text{OR}(\text{EQUAL}(X,0),\text{NOT}(\text{NUMBERP}(X)))\}$
A18. $\{\text{FIX}(X)=\text{IF}(\text{NUMBERP}(X),X,0)\}$

A19. {LESSP(X,Y)=IF(ZEROP(Y),
                   F,
                   IF(ZEROP(X),T,LESSP(SUB1(X),
                                            SUB1(Y))))}
A20. {PLUS(X,Y)=IF(ZEROP(X),
                   FIX(Y),ADD1(PLUS(SUB1(X),Y)))}

## 3. AXIOMS FOR LITERAL ATOMS

B1. {LITATOM(PACK(X1))=T}
B2. {LITATOM(NIL)=T}
B3. {LITATOM(X)=T LITATOM(X)=F}
B4. {UNPACK(PACK(X1))=X1}
B5. {LITATOM(X)=T UNPACK(X)=0}
B6. {UNPACK(NIL)=0}
B7. {LITATOM(X)=F X=NIL LESSP(COUNT(UNPACK(X)),
                                    COUNT(X))=T}
B8. {EQUAL(PACK(X1),PACK(Y1))=EQUAL(X1,Y1)}
B9. {PACK(UNPACK(X))=IF(AND(LITATOM(X),
                             NOT(EQUAL(X,NIL))),
                         X,PACK(0))}
B10. {COUNT(PACK(X1))=ADD1(COUNT(X1))}
B11. {COUNT(NIL)=0}
B12. {NIL≠PACK(X1)}
B13. {LITATOM(T)=F}
B14. {LITATOM(F)=F}
B15. {LITATOM(X)=F NUMBERP(X)=F}

## 4. AXIOMS FOR ORDERED PAIRS

C1. {LISTP(CONS(X1,X2))=T}
C2. {LISTP(X)=T LISTP(X)=F}
C3. {CAR(CONS(X1,X2))=X1}
C4. {CDR(CONS(X1,X2))=X2}
C5. {LISTP(X)=T CAR(X)=NIL}
C6. {LISTP(X)=T CDR(X)=NIL}
C7. {LISTP(X)=F LESSP(COUNT(CAR(X)),COUNT(X))=T}
C8. {LISTP(X)=F LESSP(COUNT(CDR(X)),COUNT(X))=T}

C9.  {EQUAL(CONS(X1,X2),CONS(Y1,Y2))=
            AND(EQUAL(X1,Y1),EQUAL(X2,Y2))}
C10. {CONS(CAR(X),CDR(X))=IF(LISTP(X),
                              X,CONS(NIL,NIL))}
C11. {COUNT(CONS(X1,X2))=ADD1(PLUS(COUNT(X1),
                                   COUNT(X2)))}
C12. {LISTP(T)=F}
C13. {LISTP(F)=F}
C14. {LISTP(X)=F NUMBERP(X)=F}
C15. {LISTP(X)=F LITATOM(X)=F}

## 5. A SAMPLE THEOREM IN CLAUSAL FORM

To "define" APPEND, FLATTEN, and MC.FLATTEN for a resolution theorem-prover, one could add the clauses:

{APPEND(X,Y)=IF(LISTP(X),
                CONS(CAR(X),APPEND(CDR(X),Y)),Y)},

{FLATTEN(X)=IF(LISTP(X),
               APPEND(FLATTEN(CAR(X)),FLATTEN(CDR(X))),
               CONS(X,NIL))}

{MC.FLATTEN(X,Y)=IF(LISTP(X),
                    MC.FLATTEN(CAR(X),
                               MC.FLATTEN(CDR(X),Y)),
                    CONS(X,Y))}.

When our theorem-prover is given the definitions of APPEND, FLATTEN, and MC.FLATTEN, it discovers and remembers the following theorems:

{LISTP(APPEND(X,Y))=T APPEND(X,Y)=Y},

{LISTP(FLATTEN(X))=T},

{LISTP(MC.FLATTEN(X))=T}.

One could, in principle, use a resolution theorem-prover to help perform the noninductive parts of our proofs. For example, one might ask such a theorem-prover to undertake the first inductive step of the FLATTEN.MC.FLATTEN example of Chapter II. One might therefore provide such a theorem-prover with all the foregoing clauses of this appendix and then, letting A and ANS be constants, attempt to derive a contradiction from

{LISTP(A)=T},

{MC.FLATTEN(CAR(A),MC.FLATTEN(CDR(A),ANS))
=APPEND(FLATTEN(CAR(A)),MC.FLATTEN(CDR(A),ANS))},

{MC.FLATTEN(CDR(A),ANS)=APPEND(FLATTEN(CDR(A)),ANS)},

and

{MC.FLATTEN(A,ANS)≠APPEND(FLATTEN(A),ANS)}.

Of course, the imagined resolution theorem-prover would probably fail to find a contradiction, because we know of no proof that does not depend upon the associativity of APPEND. While our theorem-prover discovered and inductively proved that APPEND is associative in the course of the FLATTEN.MC.FLATTEN proof, one would need to add the clause

{APPEND(APPEND(X,Y),Z)=APPEND(X,APPEND(Y,Z))}

to the previous collection of clauses before a resolution theorem-prover might be expected to derive a contradiction.

# References

1. J. Allen, *The Anatomy of LISP*, McGraw–Hill, New York, 1978.
2. R. Aubin, "Mechanizing Structural Induction," Ph.D. Thesis, University of Edinburgh, Edinburgh (1976).
3. W. Bledsoe, Splitting and reduction heuristics in automatic theorem-proving, *Artificial Intelligence*, **2**, 55–77 (1971).
4. W. Bledsoe, R. Boyer, and W. Henneman, Computer proofs of limit theorems, *Artificial Intelligence*, **3**, 27–60 (1972).
5. W. Bledsoe, Non-resolution theorem proving, *Artificial Intelligence*, **9**, 1–36 (1977).
6. N. Bourbaki, *Elements of Mathematics Theory of Sets*, Addison–Wesley, Reading, Massachusetts, 1968.
7. R. Boyer and J Strother Moore, Proving theorems about LISP functions, *J. Assoc. Comput. Mach.* **22**(1), 129–144 (1975).
8. R. Boyer and J Strother Moore, A lemma driven automatic theorem prover for recursive function theory, *Proc. 5th Int. Joint Conf. Artificial Intelligence*, pp. 511–520, Department of Computer Science, Carnegie-Mellon University, Pittsburgh, Pennsylvania, 1977.
9. R. Boyer and J Strother Moore, A fast string searching algorithm, *Commun. Assoc. Comput. Mach.* **20**(10), 762–772 (1977).
10. R. Boyer and J Strother Moore, "A Formal Semantics for the SRI Hierarchical Program Design Methodology," Technical Report, Computer Science Laboratory, SRI International, Menlo Park, California (1978).
11. D. Brotz, "Proving Theorems by Mathematical Induction," Ph.D. Thesis, Computer Science Department, Stanford University, Stanford (1974).
12. R. Burstall, Proving properties of programs by structural induction, *Comput. J.* **12**(1), 41–48 (1969).
13. R. Burstall and J. Darlington, A transformation system for developing recursive programs, *J. Assoc. Comput. Mach.* **21**(1), 44–67 (1977).
14. R. Cartwright, "A Practical Formal Semantic Definition and Verification System for Typed LISP," Ph. D. Thesis, Stanford University, Stanford (1976).

15. C. Chang and R. Lee, *Symbolic Logic and Mechanical Theorem Proving*, Academic Press, New York, 1973.
16. A. Church, *The Calculi of Lambda-Conversion*, Annals of Mathematical Studies No. 6, Princeton Univ. Press, Princeton, New Jersey, 1941.
17. A. Church, *Introduction to Mathematical Logic*, Princeton Univ. Press, Princeton, New Jersey, 1956.
18. R. Floyd, Assigning meanings to programs, *Mathematical Aspects of Computer Science, Proc. Symp. Appl. Math. Vol. XIX*, pp. 19–32, American Mathematical Society, Providence, Rhode Island, 1967.
19. G. Gentzen, The consistency of elementary number theory, in *The Collected Papers of Gerhard Gentzen* (M. E. Szabo, ed.), North–Holland Publ. Co., Amsterdam, 1969. pp. 132–213.
20. P. Gloess, "A Proof of the Correctness of a Simple Parser of Expressions by the Boyer–Moore System," Technical Report, Computer Science Laboratory, SRI International, Menlo Park, California (1978).
21. K. Gödel, *The Consistency of the Axiom of Choice and of the Generalized Continuum-Hypothesis with the Axioms of Set Theory*, Princeton Univ. Press, Princeton, New Jersey, 1940.
22. R. Goodstein, *Recursive Number Theory. A Development of Recursive Arithmetic in a Logic Free Equation Calculus*, North–Holland Publ. Co., Amsterdam, 1957.
23. L. Guibas and A. Odlyzko, A new proof of the linearity of the Boyer–Moore string searching algorithm, *Proc. 18th Annual IEEE Symp. Foundations Comput. Sci.*, 1977.
24. C. Hoare, An axiomatic basis for computer programming, *Commun. Assoc. Comput. Mach.* **12**(10), 576–583 (1969).
25. J. Kelley, *General Topology*, Van Nostrand, Princeton, New Jersey, 1955.
26. D. Knuth and P. Bendix, Simple word problems in universal algebras, in *Computational Problems in Abstract Algebras* (J. Leech, ed.), pp. 263–297, Pergamon Press, Oxford, 1970.
27. D. Knuth, J. Morris, and V. Pratt, Fast pattern matching in strings, *SIAM J. Comput.* **6**(2), 323–350 (1977).
28. D. Lankford and A. Ballantyne, "Decision Procedures for Simple Equational Theories with Permutative Equations: Complete Sets of Permutative Reductions," Workshop on Automatic Deduction, MIT, Cambridge, Massachusetts, 1977.
29. D. Loveland, *Automated Theorem Proving: A Logical Basis*, North–Holland Publ. Co., Amsterdam, 1978.
30. Z. Manna and A. Pnueli, Formalization of properties of functional programs, *J. Assoc. Comput. Mach.* **17**(3), 555–569 (1970).
31. Z. Manna and R. Waldinger, The logic of computer programming, *IEEE Trans. Software Eng.* **SE-4**(3), 199–229 (1978).
32. J. McCarthy, Recursive functions of symbolic expressions and their computation by machine, *Commun. Assoc. Comput. Mach.* **3**(4), 184–195 (1960).
33. J. McCarthy, Computer programs for checking mathematical proofs, *Recursive Function Theory, Proc. Symp. Pure Math., Vol. V*, pp. 219–227, American Mathematical Society, Providence, Rhode Island, 1962.
34. J. McCarthy, A basis for a mathematical theory of computation, in *Computer Programming and Formal Systems* (P. Braffort and D. Hershberg, eds.), North–Holland Publ. Co., Amsterdam, 1963.
35. J. McCarthy, P. W. Abrahams, D. J. Edwards, T. P. Hart, and M. I. Levin, *LISP 1.5 Programmer's Manual*, MIT Press, Cambridge, Massachusetts, 1965.

36. J. McCarthy and J. Painter, Correctness of a compiler for arithmetic expressions, *Mathematical Aspects of Computer Science, Proc. Symp. Appl. Math., Vol. XIX,* pp. 33–41, American Mathematical Society, Providence, Rhode Island, 1967.
37. R. Milner and R. Weyhrauch, Proving compiler correctness in a mechanized logic, in *Machine Intelligence 7* (B. Meltzer and D. Michie, eds.), pp. 51–70, Edinburgh Univ. Press, Edinburgh, 1972.
38. J Strother Moore, "Computational Logic: Structure Sharing and Proof of Program Properties," Ph.D. Thesis, University of Edinburgh, Edinburgh (1973).
39. J Strother Moore, Automatic proof of the correctness of a binary addition algorithm, *SIGART Newsletter* **52**, 13–14 (1975).
40. J Strother Moore, Introducing iteration into the pure LISP theorem prover, *IEEE Trans. Software Eng.* **1**(3), 328–338 (1975).
41. J Strother Moore, "The INTERLISP Virtual Machine Specification," CSL 76-5, Xerox Palo Alto Research Center, Palo Alto, California (1976).
42. J. Morris and B. Wegbreit, Subgoal induction, *Commun. Assoc. Comput. Mach.* **20**(4), 209–222 (1977).
43. A. P. Morse, *A Theory of Sets*, Academic Press, New York, 1965.
44. P. Naur, Proof of algorithms by general snapshots, *BIT* **6**, 310–316 (1966).
45. D. Oppen, "Reasoning about Recursively Defined Data Structures," CS Report STAN-CS-78-678, Stanford University (1978). (Appeared in *Proc. 5th ACM Symp. Principles of Programming Languages*, 1978.)
46. R. Peter, *Recursive Functions*, Academic Press, New York, 1967.
47. M. Presburger, Über die Vollständigkeit eines gewissen Systems der Arithmetik ganzer Zahlen, in welchem die Addition als einzige Operation hervortritt, *Comptes-Rendus du I Congrès de Mathematiciens des Pays Slaves*, pp. 92–101, 395, Warsaw, 1929.
48. R. Rivest, On the worst-case behavior of string-searching algorithms, *SIAM J. Comput.* **6**(4), 669–674 (1977).
49. J. A. Robinson, A machine-oriented logic based on the resolution principle, *J. Assoc. Comput. Mach.* **12**(1), 23–41 (1965).
50. J. A. Robinson, Computational logic: The unification algorithm, in *Machine Intelligence 6* (B. Meltzer and D. Michie, eds.), pp. 63–72, Edinburgh Univ. Press, Edinburgh, 1971.
51. L. Robinson and K. Levitt, Proof techniques for hierarchically structured programs, *Commun. Assoc. Comput. Mach.* **20**(4), 271–283 (1977).
52. T. Skolem, The foundations of elementary arithmetic established by means of the recursive mode of thought, without the use of apparent variables ranging over infinite domains, in *From Frege to Gödel*, (J. van Heijenoort, ed.) Harvard Univ. Press, Cambridge, Massachusetts, 1967.
53. W. Teitelman, "INTERLISP Reference Manual," Xerox Palo Alto Research Center, Palo Alto, California (1975).

# Index

## A

Accessor, 36, 37, 39, 376
ACK, 165, 170
Ackermann, W., 165, 169, 175
Add the shell, 37, 376
ADD1, 40
ADD1.EQUAL, 40, 378
ADD1.SUB1, 40, 378
ADDTOLIST, 337
ADDTOLIST2, 367
ADDTOLIST2.DELETE, 367
ADDTOLIST2.KLUDGE, 367
Alist, 60
Allen, J., 27
AND, 30
(AND a b c), 31
APPEND, 13, 203, 329
APPEND.CANCELLATION, 347
APPEND.REVERSE, 330
APPEND.RIGHT.ID, 330
Applicable induction template, 185
Applicable rewrite rule, 122
APPLY, 258, 331
Array index, 283
ASSIGNEDP, 65, 356
Assignment, 57, 60, 62
ASSIGNMENT, 60, 354
ASSIGNMENT.APPEND, 73, 356

ASSIGNMENT.IMPLIES.ASSIGNEDP, 73, 357
ASSOC, 338
ASSOC.PAIRLIST, 338
Association list, 60
ASSOCIATIVITY.OF.APPEND, 17, 330
ASSOCIATIVITY.OF.EQUAL, 339
ASSOCIATIVITY.OF.PLUS, 221, 330
ASSOCIATIVITY.OF.TIMES, 227, 331
Assume term false, 94
Assume term true, 94
ASSUME.FALSE, 65, 356
ASSUME.TRUE, 65, 356
Atom, 88
Atomic, 258
Aubin, R., xiv, 36, 152, 281

## B

Backwards chaining, 109
Ballantyne, A., 104
Bendix, P., 104
Bernays, P., 380
Big numbers, 262
BIG.PLUS, 262, 348
BIG.PLUS1, 348
Binary addition, 262
Binary tree, 42
Binding, 120

Bledsoe, W., xiv, 4, 146
Body, 46
BODY[F], 47
Boolean, 96
BOOLEAN, 339
Bottom object, 36, 37, 376
Bound, 120
Bourbaki, N., 33
Boyer, A., xiv
Boyer–Moore algorithm, 286
Brotz, D., xiv
Bugs, 279, 306
Burstall, R., 3, 16, 20, 33, 36
Buyer, 59

## C

CAAR, 43
CADDR, 43
CADR, 43
Calculator, 253
Call, 29
CAR, 42
CAR.CDRP, 42
CAR.CONS, 42, 377
CAR.LESSP, 42, 377
CAR/CDR.ELIM, 42, 378
Cartwright, R., xiv, 36, 281
CDAR, 43
CDDR, 43
CDR, 42
CDR.CONS, 42, 377
CDR.LESSP, 42, 377
CDR.NLISTP, 42, 377
CDR.NTH, 371
Chang, C., xiii
Changeables, 186
Changing variables, 190
Character position, 283
Character string, 283
Church, A., 8, 85
Clausal form, 380
Clause, 88, 124, 380
Closed, 47
CODEGEN, 263, 264, 265, 332
Coercion, 53, 310
COMMUTATIVITY.OF.APPEND.W-RT.LENGTH, 338
COMMUTATIVITY.OF.EQUAL, 339
COMMUTATIVITY.OF.EQUALP, 343
COMMUTATIVITY.OF.GCD, 351
COMMUTATIVITY.OF.PLUS, 216, 330

COMMUTATIVITY.OF.SGCD, 351
COMMUTATIVITY.OF.TIMES, 331
COMMUTATIVITY2.OF.PLUS, 211, 330
COMMUTATIVITY2.OF.TIMES, 331
COMPILE, 262, 332
Compiler, 253
Conditional rewrites, 106
CONS, 42
CONS.EQUAL, 42, 378
CONS.IF, 60, 354
Constant folding, 257
Constructor, 36, 37, 376
COPY, 342
CORRECTNESS.OF.BIG.PLUS, 262, 350
CORRECTNESS.OF.CODEGEN, 267, 269, 333
CORRECTNESS.OF.OPTIMIZE, 266, 268, 333
CORRECTNESS.OF.OPTIMIZ-ING.COMPILER, 263, 267, 333
CORRECTNESS.OF.SAMEFRINGE, 20, 360
COUNT, 53, 376
COUNT.LIST, 342
COUNT.LIST.SORT, 342
COUNT.LIST.SORT2, 368
COUNT.SORT-LP, 347
COUNT1, 45
Counterexamples, 161
COUNTING.DOWN.BY.N+1, 347
COUNTING.UP.BY.1, 168, 347
COUNTPS, 344
COUNTPS-, 344
COUNTPS-COUNTPS, 345
COUNTPS-LOOP, 344
Cross-fertilization, 20, 89, 146, 149

## D

Darlington, J., 20
Decision procedure, 61
DECREMENT, 368
DECREMENTP, 368
Default value, 37, 376
Definition type set, 100
Definition
  analysis of, 164, 180
  formal principle of, 44
  heuristic expansion of, 113
  informal discussion of principle, 13, 43
  proof of validity of, 46
  role of, xii, 85, 305
Definitions accepted, 329

DELETE, 143, 364
DELETE.ADDTOLIST2, 367
Delta1, 287
DELTA1, 288, 293, 369
DELTA1.LEMMA, 304, 373
DELTA1.LESSP.IFF.MEMBER, 304, 373
Destructive assignment, 307
Destructor, 36
Destructor terms, 133
DIFFERENCE, 233, 310, 347
DIFFERENCE.ADD1.CANCELLATION, 352
DIFFERENCE.ELIM, 353
DIFFERENCE.PLUS, 351
DIFFERENCE.PLUS2, 352
DISTRIBUTIVITY.OF.DIVIDES, 365
DISTRIBUTIVITY.OF.TIMES.O-
  VER.GCD, 309, 353
DISTRIBUTIVITY.OF.TIMES.O-
  VER.PLUS, 331
DIVIDES, 310, 352
DIVIDES.IMPLIES.TIMES, 365
DIVIDES.PLUS.REWRITE, 352
DIVIDES.PLUS.REWRITE.COM-
  MUTED, 353
DIVIDES.PLUS.REWRITE1, 352
DIVIDES.PLUS.REWRITE2, 352
DIVIDES.TIMES, 352
DIVIDES.TIMES.LIST, 364
DIVIDES.TIMES1, 363
DOUBLE, 340
DOUBLE.HALF, 340
DOUBLE.TIMES.2, 340
DSORT, 143, 366
DSORT.SORT2, 143, 368
DUPLICITY.OF.LESSEQP, 341

### E

Elaboration, 110
ELEMENT.APPEND, 351
Elementary number theory, 209, 309
Elimination lemma, 87, 133, 134, 329, 377
Elimination
  of destructors, 89, 130
  of irrelevance, 89, 159
  of nasty terms, 196
  of weak base cases, 177
ENOUGH.FACTORS, 362
Environment, 258
EQ.CHARS.AT.STRPOS, 304, 374
EQP, 334
EQUAL, 30

EQUAL.COPY, 342
EQUAL.DIFFERENCE, 352
EQUAL.EQUALP, 344
EQUAL.TIMES.O, 331
Equality, 1, 28, 89, 103, 145
EQUALP, 342
EUCLID, 353
EVAL, 258, 332
EVEN1, 340
EVEN1.DOUBLE, 340
EVEN1.EVEN2, 341
EVEN2, 340
Examples, 3, 5
EXEC, 261, 333
Existence of prime factors, 314, 318
EXP, 341
EXP.PLUS, 341
EXP.TIMES, 341
Expanding definitions, 113
Explicit value, 114
Explicit value preserving, 115
Explicit value template, 115
Extension, 47

### F

F, 30
FACT, 345
FACT-, 345
FACT-FACT, 345
FACT-LOOP, 345
FACT-LOOP.FACT, 345
FALSE, 30
FALSIFY, 81, 357
FALSIFY1, 81, 357
FALSIFY1.EXTENDS.MODELS, 83, 358
FALSIFY1.FALSIFIES, 83, 358
Fast string searching, 282
Fetch-and-execute, 260
FIDDLE, 368
FIDDLE.DOWN, 368
FIDDLE.DOWN.2, 368
FIDDLE.EQUAL, 369
FIX, 53
FLATTEN, 21, 334
FLATTEN.MC.FLATTEN, 22, 335
FLATTEN.SINGLETON, 358
FLATTEN.SWAPTREE, 344
Flawed, 194
Floyd, R., xiv, 4, 282, 296, 301, 304, 306
Forcing principle, 146
Formal parameter, 46

FORMP, 258, 331
FORMP.OPTIMIZE, 266, 268, 333
Free, 44
Free variable, 111
Fringe, 20
FSTRPOS, 288, 296
FSTRPOS.VC1, 298, 302, 370
FSTRPOS.VC2, 302, 370
FSTRPOS.VC3, 302, 371
FSTRPOS.VC4, 302, 371
FSTRPOS.VC5, 303, 372
FSTRPOS.VC6, 303, 372
FSTRPOS.VC7, 299, 303, 375
Function symbol, 29
Functional semantics, xiv, 253, 260
Fundamental theorem of arithmetic, 309

### G

GCD, 171, 309, 310, 350
GCD.DIVIDES.BOTH, 90, 309, 353
GCD.IS.THE.GREATEST, 353
General recursive, 44, 46
Generalizable, 156
Generalization lemma, 87, 157, 329, 377
Generalization and induction, 151
Generalization
  by replacing subterms, 89, 135, 151, 156
  by throwing away hypotheses, 89, 146, 149, 159
Generalized principle of induction, 33
Gentzen, G., 167
GETVALUE, 258, 331
Gloess, P., xiv, 258
Gödel, K., 41, 380
Gödelization, 41
Goodstein, R., 3
GOPHER, 359
GOPHER.PRESERVES.COUNT, 359
GOPHER.RETURNS.CORRECT.STATE, 359
GOPHER.RETURNS.LEFTMOST.ATOM, 359
Governs, 45
GREATEREQP, 337
GREATERP, 336
GREATERP.CONS, 341
GREATEST.FACTOR, 316, 360
GREATEST.FACTOR.0, 317, 361
GREATEST.FACTOR.1, 317, 361
GREATEST.FACTOR.DIVIDES, 317, 320, 361
GREATEST.FACTOR.LESSP, 317, 360

GREATEST.FACTOR.LESSP.IND, 319, 361
GT.SUB1, 374
Guibas, L., 286

### H

HACK1, 321, 322, 363
HALF, 181, 340
HALF.DOUBLE, 340
Handley, J., xiv
Henneman, W., xiv, 146
Hoare, C., 282

### I

ID, 317, 360
ID.ADD1, 362
IF, 30
IF-normal form, 58
IF.COMPLEXITY, 69, 354
IF.COMPLEXITY.GOES.DOWN1, 70, 355
IF.COMPLEXITY.GOES.DOWN2, 70, 355
IF.COMPLEXITY.NOT.0, 354
IF.COMPLEXITY.STAYS.EVEN, 70, 355
IF.DEPTH, 68, 354
IF.DEPTH.GOES.DOWN, 69, 354
IF.EXPRP, 60, 354
IF.TIMES.THEN.DIVIDES, 365
IFF, 73, 339
IFF.EQUAL.EQUAL, 339
Illegal substitution pairs, 187
IMPLIES, 31
Induced, 51
Induction
  case analysis, 175
  on conjuncts, 90
  formal principle of, 33
  heuristic application of, 89, 163, 185
  heuristic manipulation of schemes, 189
  informal discussion of principle, 15, 33
  proof of validity of, 34
  role of, xii, 301
  selection of hypotheses, 179, 186
  soundness of implementation, 188
Induction lemma, 87, 166, 180, 329, 377
Induction and recursion, 163, 171, 180, 185
Induction template, 183
Inductive assertion, xiv, 296, 306
Inductively constructed object, 11, 35
Instance, 103

Instantiation, 28
Instruction, 260
INTERLISP, 27
INTERSECT, 335
Irrelevance, 89, 159, 176, 178, 179

**K**

Kelley, J., 32, 58
KLUDGE.BRIDGE, 363
Knuth, D., 104, 284, 286

**L**

Lankford, D., 104
Laski, J., xiv
LAST, 338
LAST.APPEND, 341
LAST.REVERSE, 341
Lee, R., xiii
Left-most match, 283, 292
LEFT.BRANCH, 60, 354
LEFT.BRANCH.CONS.IF, 60, 377
LEFT.BRANCH.LESSP, 60, 377
LEFTCOUNT.GOES.DOWN, 359
Lemma, xi, 5
LENGTH, 288, 335
LENGTH.MAPCAR, 338
LENGTH.REVERSE, 335
LENGTH.SORT, 341
LENGTH.TIPCOUNT, 344
LESSEQP, 337
LESSEQP.HALF, 342
LESSEQP.NTH, 341
LESSEQP.PLUS, 337
LESSEQP.PLUS2, 338
LESSP, 52, 376
LESSP.ADD1.ID, 362
LESSP.D.V, 355
LESSP.DELETE.REWRITE, 366
LESSP.DIFFERENCE, 349
LESSP.DIFFERENCE.1, 374
LESSP.DIFFERENCE1, 349
LESSP.DIFFERENCE2, 353
LESSP.EQUAL, 334
LESSP.ID2, 360
LESSP.ID3, 361
LESSP.KLUDGE1, 372
LESSP.KLUDGE2, 372
LESSP.KLUDGE3, 374
LESSP.LENGTH, 366
LESSP.MAXIMUM.ADDTOLIST2, 367

LESSP.NOT.COMMUTATIVE, 351
LESSP.NOT.REFLEXIVE, 334
LESSP.PLUS, 373
LESSP.PLUS.CANCELLATION, 353
LESSP.PLUS.CANCELLATION2, 353
LESSP.PLUS.SUB1, 368
LESSP.PLUS.TIMES, 349
LESSP.PLUS1, 373
LESSP.QUOTIENT1, 349
LESSP.REMAINDER1, 349
LESSP.REMAINDER2, 349
LESSP.SUB1.HACK1, 375
LESSP.SUB1.PLUS.CANCELLATION, 371
LESSP.SUB1.SUB1, 374
LESSP.SUB1.SUB1.PLUS, 374
LESSP.TIMES.CANCELLATION, 353
Levitt, K., 262
Lexicographic relation, 51, 169, 182
LISP, xiii, xiv, 8, 27, 36, 60
List, 42
LISTP, 42
LISTP.CONS, 12
LISTP-GOPHER, 359
LISTP.NTH, 371
LIT, 339
LIT.APPEND, 339
LITATOM, 41
Literal, 88
Literal atoms, 41
LITTLE.STEP, 364
LOOP.ASSERT, 295, 370
Lorenzen, P., 51
Loveland, D., xiii

**M**

Machine, 165
Machine code, 256, 259
MACRO-10, 36
Manna, Z., 2, 282
MAPCAR, 338
MAPCAR.APPEND, 338
MATCH, 292, 369
MATCH.EPSILON, 374
MATCH.IMPLIES.CAR.MEMBER, 373
MATCH.IMPLIES.CAR.MEMBER1, 373
MATCH.IMPLIES.DELTA1.OK, 373
MATCH.IMPLIES.MEMBER, 373
MATCH.LENGTHS, 370
MATCH.LENGTHS1, 370
MAXIMUM, 366

# 394 / INDEX

MC.FLATTEN, 21, 22, 335
McCarthy, J., xiii, xiv, 3, 4, 8, 20, 21, 85, 253, 281, 311
Measure, 33, 172
MEASURE, 368
Measured subset, 173
Meltzer, B., xiv
MEMBER, 178, 335
MEMBER-INTERSECT, 336
MEMBER.APPEND, 335
MEMBER.APPEND2, 335
MEMBER.MAXIMUM, 366
MEMBER.OCCUR, 343
MEMBER.REVERSE, 335
MEMBER.SORT, 341
MEMBER.UNION, 336
MEMBER.UNION-, 346
Merge, 191
Metavariables, 29
Milner, R., 281
Minimal, 32
Model 0.0, 253
Morris, J., 172, 284
Morse, A. P., 32
Motivation, 2
Must be false, 94
Must be true, 94
Mutual recursion, 46

## N

Naive string searching algorithm, 283
Nasty, 196
Natural numbers, 40
Naur, P., 282
Nelson, G., xiv, 85
NEQUAL.PLUS.ADD1, 351
New, 29
NIL, 41
"NIL", 42
Nilsson, N., 6
NLISTP, 329
Noetherian induction, 33
Nonrecursive, 46
NORMALIZE, 63, 71, 355
NORMALIZE.IS.SOUND, 72, 77, 201, 358
NORMALIZE.NORMALIZES, 72, 77, 358
NORMALIZED.IF.EXPRP, 72, 356
NOT, 30
NTH, 295, 336
NTH.APPEND, 351

NTH.APPEND1, 339
NTH.MEMBER, 336
NTH.NIL, 339
NTHCHAR, 288, 291, 351
Number theory, 209, 309
NUMBER.LISTP, 342
NUMBERP, 40
NUMBERP.APPLY, 259, 331
NUMBERP.GREATEST.FACTOR, 317, 361

## O

OCCUR, 166, 343
OCCUR.SUBST, 343
ODD, 340
Odlyzko, A., 286
Opening up definitions, 113
Oppen, D., 36
Optimize, 256
OPTIMIZE, 263, 332
OR, 31
(OR a b c), 31
ORDERED, 337
Ordered pair, 42
ORDERED.ADDTOLIST, 346
ORDERED.APPEND, 342
ORDERED.SORT, 342
ORDERED.SORT-LP, 346
ORDERED2, 366
Ordinal, 170
OTHERS, 93
Overview, 6, 87

## P

(p A B), 131
PACK, 41
Painter, J., 281
Pair deletion, 187
PAIRLIST, 338
Parsing, 258
Partially correct, 47
Pattis, R., xiv
pc, 260
Peano's axioms, 40
PERM, 313, 364
Permutation, 313
Permutative, 104
Peter, R., 44, 165
PK, 368
PLISTP, 127, 204, 330

PLISTP.REVERSE, 330
PLUS, 53
"PLUS", 42
(PLUS a b c), 53
PLUS-, 346
PLUS-PLUS, 346
PLUS.2.NOT, 374
PLUS.ADD1, 330
PLUS.CANCELLATION, 369
PLUS.CANCELLATION1, 349
PLUS.DIFFERENCE.SUB1.REWRITE, 371
PLUS.DIFFERENCE3, 351
PLUS.EQUAL.0, 349
PLUS.NEQUAL.X, 351
PLUS.RIGHT.ID, 209, 330
Pnueli, A., 282
POP, 260, 331
POP-2, 36
POP.PUSH, 260, 377
POWER.EVAL, 348
POWER.EVAL.BIG.PLUS, 348
POWER.EVAL.BIG.PLUS1, 348
POWER.EVAL.POWER.REP, 350
POWER.REP, 350
Pratt, V., 284
Predicates, 10
Preprocessing, 286
Presburger, M., 307
PRIME, 312, 360
Prime factorization theorem, 309
Prime number, 312, 315
PRIME.BASIC, 315, 316, 363
PRIME.BRIDGE, 363
PRIME.FACTORIZATION.EXISTENCE, 314, 318, 319, 362
PRIME.FACTORIZATION.UNIQUENESS, 314, 323, 324, 366
PRIME.FACTORS, 314, 318, 361
PRIME.GCD, 321, 322, 364
PRIME.KEY, 321, 364
PRIME.KRUTCH, 363
PRIME.LIST, 313, 362
PRIME.LIST.APPEND, 362
PRIME.LIST.DELETE, 364
PRIME.LIST.PRIME.FACTORS, 362
PRIME.LIST.TIMES.LIST, 322, 365
PRIME.MEMBER, 324, 365
PRIME1, 312, 360
PRIME1.BASIC, 316, 363
Primitive-recursive, 196
Program counter, 260

Program output, 21, 71, 77, 160, 203, 209, 268, 324
Program semantics, xiii, xiv, 2, 298
Proof-checking, xi, 281
Proper list, 127
Propositional calculus, 1, 30, 56
Propositional IF-expression, 57
p/s, 29
Pure LISP, xiii
PUSH, 260, 331
Pushdown stack, 259
PUSH.EQUAL, 260, 378
"PUSHI", 42
"PUSHV", 42

## Q

Quantification, xiv, 3, 28, 62, 84, 282, 307, 314, 315
QUOTIENT, 242, 310, 347
QUOTIENT.DIVIDES, 364
QUOTIENT.LESSP, 362
QUOTIENT.TIMES, 365
QUOTIENT.TIMES1, 362

## R

Recognizer, 36, 37, 376
Recursion, xiv, 3, 13, 15, 319
RECURSION.BY.DIFFERENCE, 209, 233, 347
RECURSION.BY.QUOTIENT, 350
RECURSION.BY.REMAINDER, 351
Recursive, 46
REMAINDER, 242, 310, 347
REMAINDER.ADD1, 352
REMAINDER.GCD, 363
REMAINDER.GCD.1, 363
REMAINDER.QUOTIENT, 348
REMAINDER.QUOTIENT.ELIM, 209, 243, 349
REMAINDER.TIMES, 364
REMAINDER.WRT.1, 349
REMAINDER.WRT.12, 349
REMAINDER.X.X, 349
Replacement principle, 125
Resolution, xiii, 125, 146, 380
Result of substituting, 29
REVERSE, 126, 203, 329
REVERSE-, 345
REVERSE-APPEND, 346
REVERSE-LOOP, 345

REVERSE-LOOP.APPEND.REVERSE, 345
REVERSE-REVERSE, 345
REVERSE-REVERSE-, 346
REVERSE.MAPCAR, 339
REVERSE.REVERSE, 204, 334
Revised machine, 179, 182
Rewrite lemma, 87, 98, 102, 122, 329, 377
Rewrite with lemmas, 122
Rewriting, 103, 120
RIGHT.BRANCH, 60, 354
RIGHT.BRANCH.CONS.IF, 60, 377
RIGHT.BRANCH.LESSP, 60, 377
Rivest, R., 284
Robinson, J. A., 88, 125, 380
Robinson, L., 262
Rule of instantiation, 28
RUSSELL, 44

S

SAMEFRINGE, 20, 359
Schemas, 131
Score, 190
Sequence, 42
SEQUENTIAL.EXECUTION, 268, 333
SGCD, 350
SGCD.X.0.X, 351
SGCD.X.X.X, 351
Shell
    implementation of, 376
    informal discussion of principle, 35
    principle of, 37
Side-effects, 307
Simplification, 89, 124
Skolem, T., 3
Smaller than, 32
SORT, 153, 337
SORT-, 346
SORT-LP, 346
SORT.OF.ORDERED.NUMBER.LIST, 343
SORT.ORDERED, 343
SORT2, 367
SORT2.DELETE, 368
SORT2.DELETE.CONS, 367
SORT2.GEN, 367
SORT2.GEN.1, 367
SORT2.GEN.2, 367
Spitzen, J., xiv
STACKP, 260, 331
String, 291
String searching, 282

STRPOS, 283, 369
STRPOS.BOUNDARY.CONDITION, 302, 370
STRPOS.EPSILON, 374
STRPOS.EQUAL, 303, 371
STRPOS.EQUAL.0, 372
STRPOS.LESSEQP.CRUTCH, 372
STRPOS.LESSEQP.STRLEN, 372
STRPOS.LIST.APPEND, 372
Structural induction, 16
SUB1, 40
SUB1.ADD1, 40, 377
SUB1.ELIM, 40, 378
SUB1.LENGTH, 373
SUB1.LESSP, 40, 377
SUB1.LESSP.PLUS, 370
SUB1.LESSP1, 333
SUB1.TYPE.RESTRICTION, 40, 377
SUB1P, 40
Subgoal induction, 172
SUBSETP, 336
SUBSETP.CONS, 341
SUBSETP.INTERSECT, 336
SUBSETP.UNION, 336
SUBST, 343
SUBST.A.A, 343
Substitution, 29
Substitution instance, 103
Subsumption
    of clauses, 90, 125
    of induction schemes, 190
SUM, 44, 53
SWAPPED.PLUS.CANCELLATION, 371
SWAPTREE, 344
SWAPTREE.SWAPTREE, 344
Symbolic expression, 11, 42, 60, 258
Syntax, 28

T

T, 30
Tautology, 56, 57, 61
TAUTOLOGY.CHECKER, 63, 357
TAUTOLOGY.CHECKER.COMPLETENESS.BRIDGE, 83, 358
TAUTOLOGY.CHECKER.IS.COMPLETE, 62, 82, 107, 358
TAUTOLOGY.CHECKER.IS.SOUND, 62, 79, 358
TAUTOLOGY.CHECKER.SOUNDNESS.BRIDGE, 80, 358

TAUTOLOGYP, 64, 356
TAUTOLOGYP.FAILS.MEANS.FAL-
    SIFY1.WINS, 82, 83, 358
TAUTOLOGYP.IS.SOUND, 72, 74, 200, 357
Teitelman, W., 27
Term, 29
TEST, 60, 354
TEST.CONS.IF, 60, 377
TEST.LEFT.BRANCH.RIGHT.-
    BRANCHP, 60, 354
Text searching, 282
Theorem, 31
Theorems proved, 329
Theory
    formal treatment, 28
    informal sketch, 8
Tie breaking rules, 195
TIMES, 221, 309, 330
TIMES.ADD1, 223, 331
TIMES.DIFFERENCE, 352
TIMES.EQUAL.1, 365
TIMES.IDENTITY, 363
TIMES.LIST, 313, 362
TIMES.LIST.APPEND, 362
TIMES.LIST.DELETE, 365
TIMES.ZERO, 222, 330
TIPCOUNT, 344
TOP, 260, 331
TOP.ASSERT, 294, 369
TOP.POPP, 260, 331
TOP.PUSH, 260, 377
Toys, xiii, 253, 304
TRANSITIVITY.OF.EQUAL, 339
TRANSITIVITY.OF.EQUALP, 343
TRANSITIVITY.OF.GREATERP, 337
TRANSITIVITY.OF.LESSEQP, 337
TRANSITIVITY.OF.LESSP, 334
TRANSITIVITY.OF.LESSP2, 334
TRANSITIVITY.OF.LESSP3, 334
TRANSITIVITY.OF.NOT.LESSP, 334
TRANSITIVITY.OF.NOT.LESSP2, 334
TRANSITIVITY.OF.NOT.LESSP3, 334
Tree, 42
TRICHOTOMY.OF.LESSP, 337
TRUE, 30
Type, 35, 93
Type prescription, 97
Type restriction, 37, 376
Type set, 92, 93
Type set alist, 120
Type set lemma, 98, 102, 377,

## U

Unchangeables, 186
Unchanging, 190
Unflawed, 194
UNION, 336
UNION-, 346
Uniqueness of prime factors, 314, 320
UNIVERSE, 93
UNPACK, 41
UNPACKP, 41
User, 59, 67, 84, 85, 87, 90, 98, 104, 106,
    112, 132, 141, 152, 153, 161, 166, 179,
    195, 252, 281, 317, 329, 376

## V

Value, 57
VALUE, 60, 354
VALUE.CAN.IGNORE.REDUN-
    DANT.ASSIGNMENTS, 74, 357
VALUE.SHORT.CUT, 74, 357
Variable alist, 120
Variable symbol, 28
VC4.HACK.1, 371
VC5.HACK1, 372
Verification condition, xiii, 298
von Neumann, J., 380

## W

Waldinger, R., 2
Waterfall, 90
Weak base cases, 176
Wegbreit, B., 172
Well-founded, 13, 31
Weyhrauch, R., 281
WHILELOOP, 168
Worse than, 110

## X

Xi, 29
XOR, 343

## Z

ZERO, 40
0,1,2. . . , 41
ZEROP, 43
ZEROP.LENGTH, 370

## ACM MONOGRAPH SERIES

*Published under the auspices of the Association for Computing Machinery Inc.*

Editor  THOMAS A. STANDISH  *University of California at Irvine*
Former Editors
Richard W. Hamming, Herbert B. Keller, Robert L. Ashenhurst

A. FINERMAN (Ed.)  University Education in Computing Science, 1968
A. GINZBURG  Algebraic Theory of Automata, 1968
E. F. CODD  Cellular Automata, 1968
G. ERNST AND A. NEWELL  GPS: A Case Study in Generality and Problem Solving, 1969
M. A. GAVRILOV AND A. D. ZAKREVSKII (Eds.)  LYaPAS: A Programming Language for Logic and Coding Algorithms, 1969
THEODOR D. STERLING, EDGAR A. BERING, JR., SEYMOUR V. POLLACK, AND HERBERT VAUGHAN, JR. (Eds.)  Visual Prosthesis: The Interdisciplinary Dialogue, 1971
JOHN R. RICE (Ed.)  Mathematical Software, 1971
ELLIOTT I. ORGANICK  Computer System Organization: The B5700/B6700 Series, 1973
NEIL D. JONES  Computability Theory: An Introduction, 1973
ARTO SALOMAA  Formal Languages, 1973
HARVEY ABRAMSON  Theory and Application of a Bottom-Up Syntax-Directed Translator, 1973
GLEN G. LANGDON, JR.  Logic Design: A Review of Theory and Practice, 1974
MONROE NEWBORN  Computer Chess, 1975
ASHOK K. AGRAWALA AND TOMLINSON G. RAUSCHER  Foundations of Microprogramming: Architecture, Software, and Applications, 1975
P. J. COURTOIS  Decomposability: Queueing and Computer System Applications, 1977
JOHN R. METZNER AND BRUCE H. BARNES  Decision Table Languages and Systems, 1977
ANITA K. JONES (Ed.)  Perspectives on Computer Science: From the 10th Anniversary Symposium at the Computer Science Department, Carnegie-Mellon University, 1978
DAVID K. HSIAO, DOUGLAS S. KERR, AND STUART E. MADNICK  Computer Security, 1979
ROBERT S. BOYER AND J STROTHER MOORE  A Computational Logic, 1979

*In preparation*

R. L. WEXELBLAT    History of Programming Languages

J. F. TRAUB AND H. WOŹNIAKOWSKI    A General Theory of Optimal Algorithms

*Previously published and available from The Macmillan Company, New York City*

V. KRYLOV    Approximate Calculation of Integrals (Translated by A. H. Stroud), 1962